Business Statistics: Why and When

Business Statistics: Why and When

Larry E. Richards
Jerry J. LaCava
College of Business Administration
University of Oregon

McGraw-Hill Book Company

New York St. Louis San Francisco Auckland Bogotá Düsseldorf Johannesburg London Madrid Mexico Montreal New Delhi Panama Paris São Paulo Singapore Sydney Tokyo Toronto

Business Statistics: Why and When

3 4 5 6 7 8 9 0 F G R F G R 7 8 3 2 1 0 9

This book was set in Times Roman by Black Dot, Inc.
The editors were Donald E. Chatham, Jr., and Barbara Brooks;
the designer was Ben Kann;
the production supervisor was Dominick Petrellese.
The drawings were done by Fine Line Illustrations, Inc.
Fairfield Graphics was printer and binder.

Library of Congress Cataloging in Publication Data

Richards, Larry E.
Business statistics

Bibliography: p.
Includes index.
1. Statistics. 2. Business—Statistical methods.
I. LaCava, Jerry J., joint author. II. Title.
HA29.R485 519.5 77-14314
ISBN 0-07-052273-1

Contents

Preface

As there are numerous basic business statistics texts on the market today, we feel the need to explain our reasons for yet another. The authors see three interrelated situations which are believed to be generally true. Each situation, and certainly the combination, demands a change in the format of the introductory course and text on basic business statistics. These situations are discussed below.

1. *Basic business statistics texts are written with the right format for the wrong objective.* Without exception, basic business statistics texts have a format which is consistent with the "training of statisticians"—the "how to do it." Little if any effort is spent on WHY, WHERE, or WHEN—just HOW. These texts tend to follow the format of an introductory mathematical statistics text, omitting the math. It is virtually impossible to "train" a statistician in one or two terms. Our objective in the first course is *not* the training of statisticians. Most of the students going through the first course in statistics will never perform a statistical analysis outside the classroom. Rather, they will find themselves in the roles of decision makers and consumers of analyses. We would argue that it is the WHY, WHEN, and WHERE of statistics that will truly be of value to most students. No doubt the first thing the student will forget is HOW. However, in his or her eventual role as decision maker and/or consumer of statistical analyses, what is important is an awareness of and an appreciation for existing statistical techniques. Thus, we believe the objective of introductory courses should be to *introduce* statistical procedures and techniques and thereby *motivate* students to (1) become aware of statistical techniques, (2) understand the logic of statistical inference, and (3) identify applications.

2. *Undergraduate business students are not motivated by basic statistics.* The typical business student comes into the first statistics course with a relatively weak mathematical background—convinced that he or she cannot do well in a mathematically oriented course. These students are subjected to abstract concepts, notation and symbolism, and a battery of irrelevant problems, and they are asked to perform statistical analyses. The texts devote little if any effort toward motivating the students but get right to their job of "training" statisticians. The implicit assumption is that students are already aware of the value and importance of the area, and the only job remaining is to "show them how." The result is that student motivation is virtually nonexistent after a few weeks, and the course is viewed as a drudgery that someone has required. Since the problems are typically stripped of any realism, the students are unable to identify applications or to appreciate the value of the techniques. All of which leads to the question asked by the better student: "Does anyone really use this stuff?"

Clearly the general lack of student motivation is a function of text format and objective. These two situations understandably lead to the third.

3. *There is a growing dissatisfaction among the professors of introductory business statistics.* The dissatisfaction comes both from the lack of student comprehension and retention of the presented material and from the "cookbook" approach that most students adopt.

It is our feeling that if the format of the texts were changed to be consistent with WHY, WHEN, and WHERE, along with HOW, student motivation could be achieved. The objective would be to *truly* introduce and provide the much-needed motivation. We believe that such a *change in direction* of the first course would be a progressive step. It is our hope that this text represents such a change.

The specific characteristics of this text which we consider as distinguishing assets toward obtaining our objectives are:

1. Programmed mathematics review in Chapter 1
2. Introduction of each statistical technique through a realistic business situation
3. Elimination of unnecessary topics
4. Emphasis placed on identifying applications
5. Self-motivating problems
6. Cases

Chapter 1 contains a short review of the mathematical operations required both to follow the text and to solve problems. This section gives the students with a relatively weak math background a chance to "brush up" on just what is needed and introduces Σ, !, inequalities, points on a rectangular coordinate system, and how to use a square root table, without having to *stop* in the text.

Each statistical technique is introduced by describing a realistic business situation, reducing the situation to "a need to do something," and then introducing the statistical technique designed to fill the specific need. Such an approach seems both logical and more likely to "sell" the techniques. Our objective has been to present realistic situations and answer the question "Why?"—thus providing relevance.

A conscious attempt has been made to eliminate topics that are usually covered in basic statistical texts but that do not contribute or are not necessary, given our objective of introducing applications and motivating students rather than training statisticians. Such topics as average deviation and the Poisson, exponential, and rectangular distributions are not needed or used in the basic statistics course and thus only detract from the relevance of the course. Yes, the Poisson and exponential distributions are important and used in operations research. However, a student would most likely not remember them if they were introduced in a previous term or year in a statistics course, and an operations research course would certainly review or reintroduce them when needed. Our argument is that they are not needed in the basic statistics course, and thus their inclusion is not an asset in relation to motivation or relevance.

If most students in the basic statistics course eventually become decision makers in business, what they need is the ability to recognize when they have a problem capable of solution through the application of statistical techniques. Given this, it is the ability to identify applications rather then the ability to perform analyses that is important. For those chapters that introduce statistical techniques, the final section "Identification of Applications" is designed to point out the conditions under which the application of the studied technique would be of value. The idea is to focus attention on identifying applications. It seems to us that being able to identify a realistic business application for a given statistical technique is really the acid test of understanding.

We have consciously attempted to present problems which contain an explanation of why someone would want to do or solve this problem. Again, the objective is to show relevance.

Finally, to be consistent with the objective of identifying applications, a number of cases are included. The main objective for the student in relation to the cases is to identify which, if any, of the studied techniques would be appropriate and to explain why.

We would like to thank the following persons who offered comments and suggestions as our work progressed: Professor Richard Berger, Fairleigh Dickinson University, Rutherford, N.J.; Professor David A. Goodman, Southern Methodist University, Dallas; Professor John Hoftyzer, University of North Carolina, Greensboro; Professor Kenneth C. Lundahl, Jamestown (New York) Community College; Professor Zenon S. Malinowski, University of Connecticut, Storrs; Professor Paul Paschke, Oregon State University, Corvallis; Professor Grover Rodich, Portland (Oregon) State University; and Professor H. C. Rutemiller, California State University, Fullerton.

We are especially indebted to Arun Prakash for his assistance with the problems, to Arno Rethans for his numerous suggestions, and finally to Nola Ventura for her diligence in transcribing our penmanship into typed manuscript.

Larry E. Richards

Jerry J. LaCava

Chapter 1

Introduction

1.1 WHAT IS STATISTICS?

You are embarking on the study of a field of science called STATISTICS. Specifically, we will study the field of statistics as it is and can be applied to business and economics. Just what is this field? What kind of procedures will we study? What are the objectives of statistics? And, of primary importance: Why should you study statistics?

As we progress through this text, you will gain a *clear* understanding of the field of statistics, its objectives, its logic, and a variety of the more basic statistical techniques. You will find that statistics encompasses the collection, summarization, presentation, and analysis of data. These activities and their techniques are designed to aid the process of decision making—certainly an important function in business.

STATISTICS IS A FIELD OF SCIENCE ENCOMPASSING PROCEDURES DESIGNED TO AID THE PROCESS OF DECISION MAKING UNDER UNCERTAINTY.

1.2 WHY STUDY STATISTICS?

If there is one universally most important function of persons working within a business organization, it must be *decision making*. With the tremendous increase in availability of data (via computers) and the increasing complexity of business operations, extraordinary pressures are placed on the process of decision making. Thus we come to the inescapable conclusion that techniques that aid the process of decision making *are of value*. Tomorrow's decision makers will need to be sufficiently familiar with existing statistical techniques to identify *when* a problem can be analyzed through the application of statistics. True, most decision makers will not actually perform the analysis. This operation would fall in the domain of the business statistician—a staff member within the organization or an outside consultant. However, the decision maker must have the required level of statistical understanding to be an intelligent consumer of such analyses.

The need for such an understanding of statistics is not limited to the decision maker within business. The nineteenth-century English novelist H. G. Wells claimed that "statistical thinking will one day be as necessary for efficient citizenship as the ability to read and write." History has proved that Wells had an uncanny perception of the future. Today there is no way of escaping our role as consumers of statistical analyses. These analyses come to us as advertising claims, economic indicators and forecasts, results of opinion and political polls, statements of average per capita income, and so forth. Clearly this is not an exhaustive list, for such a list would be virtually endless. There is no doubt that *we are all consumers of statistical analyses*.

1.3 REQUIRED MATHEMATICS

It has been our experience that many students approach the first course in statistics with great reluctance. These students typically have limited backgrounds in mathematics. They have experienced frustration with little reward in prior mathematics courses and come to the course convinced that statistics will be another bad experience.

This text is *not* a course in mathematics. Yes, because of the nature of statistics, we will be involved with some mathematical operations. However, the level of mathematics required in this text is minimal.

The remainder of this chapter contains *all* the mathematical operations required to follow the material presented. If you study this material (most of it will certainly be a review) and can solve the problems at the end of the chapter, your background is sufficient, and mathematics will not impede your study and understanding.

The last part of this chapter is in the *programmed text format*. The reasons for this format selection are: (1) the programmed text format is particularly well suited for self-study and (2) many instructors presume the student has previously acquired this background and thus it is not covered in class as an integral part of the course.

The following material is arranged in a vertical series of frames along the right-hand side of each page. The correct response to a frame is located in the left-hand side of the following frame. We encourage you to cover the following frame with a piece of paper, write your answer in the blank, and then check it.

This is not a comprehensive review of mathematics. However, it *is* a review of the mathematical operations which you will need to follow the material presented in this text.

1.3a Algebra of Fractions

1. Integers

To begin we must be familiar with integers. . . . , −5, −4, −3, −2, −1, 0, +1, +2, +3, +4, +5, . . . are called *integers*. Note that all integers, except ________, have associated signs (+ or −).

---

0 or zero

2. +1, +2, +3, . . . are positive integers and −1, −2, −3, . . . are negative integers. Which of the following are integers: −17, +2.45, +6, −0.01, +100.3? ________

−17, +6

3. Fractions

If a and b are both integers and b is not equal to ($\neq$) zero, then a/b is called a *fraction*. a is called the *numerator* and b is called the *denominator*. In the fraction 2/10, the integer 2 is the ________ and 10 is the ________.

---

numerator, denominator

4. Equality of Fractions

We define $a/b = c/d$ if $ad = bc$. Thus, $^2/_3 = {}^6/_9$ as (2)(9) = (3)(6). Which of the following fractions are equal to $^2/_5$?

(a) $^7/_{15}$ **(b)** $^1/_3$ **(c)** $^8/_{20}$

____

---

only **(c)** $8/20$:
$(2)(20) = (5)(8)$

5. Multiplying or dividing both numerator *and* denominator of a fraction by a non-zero value does not change the value of fraction.

$a/b = ac/bc$

and

$$a/b = \frac{a/c}{b/c}$$

where $c \neq 0$. c is known as a common factor of ac and bc. $1/c$ is then a ________ ________ of a/c and b/c.

---

common factor

6. Frame 5 points out the important fact that a fraction may be reduced by canceling common factors in the numerator and denominator. Reduce the following fractions:

(a) $3/9$ ________

(b) $\frac{2x}{7x}$ ________

(c) $\frac{2 + x}{7 + x}$ ________

(a) $3/9 = \frac{(3)(1)}{(3)(3)} = 1/3$

(b) $2x/7x = 2/7$
x is a common factor of both numerator and denominator.

(c) This cannot be reduced; x is not a common factor.

7. Addition of Fractions

In order to add two fractions they must have equal denominators. Thus,

$$\frac{a}{b} + \frac{c}{d} = \frac{ad + bc}{bd}$$

where b and $d \neq 0$. If $b = d$, we can add the two fractions directly by adding the numerators.

$1/10 + 3/10 = 4/10$

$3/4 + 2/5 =$ ________

$$\frac{(3)(5) + (4)(2)}{(4)(5)} = 23/20$$

8. Subtraction of Fractions

As with addition, in order to subtract two fractions they must have equal denominators.

$$\frac{a}{b} - \frac{c}{d} = \frac{ad - bc}{bd}$$

where b and $d \neq 0$.

3/8 − 1/8 = 2/8, which can be reduced to 1/4 because 2 is a common factor of both numerator and denominator.

4/5 − 1/3 = ________

$$\frac{(3)(4) - (5)(1)}{(5)(3)} = 7/15$$

9. Multiplication of Fractions

$$\frac{a}{b} \cdot \frac{c}{d} = \frac{ac}{bd}$$

where b and $d \neq 0$. We simply multiply both numerators to obtain the numerator of this product and multiply both denominators to obtain the denominator of the product.

3/8 · 1/5 = 3/40

2/3 · 3/5 = ________

$$\frac{(2)(3)}{(3)(5)} = 2/5$$

Note that 3 is in the numerator of one of our fractions and also in the denominator of the other. Thus, 3 will be a common factor of the product, and we could cancel the 3s before multiplying.

10. The result in Frame 9 leads us to the realization that

$a/b \cdot c/a = c/b$

$3x/7 \cdot 3/2x = 9/14$

$$\frac{(a + b)}{c} \cdot \frac{3c}{a} = \text{________}$$

$$\frac{3(a + b)}{a} = \frac{3a + 3b}{a}$$

11. Division of Fractions

$a/b \div c/d = a/b \cdot d/c = ad/bc$

Note that we divide one fraction by another by inverting the *divisor* fraction and multiplying.

2/5 ÷ 1/3 = 2/5 · 3/1 = 6/5

2/7 ÷ 3/4 = ________

2/7 · 4/3 = 8/21

12. It is important to realize that an integer can be expressed as a fraction.

7 = 7/1

Thus $2/9 \div 4 = 2/9 \div 4/1$ which is equal to $2/9 \cdot 1/4 = 2/36$ or $1/18$.

$$\frac{(n-1)s^2}{\sigma^2} \div (n-1) = \_\_\_\_\_\_\_\_$$

$$\frac{(n-1)s^2}{\sigma^2} \div \frac{(n-1)}{1} =$$

$$\frac{(n-1)s^2}{\sigma^2} \cdot \frac{1}{(n-1)} = \frac{s^2}{\sigma^2}$$

$(n - 1)$ will be a common factor of this product.

1.3b Operations with Equations

13. Grouping Symbols

Quite often we need to be able to indicate that several values or terms in an equation should be treated as a unit. This is accomplished by using parentheses ( ), brackets [ ], and/or braces { }. For example,

$$8 = 2 + 3(4 - 2)$$

but

$$8 \neq 2 + 3 \cdot 4 - 2$$
$$12 = 2 + 3 \cdot 4 - 2$$

If the parentheses, brackets, or braces are preceded by a negative sign, *all* signs within the parentheses, brackets, or braces are changed.

$$3 + (a - 4) = a - 1$$
$$3 - (a - 4) = 3 + (-a + 4)$$
$$= 7 - a$$

$5x - (y + 2x) = \_\_\_\_\_\_\_\_$

$5x - y - 2x = 3x - y$

14. The equality is preserved when the same value is added to or subtracted from both sides of an equation.

If $a = b$

$a + c = b + c$

and $a - c = b - c$

If $x = 4$

$x + 7 = \_\_\_\_\_\_\_\_$

and if $\mu = 38$

$\mu - 2\sigma = \_\_\_\_\_\_\_\_$

$x + 7 = 11$

$\mu - 2\sigma = 38 - 2\sigma$

15. Multiplying or dividing both sides of an equation by a constant preserves the equality.

If $a = b$

then $ac = bc$

and $a/c = b/c$

where $c \neq 0$.

If $3x - 2 = 28$

$4(3x - 2) =$ ________

$(4)(28) = 112$

16. If $7z = \sigma$

$7z/3 =$ ________

$\sigma/3$

17. With these operations we are able to solve an equation for an unknown. For example,

$6z + 5 = -3z - 2$

We may subtract 5 from both sides of this equation, giving $6z = -3z - 7$. Now, we may add $3z$ to both sides, resulting in the equation $9z = -7$. Finally, we can solve this equation for z by dividing both sides by 9.

$z = -7/9$

Solve the following equation for $\bar{y}$.

$$z = \frac{\bar{y} - \mu}{\sigma/\sqrt{n}}$$

$\bar{y} =$ ________

$z(\sigma/\sqrt{n}) = \bar{y} - \mu$

$\bar{y} = \mu + z(\sigma/\sqrt{n})$

1.3c Inequalities

18. Frequently we need to indicate the relative size or value of one expression compared with another. For this reason we have defined the following symbols.

$>$ is read " is greater than"

$<$ is read "is less than"

$\geq$ is read "is greater than or equal to"

$\leq$ is read "is less than or equal to"

Thus, $7 < 10$

$4 > 3$

Insert the proper inequality sign for each of the following:

(a) 1.36 ____ 2.41

(b) 3.01 ____ 2.62

(c) −7 ____ −3

(a) $<$
(b) $>$
(c) $<$

19. For the following values of x, is the inequality $x \leq 32$ true or false?

(a) 24 ______

(b) 0 ______

(c) 41 ______

(d) 32 ______

(a) true
(b) true
(c) false
(d) true

20. A constant can be added to or subtracted from both sides of an inequality and the inequality is preserved.

If	a	$<$	b
	$a + c$	$<$	$b + c$
and	$a - c$	$<$	$b - c$
If	6	$<$	9
	$6 + 2$	$<$	$9 + 2$
and	$6 - 4$	$<$	$9 - 4$
If	28	$>$	14
	$28 + 6$	___	$14 + 6$
If	$3x$	$\leq$	12
	$3x - 4$	___	$12 - 4$

$>$
$\leq$

21. We can multiply or divide both sides of an inequality by a *positive* (nonzero) number and the inequality is preserved.

If	a	$<$	b
	$3a$	$<$	$3b$
and	$a/2$	$<$	$b/2$
Since	16	___	4
	$16/3$	___	$4/3$

$>$
$>$

22. If we multiply or divide both sides of an inequality by a *negative* number, the direction of the inequality is *reversed*.

If	a	$<$	b
	$-3a$	$>$	$-3b$
Since	4	$<$	7
	$(-1)(4)$	___	$(-1)(7)$
and	$4/-3$	___	$7/-3$

1.3d Subscripts and Summation

> >

23. Virtually *all* statistical analyses deal with collections of data. In such a collection we might have numeric values for the same characteristic—but from different units. For example, we may have measures of dollar sales (characteristic) from different invoices (unit). We find it convenient to express the observations on a characteristic in compact notational form. In order to accomplish this, we adopt *subscripts*. In the expression x_i, i is the subscript and x_i can be read as the ith observation or measurement of characteristic x. The capital Greek letter sigma (Σ) is used to indicate *summation*. Thus, the expression

$$\sum_{i=1}^{n} x_i$$

is a compact notation for $x_1 + x_2 + x_3 + \cdots + x_n$.

If $x_1 = 2$
$x_2 = 4$
$x_3 = 1$
$x_4 = 8$

$$\sum_{i=1}^{4} x_i = \_\_\_\_\_\_\_\_$$

15

24. There are three very important sums in regard to your study and understanding of statistics. These are:

$$\sum_{i=1}^{n} x_i \qquad \sum_{i=1}^{n} x_i^2 \qquad \text{and} \qquad \left(\sum_{i=1}^{n} x_i\right)^2$$

$$\sum_{i=1}^{n} x_i = x_1 + x_2 + \cdots + x_n$$

$$\sum_{i=1}^{n} x_i^2 = x_1^2 + x_2^2 + \cdots + x_n^2$$

$$\left(\sum_{i=1}^{n} x_i\right)^2 = (x_1 + x_2 + \cdots + x_n)^2$$

Note that

$$\sum_{i=1}^{n} x_i^2 \neq \left(\sum_{i=1}^{n} x_i\right)^2$$

This is an important distinction.

Let $x_1 = 2$
$x_2 = 7$
$x_3 = 4$
$x_4 = 6$
$x_5 = 9$

Find:

(a) $\sum_{i=1}^{5} x_i =$ ________

(b) $\sum_{i=1}^{5} x_i^2 =$ ________

(c) $\left(\sum_{i=1}^{5} x_i\right)^2 =$ ________

(a) $\Sigma x_i = 28$
(b) $\Sigma x_i^2 = 186$
(c) $(\Sigma x_i)^2 = 28^2 = 784$

1.3e Factorial

25. In the approach and solution of certain probability problems we find it necessary to enumerate or at least determine how many ways a given outcome could occur. Thus we are interested in ways of facilitating the counting or enumeration process. One such aid is the concept and notation of *factorials*. 6! is read "six factorial" and is a convenient method for representing a product of descending integers.

$6! = 6 \cdot 5 \cdot 4 \cdot 3 \cdot 2 \cdot 1$

In general terms,

$n! = n(n - 1)(n - 2) \cdots (2)(1)$

$8! =$ ________

---

(8)(7)(6)(5)(4)(3)(2)(1) = 40,320

26. We also need to note that

$1! = 1$

and by definition

$0! = 1$

Finally, notice that

$$8! = (8)(7)(6!)$$

$$\frac{8!}{(6!)(2!)} = \frac{(8)(7)(6!)}{(6!)(2)(1)}$$

As 6! is a factor of both numerator and denominator, we can reduce this expression as follows:

$$\frac{(8)(7)(\cancel{6!})}{(\cancel{6!})(2)(1)} = \frac{(\overset{4}{\cancel{8}})(7)}{(\cancel{2})(1)} = \frac{(4)(7)}{1} = 28$$

$$\frac{10!}{(3!)(7!)} = \underline{\qquad\qquad}$$

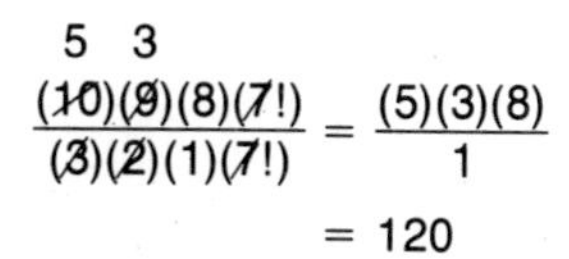

$$\frac{(\overset{5}{\cancel{10}})(\overset{3}{\cancel{9}})(8)(\cancel{7!})}{(\cancel{3})(\cancel{2})(1)(\cancel{7!})} = \frac{(5)(3)(8)}{1} = 120$$

1.3f Rectangular Coordinate System

27. The following diagram represents a rectangular coordinate system.

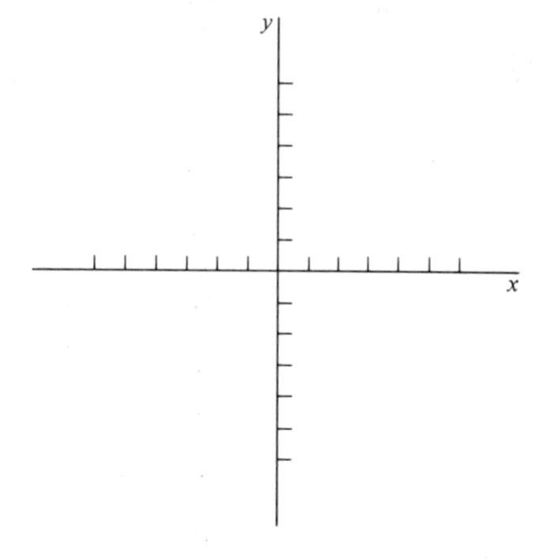

We have labeled the horizontal axis x and the vertical axis y. The intersection of these two axes is called the *origin*. At the origin both x and $y = 0$.

By convention we will indicate a point in the rectangular coordinate system by an ordered pair of coordinates

(x, y). Thus, the origin would correspond to the point with coordinates (0, 0). We can plot any point by locating the x coordinate on the horizontal axis and drawing a vertical line through it. Next we find the y coordinate on the vertical axis and draw a horizontal line through it. The intersection of the two lines will correspond to (x, y) and this is the location of the point in question.

The following shows the location of the point (−3, 5).

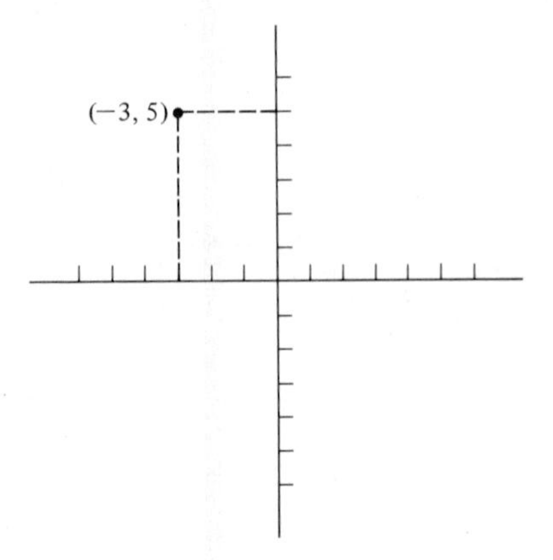

Draw your own rectangular coordinate system and plot the points:

(a) (2, 1)
(b) (5, 7)
(c) (−4, 2)
(d) (3, −3)
(e) (−2, −4)

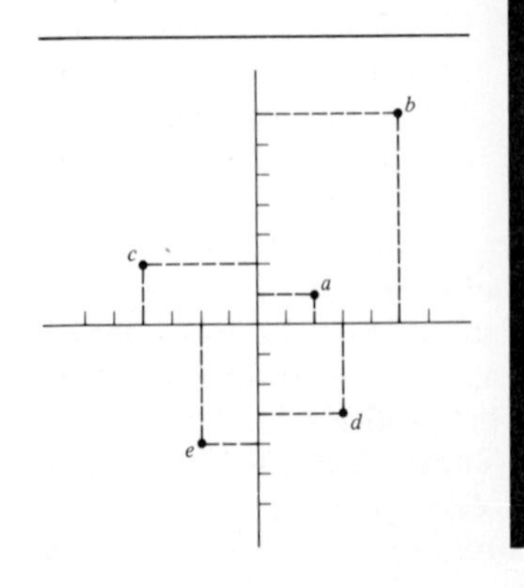

1.3g Squares

28. Algebraic manipulation of expressions containing squares and square roots is quite common in statistics. The square of a number is equal to the number multiplied by itself. That is, $8^2 = (8)(8) = 64$. We can also talk of the squares of symbols or expressions. For example,

$$\sigma^2 = (\sigma)(\sigma)$$

or $$(x - \mu)^2 = (x - \mu)(x - \mu)$$

Find the following, under the assumption that $x = 4$, $\sigma = 2$, and $\mu = 4.5$.

(a) 1.96^2 ________

(b) $(x - \mu)^2$ ________

(c) $(\sqrt{\sigma})^2$ ________

(a) 3.8416
(b) 0.25
(c) $\sigma = 2$

1.3h Square Roots

29. We can find the square root of any positive number rounded to three digits by using Table A-1 in the Appendix. This can be accomplished by using the following rule:

Step 1. Move the decimal point two places at a time to the right or left until we obtain a number greater than or equal to 1 but less than 100. If the resulting number is less than 10, read the table entry from the $\sqrt{n}$ column; otherwise the proper table entry is read from the $\sqrt{10n}$ column.

Step 2. The final step is to locate the decimal. The decimal point in the tabulated entry is shifted half as many places as in step 1, and in the opposite direction.

Example: To find $\sqrt{1380}$, step 1 directs us to shift the decimal point two places to the left, giving us 13.80. As $13.80 \geq 10$, we read the table value in column $\sqrt{10n}$. This gives us 3.7148. Since we shifted the decimal two places to the left in step 1, step 2 indicates that we must now move the decimal in the tabulated value one place to the right. Thus, $\sqrt{1380} = 37.148$.

Similarly, we find the square root of 0.00026 by shifting the decimal four places to the right, which gives us 2.6. As $2.6 < 10$, we read the tabulated value from column $\sqrt{n}$. This gives us the value 1.6125. By shifting the decimal two digits to the left, we obtain $\sqrt{0.00026} = 0.016125$.

Find the square roots for the following values:

(a) $\sqrt{384}$ ________

(b) $\sqrt{2.64}$ ________

(c) $\sqrt{0.726}$ ________

(d) $\sqrt{0.0000039}$ ________

(a) 19.596
(b) 1.6248
(c) 0.85206
(d) 0.0019748

OK—that's it! Virtually *all* the mathematical operations you will need to master this course are presented in the preceding 29 frames. You have an opportunity to begin your study of statistics without a mathematical handicap. Seize this opportunity by reviewing any of the frames you may have found difficult. We also recommend that you work through the problems that follow.

PROBLEMS

1. Identify the integers:

   a. +7, −3, +2.1, +21, −.1, +1/10, +10/1, −27/3, −19/9, +4.76

   b. 12; −12/1; 139,056; −8,742,278; −0.2359144; 121/11

2. Construct a fraction with the indicated property:

   a. A denominator which is twice the numerator

   b. A numerator which is one-third the denominator

   c. Equal to 4

   d. Equal to 3/10

   e. Equal to 0.8

   f. Equal to 2.47

3. Reduce the following fractions:

   a. $\frac{9}{18}$ *d.* $\frac{48}{36}$ *g.* $\frac{11a}{66a}$ *j.* $\frac{6+6}{12+6}$

   b. $\frac{-25}{15}$ *e.* $\frac{24x}{11x}$ *h.* $\frac{13a}{19b}$ *k.* $\frac{11+7}{9}$

   c. $\frac{5}{6}$ *f.* $\frac{12y}{3y}$ *i.* $\frac{ab}{a^2}$ *l.* $\frac{14+15}{3+1}$

4. Perform the indicated operation:

a. $\frac{4}{5} + \frac{3}{5}$ *d.* $\frac{1}{7} - \frac{6}{7}$ *g.* $\frac{x}{\sigma} - \frac{\mu}{\sigma}$ *j.* $\frac{5}{6} \cdot \frac{4}{15}$

b. $\frac{5}{3} + \frac{4}{3}$ *e.* $\frac{7}{2} + \frac{2}{3}$ *h.* $\frac{x - \mu}{\sigma} + \frac{\mu - \bar{x}}{\sigma}$ *k.* $\frac{8}{7} \div \frac{9}{8}$

c. $\frac{8}{9} - \frac{2}{9}$ *f.* $\frac{1}{4} - \frac{5}{6}$ *i.* $\frac{2}{3} \cdot \frac{4}{9}$ *l.* $\frac{12}{5} \div \frac{6}{25}$

5. Remove the parentheses:

a. $4 + (3 - 1)$ *c.* $(x + 3) - (x - 2)$

b. $(14 + 1 - 2) + 11$ *d.* $(y + 5x) + 3 - (2x - 2y)$

6. Solve the following equations for x:

a. $x + 4 = 7$ *f.* $(2x + 3) - (5 + 6x) = 14$

b. $2x = 9$ *g.* $4x - 7 = 3 + 2x$

c. $3x - 5 = 8$ *h.* $\frac{x - 100}{10} = x$

d. $6 - 2x = 10$ *i.* $\frac{x - \mu}{\sigma} = 1.645$, when $\mu = 10$, $\sigma = 1$

e. $-1 + (1/2)x = -3$ *j.* $\frac{x - \mu}{\sigma} = 1.96$, when $\mu = 5.1$, $\sigma = 3$

7. Identify the appropriate sign (>, <):

a. 4 ____ 1.6 *c.* 7 + 5 ____ 3^2

b. 3 ____ −6 *d.* 1 ____ $\frac{8 + b}{8}$, $b > 0$

8. Given the following, which is larger, $\frac{x - \mu}{\sigma}$ or 1.96?

a. $x = 2.0$ $\mu = 1.3$ $\sigma = 1$

b. $x = 2.0$ $\mu = 1.3$ $\sigma = 0.1$

c. $x = 107$ $\mu = 100$ $\sigma = 3$

d. $x = 64$ $\mu = 74$ $\sigma^2 = 25$

9. Given the following, which is smaller, $\frac{x - \mu}{\sigma}$ or 1.645?

a. $x = 0.56$ $\mu = 0.57$ $\sigma = 0.01$

b. $x = 3.6$ $\mu = 3.6$ $\sigma = 0.7$

c. $x = 0.1$ $\mu = 0.09$ $\sigma = 0.004$

d. $x = 18$ $\mu = 25$ $\sigma^2 = 16$

10. Identify the appropriate sign (>, <):

a. If $a > b$, then $1/2a$ ____ $1/2b$

b. If $c < d$, then $-c$ ____ $-d$

c. If $c > 5.6$, then $c - 2$ ____ 3.6

d. If $a < 1$, then $2a$ ____ $1 + a$

e. If $x > 11.3$, then $x - \mu$ ____ $11.3 - \mu$

f. If $x - \mu < 3 - \mu$, then x ____ 3

g. If $x \geq 5$, $\sigma > 0$, then $\dfrac{x - \mu}{\sigma}$ ____ $\dfrac{5 - \mu}{\sigma}$

h. If $x < \mu$, $\sigma > 0$, then $\dfrac{x - \mu}{\sigma}$ ____ 0

i. If $\dfrac{x}{\sigma} < 2$, $\sigma > 0$, then x ____ 2σ

j. If $\dfrac{x}{\sigma} < -1$, $\sigma > 0$, then x ____ $-\sigma$

k. If $\dfrac{x - \mu}{\sigma} > 1.645$, $\sigma > 0$, then x ____ $\mu + 1.645\sigma$

l. If $\dfrac{x - \mu}{\sigma} < -1.96$, $\sigma > 0$, then x ____ $\mu - 1.96\sigma$

m. If $\dfrac{x - \mu}{\sigma} > -1.645$, $\sigma > 0$, then μ ____ $x + 1.645\sigma$

n. If $\dfrac{x - \mu}{\sigma} < 1.96$, $\sigma > 0$, then μ ____ $x - 1.96\sigma$

11. For

i	1	2	3	4	5	6	7	8	9	10
x_i	18	−7	−30	25	8	11	−3	−12	1	9

Evaluate

a. $x_1 + x_4 - x_7$

b. $x_2 + x_3$

c. $x_6 \cdot x_{10}$

d. $\dfrac{(x_5 \cdot x_8)}{x_9}$

e. $\sum_{i=1}^{6} x_i$

f. $\dfrac{1}{10} \sum_{i=1}^{10} x_i$

g. $\dfrac{1}{5} \sum_{i=5}^{8} x_i$

h. $\sum_{i=1}^{10} x_i^2$

i. $\left(\sum_{i=1}^{10} x_i \right)^2$

12. Evaluate:

a. $4!$

b. $\dfrac{7!}{3!4!}$

c. $\dfrac{5!}{4!1!}$

d. $\dfrac{6!}{0!6!}$

e. $\dfrac{9!}{7!2!} \cdot \dfrac{11!}{9!2!}$

f. $\dfrac{8!}{2!6!} + \dfrac{8!}{1!7!} + \dfrac{8!}{0!8!}$

13. Plot the following points on a rectangular coordinate system:

    a. (2, 3.5), (5, 5), (4, 4.5), (6, 4.5), (3, 3)

    b. All points (x, y) which satisfy $6x + 4y = 8$

14. Given the following, plot the points (x_i, y_i) on a rectangular coordinate system:

i	1	2	3	4	5	6
x_i	10	3	−6	−1	2	7
y_i	2	8	4	9	9	3

15. Compute $(x - \mu)^2$ when:

    a. $x = 3$ $\mu = 2.5$

    b. $x = 12.1$ $\mu = 10$

    c. $x = -0.2$ $\mu = 0.15$

    d. $x = 0.2$ $\mu = 0.15$

    e. $x = 10{,}374$ $\mu = 10{,}500$

    f. $x = 10.734$ $\mu = 10.5$

16. Let $\mu = 10$. Compute $\sum_{i=1}^{5} (x_i - \mu)^2$ for:

    a. $x_1 = 3,\ x_2 = 9,\ x_3 = 12,\ x_4 = 15,\ x_5 = 9$

    b. $x_1 = 18,\ x_2 = 11,\ x_3 = 14,\ x_4 = 17,\ x_5 = 16$

17. For each set of numbers, let $\bar{x}$ denote $(1/4)\Sigma x_i$, the average of four numbers. Compute $\Sigma(x_i - \bar{x})^2$ for each set:

    a. 1, 3, 5, 7

    b. 22, 24, 26, 28

    c. 9, 10, 20, 11

18. Find the square root of:

    a. 21.7

    b. 0.056

    c. 489

    d. 489,000

    e. 0.00033

    f. 0.0062

Chapter 2

Descriptive Statistics

2.1 DATA AND THE PARTITION OF VARIABLES

Throughout our study of statistics we will be interested in data: observations of characteristics. If the characteristic of interest can assume different values or outcomes, it is called a *variable*.

Variables can be partitioned into two types, either *quantitative* or *qualitative*. If the possible outcomes of a variable can be expressed numerically, it is known as a *quantitative variable*. Perhaps the easiest way of making this distinction is to equate numeric with quantitative. Examples of quantitative variables could be such characteristics as the price-earnings ratio of a stock, the proportion of defective transistors in a shipment, the length of a log, the net weight of the contents of a box of cereal, or the number of televisions sold by a firm during a given week. In all cases, the value of the quantitative variable can be expressed *numerically*.

Quantitative variables can be further partitioned into *discrete* or *continuous*. If the quantitative variable can assume any value in an interval (a range of values), it is continuous. An example of a continuous quantitative variable would be the weight of the contents of a 3-pound tin of coffee. We do not expect the net weight to be exactly 3 pounds. It would not surprise us to find the net weight for a 3-pound tin to be 3.001 or 2.984 pounds. Actually these weights are rounded to the nearest thousandth of a pound. Theoretically we could refine our measurement forever. Thus such a variable could assume any value within an interval.

On the other hand, some quantitative variables can assume only specific values and are referred to as *discrete*. The number of refrigerators sold per week by a specific Sears store would be a *discrete quantitative variable*. The variable (number of units sold) could assume only nonnegative integer values. It would not be possible to sell 7.25, 23.62, or even 4.5 units.

The other classification or type of variable is *qualitative*. A qualitative variable is one that cannot be expressed numerically but is categorical in nature. That is, instead of having numeric values, the possible outcomes belong to classes or categories. Examples of qualitative variables could be sex, nationality, brand preference, political affiliation, package design, or type of advertising media. Let's take the qualitative variable political affiliation. We do not have a universally accepted, logical way to assign numeric values to the categories Democrat, Independent, and Republican. Thus, a person's political affiliation would be a qualitative variable.

A summary of the partitioning of variables is shown in Figure 2-1. It is important to be able to identify whether a variable of interest is quantitative or qualitative. And, if quantitative, whether it is discrete or continuous. The appropriate statistical analysis depends upon the type of variable being studied. For this reason, such an ability is a prerequisite for identifying applications.

2.2 SITUATION I: WHY?

Suppose you are an executive vice president of a sizable commercial bank. At a recent board meeting a policy decision was made to expand the bank's commitment to personal unsecured loans by $10 million. However, your bank has not been able to attract a sufficient number of "acceptable" loan applications to support such an expansion.

The night after the board meeting you happened to read an article entitled "The Profile of the Personal Loan Applicant" in which the "typical" individual

Figure 2-1 Partitioning of Variables

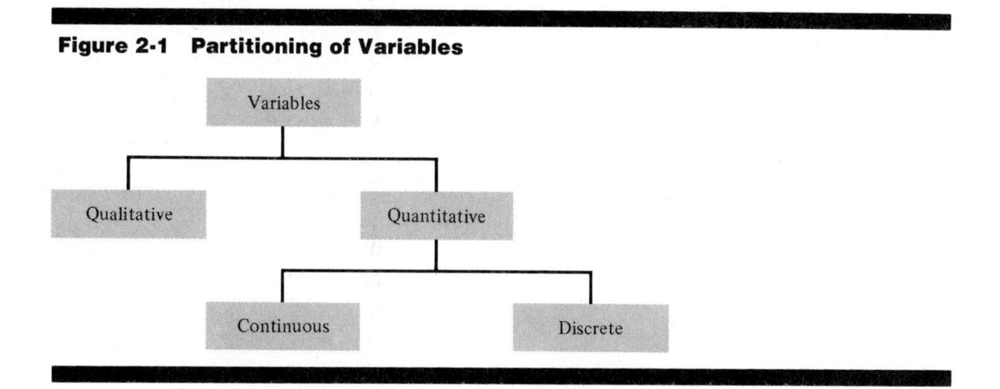

who borrows from consumer finance or small loan companies was described relative to age and marital status. The article indicated that the typical borrower from small loan companies is single and in his or her twenties or thirties. It occurred to you that the individuals seeking loans at your bank may differ in one or both of these characteristics from the applicants for consumer finance or small loan companies. If such a difference exists and can be detected, your bank may be able to carry on an effective advertising campaign to attract the individuals seeking loans elsewhere.

Realizing the need for an accurate comparable description (profile) of loan applicants at your bank, you ask a subordinate to obtain data on your loan applicants during the past three months. Specifically, you ask for data on the characteristics age and marital status.

One week has gone by and your subordinate has just come through your office door and presented you with the requested data: Table 2-1.

Well—what do you think? Do you have a feel for the profile of your bank's loan applicants? Can you detect any differences between the profile cited in the magazine article and your bank's applicants? Probably not. The problem is that it is difficult to digest or assimilate a large amount of data. What is needed is some means of summarizing and presenting these data so that the information present can be extracted and understood.

THE OBJECTIVE OF DESCRIPTIVE STATISTICS IS TO *SUMMARIZE* AND TO *PRESENT* DATA IN ORDER TO FACILITATE THEIR USE.

2.3 DESCRIPTIVE STATISTICS: HOW?

There are a variety of procedures and techniques which come under the heading of descriptive statistics. Such procedures and techniques can be classified as *graphic* with the emphasis on *presentation*, or as *numeric* with the emphasis on *summarization*. This is *not* to say that graphic techniques only present and numeric techniques only summarize. The processes of presenting and summarizing are interrelated and any technique which comes under the heading of descriptive statistics can both present *and* summarize data. However, the graphic techniques tend to emphasize presentation, whereas the numeric techniques emphasize summarization.

In order to more easily introduce the basic graphic and numeric techniques of descriptive statistics, let us deal with smaller sets of data than those presented in Situation I. However, keep in mind that *the larger the set of data, the greater the need for descriptive statistical techniques.* After illustrating the techniques with smaller data sets, we will return to the problem in Situation I.

2.3a Graphic Techniques

Let us suppose that Wilson Electric is a retailer of major electrical appliances. Jim Wilson, the owner and manager, has decided to "try" a number of spot

Table 2-1a Age of Loan Applicants

41	56	45	52	52	33	45	49	39	49
32	37	51	38	47	40	49	48	50	73
44	51	54	35	52	55	39	52	45	51
49	50	60	40	49	37	34	64	37	28
23	24	25	18	27	20	24	24	28	40
40	35	23	56	21	25	32	42	48	23
34	34	25	25	62	50	35	53	25	20
62	24	21	21	24	51	43	41	26	21
21	22	81	46	41	33	25	54	47	33
44	50	32	24	34	24	38	23	38	58
38	28	39	42	37	29	36	27	33	29
39	41	43	39	30	44	44	48	34	48
32	36	36	44	38	40	24	24	37	54
46	60	42	54	37	52	38	33	40	38
47	51	38	32	44	39	43	51	38	56
42	27	42	38	44	28	48	37	35	51
30	31	31	43	32	32	45	44	48	45
38	38	30	39	46	31	49	49	31	44
41	63	40	46	39	21	55	40	48	49
34	31	20	25	34	27	20	32	32	23
34	37	22	30	23	39	22	25	24	57
25	36	32	30	44	48	40	41	30	24
25	35	35	23	24	28	38	35	32	25
47	53	34	24	25	19	41	27	25	23
48	35	42	29	38	27	23	37	35	37
47	49	30	38	35	27	33	20	37	48
43	25	39	64	25	28	30	23	42	28
35	27	44	25	41	37	63	22	39	39
28	24	41	28	56	39	33	26	22	60
30	25	28	38	25	24	50	25	33	24
33	40	50	37	32	26	22	26	37	26
34	43	27	38	24	33	23	22	60	51
33	44	52	21	34	24	22	43	40	33
30	37	50	41	30	27	72	31	47	27
35	42	27	23	61	23	24	21	34	21
18	24	31	21	28	27	32	26	63	21
38	28	27	28	33	26	27	29	49	33
37	54	30	30	61	21	24	52	35	41
42	45	50	43	63	16	54	47	42	38
30	44	49	28	31	38	27	36	41	30
56	47	33	33	31	38	24	42	35	31
42	45	36	28	49	47	33	39	48	49
32	51	23	46	36	40	33	76	25	29
27	47	29	22	60	36	62	52	60	37
44	50	32	51	55	42	31	26	68	47
38	30	55	51	63	31	33	53	23	51
44	37	37	41	40	54	41	39	41	34
28	48	43	27	30	28	65	48	37	49
31	42	26	23	42	25	34	45	36	32
45	36	41	49	40	36	52	34	34	48
45	69	19	24	22	59	50	51	26	38
24	42	42	30	42	60	35	32	23	24
29	22	33	47	48	33	35	28	30	39
21	61	47	62	34	31	30	38	28	34
19	74	54	47	24	35	44	45	21	40
32	58	38	34	28	35	36	23	23	27
23	34	20	34	36	32	39	27	47	23
21	31	51	31	43	25	37	23	26	26
42	42	43	24	30	48	18	30	21	21
41	22	31	37	24	31	45	27	45	25
40	50	27	34	24	25	32	33	44	33
26	31	40	34	26	47	42	41	21	42
23	36	33	30	36	31	36	40	23	24
34	38	37	45	38	44	36	27	26	21
23	28	24	58	51	24	22	24	21	52
30	44	29	34	32	48	36	25	46	48
24	29	36	35	33	38	41	38	34	35
36	29	25	37	58	27	63	24	27	36
26	45	22	25	49	28	50	52	58	30
42	33	44	28	30	36	33	39	52	56
39	40	25	38	39	39	24	30	24	24
21	32	28	23	42	54	43	24	29	48
23	19	23	46	43	45	29	39	20	34
29	30	29	40	32	59	35	28	26	42
31	43	49	42	55	38	28	42	38	24
49	47	28	23	32	32	39	58	26	36
55	41	43	53	55	52	37	33	29	50
41	46	32	57	60	43	26	34	42	30
51	32	40	32	41	57	66	30	44	58
58	23	37	31	38	36	39	52	32	32
25	26	26	43	43	46	65	47	47	24
24	27	23	59	27	37	35	49	29	70
43	40	21	26	30	21	48	37	29	22
41	36	35	33	28	42	40	37	38	40
47	37	28	52	24	30	61	28	38	24
53	25	27	52	30	30	39	62	49	52
34	29	33	34	32	28	29	28	32	30
35	40	30	42	51	35	46	29	45	53
31	34	24	34	34	29	35	40	30	28
37	42	26	26	52	39	38	27	28	47
36	35	25	23	25	41	33	30	47	39
32	25	28	57	20	21	32	49	21	42
36	63	32	24	54	51	51	42	41	40
43	26	31	41	39	38	32	30	28	26
24	24	41	36	23	27	29	35	29	36
47	21	26	39	41	25	23	27	38	40
22	43	31	50	43	36	22	39	55	59
31	21	42	29	28	32	53	22	51	53
46	44	43	29	25	30	35	29	41	28
45	39	42	30	24	28	46	46	26	31

Table 2-1b Marital Status of Loan Applicants

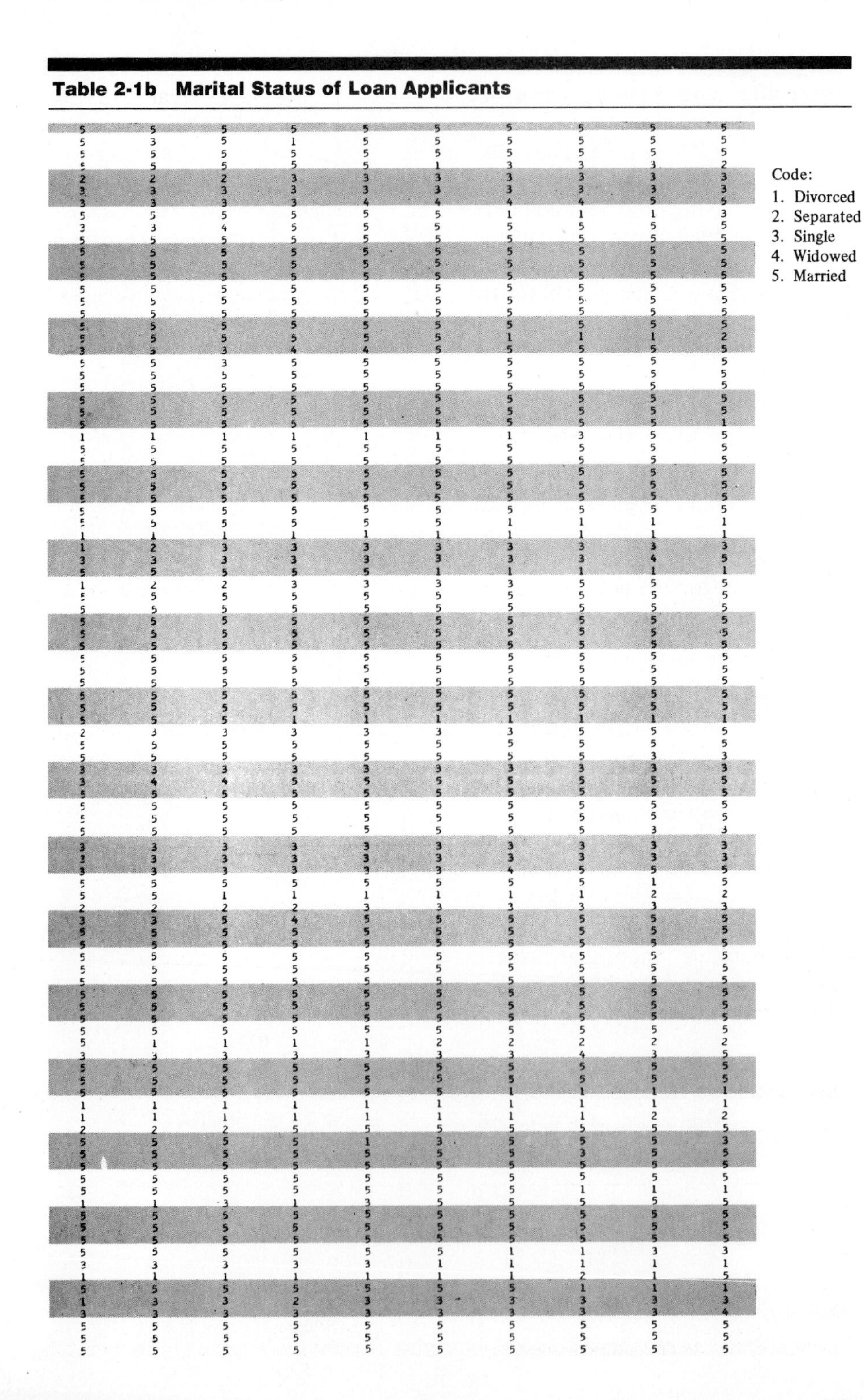

5	5	5	5	5	5	5	5	5	5
5	3	5	1	5	5	5	5	5	5
5	5	5	5	5	5	5	5	5	5
5	5	5	5	5	1	3	3	3	2
2	2	2	3	3	3	3	3	3	3
3	3	3	3	3	3	3	3	3	3
3	3	3	3	4	4	4	4	5	5
5	5	5	5	5	5	1	1	1	3
3	3	4	5	5	5	5	5	5	5
5	5	5	5	5	5	5	5	5	5
5	5	5	5	5	5	5	5	5	5
5	5	5	5	5	5	5	5	5	5
5	5	5	5	5	5	5	5	5	5
5	5	5	5	5	5	5	5	5	5
5	5	5	5	5	5	5	5	5	5
5	5	5	5	5	5	5	5	5	5
5	5	5	5	5	5	5	5	5	5
5	5	5	5	5	5	1	1	1	2
3	3	3	4	4	5	5	5	5	5
5	5	3	5	5	5	5	5	5	5
5	5	5	5	5	5	5	5	5	5
5	5	5	5	5	5	5	5	5	5
5	5	5	5	5	5	5	5	5	5
5	5	5	5	5	5	5	5	5	5
5	5	5	5	5	5	5	5	5	1
1	1	1	1	1	1	1	3	5	5
5	5	5	5	5	5	5	5	5	5
5	5	5	5	5	5	5	5	5	5
5	5	5	5	5	5	5	5	5	5
5	5	5	5	5	5	5	5	5	5
5	5	5	5	5	5	5	5	5	5
5	5	5	5	5	5	5	5	5	5
5	5	5	5	5	5	1	1	1	1
1	1	1	1	1	1	1	1	1	1
1	2	3	3	3	3	3	3	3	3
3	3	3	3	3	3	3	3	4	5
5	5	5	5	5	1	1	1	1	1
1	2	2	3	3	3	3	5	5	5
5	5	5	5	5	5	5	5	5	5
5	5	5	5	5	5	5	5	5	5
5	5	5	5	5	5	5	5	5	5
5	5	5	5	5	5	5	5	5	5
5	5	5	5	5	5	5	5	5	5
5	5	5	5	5	5	5	5	5	5
5	5	5	5	5	5	5	5	5	5
5	5	5	5	5	5	5	5	5	5
5	5	5	5	5	5	5	5	5	5
5	5	5	5	5	5	5	5	5	5
5	5	5	1	1	1	1	1	1	1
2	3	3	3	3	3	3	5	5	5
5	5	5	5	5	5	5	5	5	5
5	5	5	5	5	5	5	5	3	3
3	3	3	3	3	3	3	3	3	3
3	4	4	5	5	5	5	5	5	5
5	5	5	5	5	5	5	5	5	5
5	5	5	5	5	5	5	5	5	5
5	5	5	5	5	5	5	5	5	5
5	5	5	5	5	5	5	5	3	3
3	3	3	3	3	3	3	3	3	3
3	3	3	3	3	3	3	3	3	3
3	3	3	3	3	3	4	5	5	5
5	5	5	5	5	5	5	5	1	5
5	5	1	1	1	1	1	1	2	2
2	2	2	2	3	3	3	3	3	3
3	3	5	4	5	5	5	5	5	5
5	5	5	5	5	5	5	5	5	5
5	5	5	5	5	5	5	5	5	5
5	5	5	5	5	5	5	5	5	5
5	5	5	5	5	5	5	5	5	5
5	5	5	5	5	5	5	5	5	5
5	5	5	5	5	5	5	5	5	5
5	5	5	5	5	5	5	5	5	5
5	5	5	5	5	5	5	5	5	5
5	5	5	5	5	5	5	5	5	5
5	1	1	1	1	2	2	2	2	2
3	3	3	3	3	3	3	4	3	5
5	5	5	5	5	5	5	5	5	5
5	5	5	5	5	5	5	5	5	5
5	5	5	5	5	5	1	1	1	1
1	1	1	1	1	1	1	1	1	1
1	1	1	1	1	1	1	1	2	2
2	2	2	5	5	5	5	5	5	5
5	5	5	5	1	3	5	5	5	3
5	5	5	5	5	5	5	3	5	5
5	5	5	5	5	5	5	5	5	5
5	5	5	5	5	5	5	5	5	5
5	5	5	5	5	5	5	1	1	1
1	1	3	1	3	5	5	5	5	5
5	5	5	5	5	5	5	5	5	5
5	5	5	5	5	5	5	5	5	5
5	5	5	5	5	5	5	5	5	5
5	5	5	5	5	5	1	1	3	3
3	3	3	3	3	1	1	1	1	1
1	1	1	1	1	1	1	2	1	5
5	5	5	5	5	5	5	1	1	1
1	3	3	2	3	3	3	3	3	3
3	3	3	3	3	3	3	3	3	4
5	5	5	5	5	5	5	5	5	5
5	5	5	5	5	5	5	5	5	5
5	5	5	5	5	5	5	5	5	5

Code:
1. Divorced
2. Separated
3. Single
4. Widowed
5. Married

advertisements on the local television stations. Wilson's hope is that the cost of the spot ads will be more than offset by increased unit sales. A local advertising agency has been retained to develop the advertisements and the campaign is ready for initiation. The spot ads will be shown twice daily for a one-week period. The total cost to Wilson Electric is set at $2500.

Not having used spot ads before, Wilson is uncertain about the effectiveness of such an advertising campaign. If the campaign turns out to be successful, he may choose to repeat the ads at a later date. Therefore, Wilson is quite interested in obtaining a feel for the effect on unit sales of the advertisements. Specifically he would like to know whether unit sales increased as a result of the ads. If indeed sales did increase, Wilson would like answers to these questions:

1. How large an increase was realized?
2. How soon after the first ad was shown did sales increase?
3. How long did the effect of the advertisements last?

In order to determine whether sales actually increased after running the ads, Wilson realized the need for a description of unit sales prior to placing the ads. Thus, to obtain a "standard" with which to compare, Wilson Electric has recorded its unit sales by day for the past month. These data are given in Table 2-2.

Our first step in describing (presenting and/or summarizing) these data would be to tabulate the observations. A quick look at the data reveals that our observations range from a low of 11 to a high of 17 units per day. With this we can easily tabulate our data as shown in Table 2.3.

The frequency, given in column 3, is simply the number of times a particular number of sales appears in the data in Table 2-2. Columns 1 and 3 constitute what is called a *frequency distribution*. A frequency distribution gives both the values or classes for the variable and their corresponding frequencies.

In some situations the interest is in the number of observations that are "less than" or "more than" specific values. In such cases, a *cumulative frequency distribution* would be called for. Columns 1 and 4 constitute a cumulative (or less than) frequency distribution. You will note that column 4 shows the

Table 2-2 Unit Sales by Day for Wilson Electric, September 197X

Sun.	Mon.	Tues.	Wed.	Thurs.	Fri.	Sat.
				14	12	
	13	11	12	16	14	
	14	15	17	13	11	
	13	14	15	13	14	
	14	13	14	13	12	

Table 2-3 Tabulation of Frequencies and Cumulative Frequencies for the Number of Units Sold per Day for Wilson Electric, September 197X

(1) Unit Sales (x)	(2) Tally	(3) Frequency (f)	(4) Cumulative Frequency
11	||	2	2
12	|||	3	5
13	~~||||~~ |	6	11
14	~~||||~~ ||	7	18
15	||	2	20
16	|	1	21
17	|	1	22
		$N = 22$	

number of observations (frequencies) that are less than or equal to the corresponding value of x.

The tabulation of data is in effect a sorting or grouping of observations into, it is hoped, a more meaningful form. Frequency distributions tend to exhibit how the data are distributed.

Once we have constructed a frequency distribution, we can extend our description *graphically* by means of a *histogram*. Histograms are constructed by placing rectangles or bars over each value or interval for the quantitative variable of interest (in our example, daily unit sales). The height of a specific rectangle is measured on the vertical axis and represents frequency or cumulative frequency, depending on the type of histogram. Figure 2-2 shows the histograms for both the frequency and cumulative frequency distributions relative to daily unit sales for Wilson Electric.

From Figure 2-2, it is clear that histograms can be effectively used to concisely present and summarize quantities of data. Jim Wilson can get an appreciation for the data with a quick glance at Figure 2-2. This was not possible with Table 2-2. You will notice that the variable, number of units sold per day, is a discrete quantitative variable. If the number of observed values were much larger, say ranging from 0 to 50, the histogram would appear rather cluttered and thus less effective. A similar problem would result if the quantitative variable of interest was continuous. In these cases, it becomes necessary to establish classes (intervals) for the variable if we are to either tabulate or graphically describe the data by means of a histogram.

The following are general guidelines for the categorization (dividing into intervals) of a continuous variable.

1. *Ordinarily* we choose to establish between 5 and 15 classes. This choice is a balance between the amount of summarization and lost information. As we increase

the degree of summarization (by forming fewer classes), we lose detail. However, the more classes we establish, the more difficult it is to extract useful information. We are typically willing to sacrifice some detail in order to facilitate use.

2. *Ordinarily* we form equal-width classes. Intervals that are unequal in width tend to distort comparisons. In comparing classes shown in histograms, we focus on area. If the classes are of equal width, we need only examine the heights of rectangles to compare classes. However, if they are formed with unequal widths, the variation in widths must also be taken into account. As our objective is to summarize and present data to *facilitate* their use, unequal intervals tend to be counterproductive.
3. We *always* form nonoverlapping classes. In order to eliminate any possible ambiguity about which class an observation belongs to, we form nonoverlapping classes.

Figure 2-2 Histograms for (*a*) the Frequency Distribution and (*b*) the Cumulative (or Less Than) Frequency Distribution

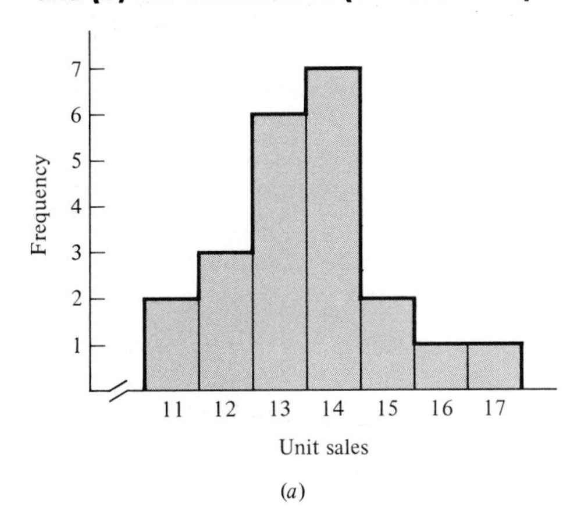

(*a*)

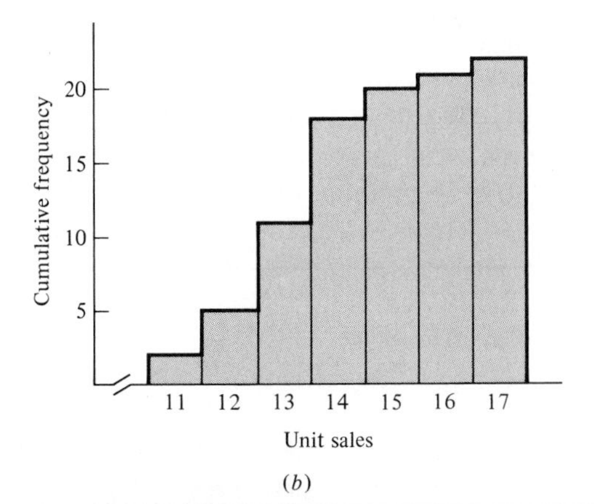

(*b*)

Table 2-4 Inside Diameters of Bearings Produced on Machine A, in Inches, 8/7/7X

5.02	5.06	4.99
5.00	5.01	4.97
4.96	5.00	5.01
4.96	5.02	5.04
5.01	4.98	5.02
5.04	5.04	4.98
4.95	4.95	5.01
4.99	4.97	5.03
4.98	4.99	4.98
5.01	5.00	5.00
5.00	5.00	5.03
4.96	4.98	5.01
5.01	5.03	5.02
4.94	5.00	5.01
4.97	5.02	4.99

Actually, the categorization of a continuous quantitative variable, including size and number of classes, should depend upon the objective of gathering the data. We do not gather data without purpose. Clearly, the purpose may dictate the appropriate number and size of intervals.

As an example of the categorization of a quantitative variable, consider the following. Rutherford Equipment has just installed a new machine for manufacturing bearings. In order to establish a standard of performance and determine the precision of this new machine, *all* the bearings produced on a given day are measured. The characteristic of interest is inside diameter, which is a continuous quantitative variable. The measurements on inside diameters are given in Table 2-4.

We see that the quantitative values range from 4.94 to 5.06 inches and that all observations are measured to the nearest hundredth of an inch. One possible categorization of our continuous quantitative variable might be the following:

4.94″–4.95″
4.96″–4.97″
4.98″–4.99″
5.00″–5.01″
5.02″–5.03″
5.04″–5.05″
5.06″–5.07″

You will note that the class intervals are of equal width and nonoverlapping and

Table 2-5 Frequency Distribution of Inside Diameters of Bearings

Inside Diameter (Inches)	Class Marks	Frequency
4.94–4.95	4.945	3
4.96–4.97	4.965	6
4.98–4.99	4.985	9
5.00–5.01	5.005	15
5.02–5.03	5.025	8
5.04–5.05	5.045	3
5.06–5.07	5.065	1
		45

that the limits for the classes are measured to the same degree of accuracy as our data (nearest hundredth of an inch). The value of 4.94 inches is referred to as the lower class limit for the first class. Conversely, 4.95 inches is the upper class limit for this class. Midway between the lower and upper class limits is known as the *class mark*. The class mark is used to numerically represent its class and can easily be obtained by adding the lower and upper class limits and dividing this sum by 2. Thus, the class mark for our first class would be (4.94 + 4.95)/2 = 4.945.

As our intervals are nonoverlapping, we have no difficulty determining into which class an observation belongs. The frequency distribution for these data is given in Table 2-5 and is graphically presented in Figure 2-3.

Figure 2-3

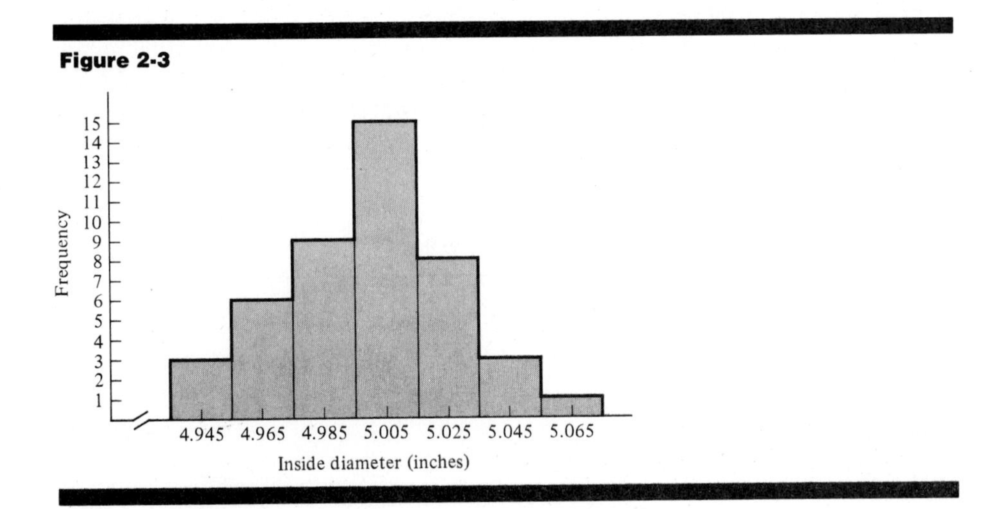

These tabulations and graphic presentations have all been for quantitative variables. The only difficulty we find in the tabulation and graphic presentation of qualitative variables is that there may be no apparent logical ordering or arrangement of classes or categories. For this reason we are reluctant to place the classes of a qualitative variable on an axis, as we have come to interpret relative location on an axis with magnitude.

Ordinarily when we are interested in a qualitative variable, it is the percentage breakdown into the various categories that is most important. For example, Bi-Mart is a chain of discount stores in Oregon. Suppose there are nine separate stores and all nine are virtually identical. That is, the stores are the same size, have the same layout, and carry identical product lines, as buying for all stores is centralized. Each store is divided into the following six departments:

Department

1. Appliances
2. Drugs and Cosmetics
3. Hardware
4. Household Supplies
5. Sporting Goods
6. Toys

Suppose that every January, company decisions are made relative to allocation of both floor space and inventory investment for the six departments. As competitors come and go, the profitability of the various departments changes. If a department becomes more profitable than others, Bi-Mart would like to increase that department's allocated floor space and inventory. Let us also assume that Bi-Mart's pricing is "cost plus X percent." Thus, total revenue for a department is directly related to its profit.

In order to make the decision on allocation of floor space, the directors of Bi-Mart examine many factors—one of which is the distribution of total revenue by department. In order to obtain the necessary data, the directors ask each store manager to provide a departmental breakdown of total store revenue.

Let's see how we, as one of the managers, might present our data. Suppose that after sorting receipts by departments we arrive at the following:

Department	Total Receipts for 197X
Appliances	$143,680
Drugs and Cosmetics	$138,422
Hardware	$234,714
Household Supplies	$159,307
Sporting Goods	$ 34,698
Toys	$ 82,015
	$792,836

When we examine a given receipt, we are interested in both dollar amount and departmental designation. Our objective is to summarize and present the distribution of total receipts by *department*. The departmental designation is then our variable and it is *qualitative*.

The percentage breakdown of a qualitative variable can effectively be presented by means of a *pie chart*. A pie chart is basically a circle partitioned into sections. As there are 360 degrees in a circle, 3.6 degrees correspond to 1 percent of the central angle. Thus, if one category of a qualitative variable contained 15 percent of the total, it could be represented by a section of a circle whose central angle was $(3.6)(15) = 54$ degrees.

By dividing each department's total revenue by $792,836 (the total of all departmental receipts), we can convert our departmental revenue figures into percentages. The resulting percentage distribution is:

Department	Percent of Total Receipts, 197X
Appliances	18.1
Drugs and Cosmetics	17.5
Hardware	29.6
Household Supplies	20.1
Sporting Goods	4.4
Toys	10.3
	100.0

Multiplying each department percentage by 3.6 degrees, we arrive at the desired partitioning of the central angle. For example, the section of the circle representing the appliance department would have a central angle of 65 degrees—which is (18.1)(3.6) rounded to the nearest degree. The complete pie chart depicting the percentage breakdown of total store revenue to the six departments is shown in Figure 2-4.

Figure 2-4 Pie Chart for Proportional Distribution of Total Receipts by Department

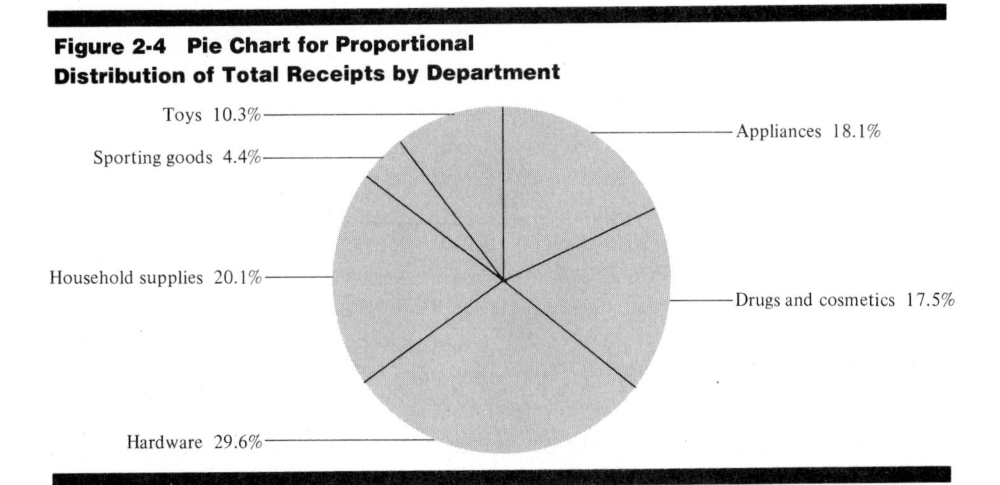

Histograms and pie charts are by no means the only forms of graphical presentation of data. You have no doubt been exposed to a number of different types and variations of such presentations through television advertisements, magazines, newspapers, and other textbooks. It appears as though the only limits to ways of graphically presenting data are the imagination and artistic ability of those preparing the presentations.

There is no doubt that graphic statistical techniques can be effectively used to both summarize and present data. However, there are situations where our interest extends beyond mere presentation and summarization to *analysis*. In such cases we will be interested in more precise analytic descriptors than those typically afforded by graphic techniques. Such analytic descriptors are introduced under the heading of numeric techniques.

2.3b Numeric Techniques

There are a variety of ways we could numerically describe a set of data. However, there are two categories of numerical descriptors which are of particular interest in applied statistics. The categories are:

1. Measures of Location
2. Measures of Dispersion

Measures of Location

We will examine three measures of location. Each one to some extent describes the "typical" observation or measurement.

Mean The first such value is the *arithmetic mean*, hereafter referred to as the *mean*. The mean of a set of measured observations is defined as the sum of the measurements divided by the number of measurements. Quite simply, the mean is what most individuals refer to as the "average."

Symbolically, if we have N measurements, say $x_1, x_2, \ldots, x_N$, then

$$\mu = \sum_{i=1}^{N} x_i/N \qquad (2\text{-}1)$$

where μ (mu) is the symbol we adopt to denote the mean. For the Wilson Electric data on units sold, Table 2-2, we find the sum of the 22 observations to be 297.

$$\sum_{i=1}^{22} x_i = 14 + 12 + 13 + \cdots + 12 = 297$$

Thus, $\mu = 297/22 = 13.5$.

Suppose that instead of having access to the actual measurements, we were given a frequency distribution like that of Table 2-5. As the measurements have been grouped, we could not merely apply Equation (2-1) to determine the mean. Without recourse to the actual measurements, the mean can only be approximated by Equation (2-2).

$$\mu \approx \sum_{i=1}^{k} m_i f_i / N \tag{2-2}$$

where k = the number of classes
m_i = the class mark for the ith class
f_i = the frequency for the ith class
N = the total number of observations

The reason Equation (2-2) gives only an approximation is that we represent the actual numeric values of the observations within a given class with the class mark. As it would be rare for $m_i f_i$ to equal the sum of the actual observation within the ith class, $\sum_{i=1}^{k} m_i f_i$ would rarely equal the true sum of the N observations. Thus, Equation (2-2) yields an approximation. Although we might prefer to use the actual values, there are situations where the data have been collected by an outside source, such as a governmental agency, and we have access only to a published frequency distribution.

Applying Equation (2-2) to the frequency distribution shown in Table 2-5, we obtain the following:

Inside Diameter (Inches)	Class Mark m	Frequency f	mf
4.94–4.95	4.945	3	14.835
4.96–4.97	4.965	6	29.790
4.98–4.99	4.985	9	44.865
5.00–5.01	5.005	15	75.075
5.02–5.03	5.025	8	40.200
5.04–5.05	5.045	3	15.135
5.06–5.07	5.065	1	5.065
		45	224.965

$$\mu \approx \frac{224.965}{45} = 4.9992$$

If we sum the actual values, given in Table 2-4, we obtain 224.940 rather than the approximation 224.965. Thus, $\mu = 224.940/45 = 4.9986$.

Median The second measure of location which we will look at is the *median*. The median is defined as the value of the middle observation in a series which has been arranged in either ascending or descending order of magnitude. If the series contains an odd number of observations, the median is the middle observation. If the number of observations is even, the median is the mean of the two middle values.

Again referring to the Wilson Electric data, Table 2-2, if we arrange the values in ascending order, we obtain the following:

11
11
12
12
12
13
13
13
13
13
13

$\longleftarrow$ MEDIAN $= \dfrac{13 + 14}{2} = 13.5$

14
14
14
14
14
14
14
15
15
16
17

With 22 values, the median is the value halfway between (the mean of) the eleventh and twelfth observations. Our median is therefore 13.5.

If one of the observations with value 13 had actually been 14, then both the eleventh and twelfth observations would have been 14s. In this case the mean of the two middle values would obviously be 14. Thus, our median would then be 14.0.

The median is also referred to as the *50 percentile* and the *second quartile*. The X percentile is the value of the variable such that X percent of all the observations are less than the specific value. As the median is the middle value, 50 percent of the values are less than the median; thus the median is the 50 percentile. For the Wilson Electric data, we can see that the value 16 is the 20/22 = 90.90909 percentile. The 25 percentile is called the *first quartile* and the 75 percentile is the *third quartile*. Percentiles and quartiles are common additional descriptive measures of location.

Mode The final measure of location is the *mode*. The mode is defined as the most frequently occurring value. Table 2-3 shows that the value 14 has the greatest frequency and is therefore the mode for this set of data.

Realize we have introduced three basic measures of location: *mean*, *me-*

dian, and *mode*. Each of these values has been determined for the same data set, Table 2-2. Let's look at our three values.

Mean = 13.5

Median = 13.5

Mode = 14

Although these measures of location tend to indicate the central or typical value of a variable, they are quite distinct. The mean takes into account all observations and is what most people refer to as the "average." The median is affected only by the middle value or values *after* the data have been arranged in order of magnitude. The mode is merely the most frequent value. In the special case where the frequency distribution is unimodal (the property of having only one mode) and symmetric, all three measures of location will be equal. Otherwise, they will ordinarily not be the same—although they could be.

Which one of the three is the "best" measure of location? We really cannot answer this question, as the appropriateness depends upon the purpose. We do know that the mean is affected by extremes. For example, suppose we change the value 17 in Table 2-2 to 170—clearly an extreme value relative to the other 21 observations. The sum of these 22 observations would then be 450, thus $\mu = 450/22 = 20.45$. Verify that the median and mode are unaffected by this change. As 20.45 is greater than all observations except one, it would be a poor summarization or representation of these data. Our conclusion is that if extreme values occur, the median may more accurately represent the quantitative observations. However, you will see that the mean is by far the more important measure when we get to statistical inference.

Measures of Dispersion

When we describe a set of measurements by using numeric descriptors, we are often interested in the degree of *variation* or *dispersion* among observations as well as measures of location.

Range No doubt the simplest measure of the variation is the *range*. The range is defined as the difference between the largest and smallest measurements. For our Wilson Electric data, Table 2-2, we find:

Range $= 17 - 11 = 6$

Although the range is easy to compute and to some extent does measure variation, it has two important drawbacks. First, only the extremes (largest and smallest values) are considered. Thus, the range contains no information about the variability of all the intervening values. We can foresee two different sets of measurements producing the same range but having very distinct degrees of variation. Second, although the range is of value, there is one other measure of dispersion that considers all values—not just the extremes—and is also an important value for further analysis.

Variance This other important measure of dispersion is called *variance*. Variance is defined as the average or the mean of the squared deviations of the measurements about their mean.

Symbolically, if the N measurements, $x_1, x_2, \ldots, x_N$, have mean μ, then

$$\sigma^2 = \frac{\sum_{i=1}^{N} (x_i - \mu)^2}{N} \tag{2-3}$$

where σ^2 (lowercase sigma) is the symbol we adopt to denote the variance. The following shows the computation of variance for the Wilson Electric data, Table 2-2. Recall the mean for these data was $\mu = 13.5$.

x_i	$x_i - \mu$	$(x_i - \mu)^2$
11	−2.5	6.25
11	−2.5	6.25
12	−1.5	2.25
12	−1.5	2.25
12	−1.5	2.25
13	−0.5	0.25
13	−0.5	0.25
13	−0.5	0.25
13	−0.5	0.25
13	−0.5	0.25
13	−0.5	0.25
14	+0.5	0.25
14	+0.5	0.25
14	+0.5	0.25
14	+0.5	0.25
14	+0.5	0.25
14	+0.5	0.25
14	+0.5	0.25
15	+1.5	2.25
15	+1.5	2.25
16	+2.5	6.25
17	+3.5	12.25
	0.0	45.50

$\sigma^2 = \Sigma(x_i - \mu)^2/N = 45.50/22 = 2.068$ population variance

Note the necessary fact that the sum of the deviations of the observations about their mean (second column) is zero. This will always be true.

There is one problem in using variance as a numeric descriptor of dispersion. Variance is in terms of square units. Say a characteristic of interest is measured in dollars. The variance of our measurements would be in *square*

For sample variance

$$s^2 = \frac{\Sigma (x_i - \bar{x})^2}{n-1}$$

dollars. Alternatively, we define the positive square root of the variance as the *standard deviation*:

$$\sqrt{\sigma^2} = \sigma = \text{standard deviation}$$

The standard deviation is in the same units as the original measurements.

Although variance and standard deviation may not have a great deal of intuitive appeal, they *do* measure variation or dispersion among measurements, and they do consider all the observations. If all the observations were equal, σ^2 would be zero. Conversely, the greater the degree of dispersion, the larger σ^2 will be. Suffice it to say that as we progress through this text you will see the importance of variance and standard deviation as measures of dispersion.

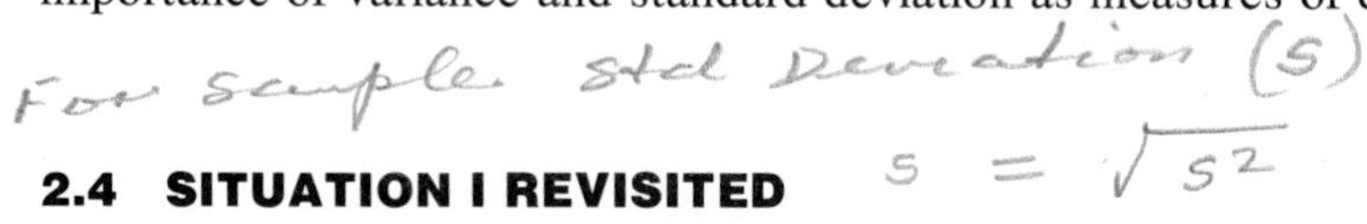

2.4 SITUATION I REVISITED

Now that we have seen *how* data can be summarized and presented through both graphic and numeric techniques, let us return to our original problem presented in Situation I. Recall that as an executive vice president of a commercial bank, you were trying to determine whether there was a difference between your loan applicants and the individuals who seek loans from consumer finance or small loan companies. The "typical" borrower was reported to be single and in his or her twenties or thirties.

Figures 2-5 and 2-6 are graphic presentations of the data in Tables 2-1*a* and 2-1*b*. The histogram for age of loan applicant (Figure 2-5) clearly indicates that your bank has appealed to the same age group as reported in the article. That is, a substantial number of your applicants appear to be in their twenties or thirties. However, the pie chart for marital status (Figure 2-6) indicates that your bank's appeal has been predominately to married individuals. Note that slightly

Figure 2-5 Age of Loan Applicants

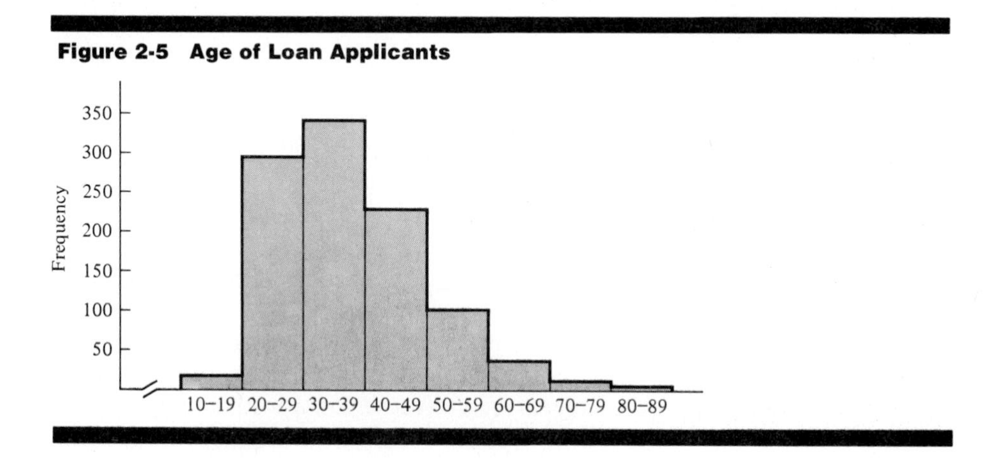

Figure 2-6 Marital Status of Loan Applicants

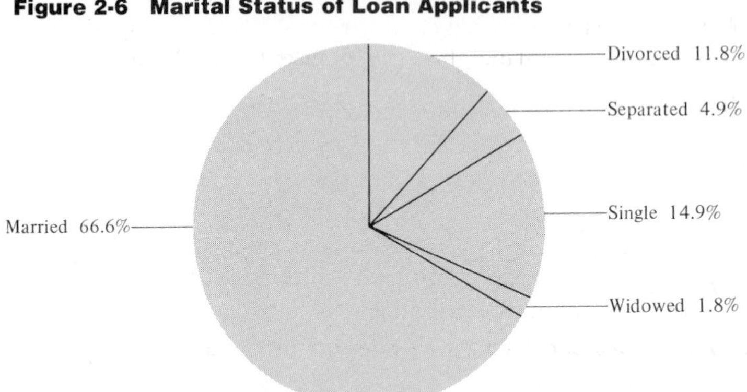

less than 15 percent of your applicants were single. SUCCESS—this is just what you were looking for! You may be able to increase your bank's demand for personal unsecured loans through an advertising campaign. The advertisement would be designed to appeal to single individuals, in an attempt to draw new applicants away from competitors.

2.5 IDENTIFICATION OF APPLICATIONS: WHEN?

What are the characteristics of a situation in which descriptive statistical techniques would be of value? Look again at the objective of descriptive statistics—TO SUMMARIZE AND TO PRESENT DATA IN ORDER TO FACILITATE THEIR USE. The key words are *summarize* and *present*.

Summarization

The application of descriptive statistical techniques could and would be of value in any situation where one is the consumer of *mass* data. Typically, the larger the data set is, the greater the need for summarization. Certainly, the advent of the computer, with the tremendous increase in capability to store large amounts of data and produce reports, has in turn increased the need and demand for data summarization techniques.

Presentation

The objective of data presentation is communication. Recall the adage "a picture is worth a thousand words." This is our point. The information contained in a set of data may effectively be communicated by means of graphical descriptive statistical techniques. Therefore, the use of such graphic techniques could be of value in a situation where the objective is to communicate the information content of data.

PROBLEMS

1. Identify an application of descriptive statistics in your major field of study. Point out the conditions present in your situation that would make this application both appropriate and of value.

2. As a new accountant for a medium-sized department store, you have been asked to "prepare something that will show our stockholders what happens to our sales revenue after we receive it." The stock in the firm is held by approximately 125 members of the local community. While the shareholders are, for the most part, not business persons, a substantial number of them are interested in the operation of the firm. Furthermore, many are aware that the firm utilizes a relatively high markup. Management anticipates that the annual stockholders' meeting, one week away, will see a number of questions concerning why after-tax profits are not higher than currently reported. An examination of last year's sales and cost data reveals information shown below. Prepare a pie chart of this information for presentation to the stockholders.

Net sales	3,500,000
Cost of goods sold	1,400,000
Sales force compensation	350,000
Administrative salaries	525,000
Overhead (rent, insurance, etc.)	525,000
Inventory shrinkage (spoilage, theft, etc.)	175,000
Advertising	70,000
Taxes (local, state, and federal)	350,000
After-tax profit	105,000

3. As a new trainee in financial management, you have been asked to analyze the variability in the prices of three different stocks to assist the bank manager in making an investment for a trust fund. These three corporations are in the same industry and have been affected by a recent change in government regulations. For this reason, only the past 16 weeks are indicative of future performance. The stocks have paid similar dividends in the past, which is the manager's primary criterion. However, to avoid speculation, the manager also prefers stocks that do not fluctuate widely in price. Weekly closing prices for the three stocks for the past 16 weeks are shown below.

Stock A		Stock B		Stock C	
90	94	94	97	98	96
100	102	97	106	93	105
110	101	112	94	115	94
105	106	96	97	102	101
98	98	106	99	82	102
97	99	113	110	103	105
98	97	92	96	101	98
103	102	95	96	100	105

a. Compute the mean for each stock.

b. Compute the variance for each stock.

c. Give the standard deviation for each stock.

d. What is your recommendation?

4. As personnel director for a medium-sized manufacturing firm, you are concerned about the upcoming contract negotiations with the union. A recent survey of the work force indicates the existence of three distinct groups, each with its own unique set of desires for the new contract. The younger workers (20–34) want more pay. The middle-aged workers (35–49) are concerned primarily with vacation time. The older workers (50–64) want increased retirement benefits. The firm cannot afford to meet all three demands. Your new assistant has just completed a study of the ages of the work force. The study shows that the average age is 41.85 years and the ages are distributed as follows:

Age	Number	Age	Number
20–24	873	45–49	369
25–29	1122	50–54	392
30–34	414	55–59	915
35–39	337	60–64	989
40–44	287		

Your assistant recommends that the company offer increased vacation time since the "average worker is 41.85 years old and our survey shows that workers this age are primarily concerned with vacation time."

a. List the class marks.

b. Approximate the mean age of your employees.

c. Present these data in a histogram.

d. Comment on the recommendation.

5. In designing a birth-control campaign for a developing country, a consulting firm found that the length and complexity of its messages must be based on the education level of the population. Messages effective with individuals with limited educations were not effective with those with more education and vice versa. This relationship held even with only one year's difference in education. Research found that the mean education level in this country was 10 years, the median 8 years, and the mode 4 years. In order to reach the most people, at what education level should the message be aimed? Why?

6. Your firm is comparing two brands of an automatic packaging machine. The function of the machine is to insert 100 nails of a specified size into a plastic container. In order to label the package as containing 100 nails, regulations require that an average of 100 nails be placed in each package and the standard deviation be less than 2. The firm wishes to minimize any variation around the average of 100. Since the firm sells almost a million packages of nails a year, a slight overage costs a considerable amount. In addition, consumer groups frequently count the contents of packages such as this, using only a few packages as their sample. If there is a shortage, the firm involved receives substantial negative publicity, and it is often subjected to

costly governmental investigations. Therefore, the firm wants to select a machine that will package an average of 100 nails with a minimum amount of variation. One thousand trials have been run with each of the two machines and the results are presented below.

Number of Nails per Package	Machine A	Machine B
97	35	10
98	50	45
99	60	120
100	700	655
101	75	130
102	50	10
103	30	30

As a preliminary step in forming your recommendation, describe the performances of the two machines by:

a. Histograms

b. Means

c. Variances

7. The number and characteristics of new employees of a medium-sized Chicago firm for 1974–1977 are presented below.

	1974	1975	1976	1977
Total	200	230	220	300
College graduates	80	88	84	116
Male	74	80	78	96
Female	6	8	6	20
High school graduates	120	142	136	184
Male	100	116	112	138
Female	20	26	24	46

a. Assume that your purpose is to inform the stockholders of the nature of the new employees entering the firm. How would you present this information?

b. How would you present these data if you were an attorney working for a group charging the firm with a failure to actively recruit female employees? (Assume that your goal is to convince a jury of the accuracy of your client's charge.)

c. How would you present these data if you were an attorney for the firm responding to the charge stated in part *b* above?

8. Management is about to have its annual managerial performance evaluations. You have been manager of the appliance department since 1967. During that time, the store's overall sales have shown a steady but slow upward trend, as have profits. Your department, on the other hand, has shown substantial increases in both sales and profits. If you can impress this fact on management, you stand a good chance for a promotion. The basic data are presented below. How would you present them to management?

	Sales		Profits	
	Store	Department	Store	Department
1967	4,500,000	675,000	495,000	70,900
1968	4,770,000	743,000	520,000	78,700
1969	5,056,000	817,000	546,000	87,400
1970	5,360,000	898,000	573,000	96,700
1971	5,681,000	988,000	602,000	107,600
1972	6,022,000	1,087,000	632,000	119,500
1973	6,383,000	1,196,000	663,000	132,600
1974	6,766,000	1,315,000	697,000	147,200

9. The management staff of a regional chain of supermarkets is trying to decide on the appropriate number of checkout counters to keep in operation during the slack times of the stores' operations. All the stores currently utilize two checkout lines during slack times. If customers are kept waiting in line, they may take their business elsewhere. However, given the spiraling costs of grocery items, it is essential to minimize costs. Having an underutilized checkout line in operation raises costs without offsetting benefits. You have been asked to gather information to help make this decision. You decide to visit five of the chain's outlets during the slack time of day and measure the waiting time of 20 customers in each store. The results are shown below in minutes and seconds. How would you present these data to management?

Elapsed Time	Elapsed Time	Elapsed Time	Elapsed Time	Elapsed Time	Elapsed Time	Elapsed Time
3:20	2:15	4:10	1:50	6:20	0	1:15
1:15	1:20	5:35	7:30	1:10	0	:45
0	:40	2:20	2:10	:15	0	2:30
:45	0	2:10	:55	3:35	4:00	1:40
8:40	0	3:30	1:40	:25	1:55	3:10
2:00	9:30	0	4:00	1:45	3:10	2:25
0	4:15	:55	5:10	2:35	2:20	1:05
1:30	0	3:15	0	5:05	:40	0
6:50	8:00	0	1:10	1:20	1:40	0
11:35	4:55	:40	1:30	2:30	8:15	1:20
0	0	1:25	1:15	0	9:10	
1:10	6:10	2:45	0	0	3:20	
0	:20	:45	4:30	3:30	1:30	
6:00	1:30	3:10	0	1:10	0	
:15	:45	10:20	5:20	1:45	2:10	

10. International Expansion, Inc., has developed a consumer product which is particularly appropriate for developing countries. In light of legal, economic, and social responsibility considerations, the firm has a policy of manufacturing its products in the countries which are the primary target markets. In order to meet the firm's investment criteria, the current product will be introduced only in those countries with 1 million or more households with an annual income of $400 or more. As a

newly hired marketing research assistant, your first assignment is to gather initial data on five countries to determine which of them, if any, meet the minimum requirements specified above. Those meeting these requirements will then be subjected to extensive analysis to determine the feasibility of introducing the product in them. Your initial research uncovers the data shown below. Which country or countries, if any, should you recommend for additional analysis? Why?

Country	Mean	Median	Mode	Variance	Number of Households
A	$500	$137	$125	5625	5,000,000
B	435	159	147	2704	3,500,000
C	403	402	398	100	2,700,000
D	415	390	360	400	2,000,000
E	417	410	408	324	1,900,000

11. As a new management trainee, you have been asked to prepare a report for your immediate supervisor on the growing importance of foreign sales to the firm. Your supervisor hopes to use the report as evidence for the need to reorganize the firm with greater emphasis on international operations. The first set of relevant data that you can find is shown below. How would you present these data to management?

Year	Domestic Sales (In Thousands of Dollars)	Foreign Sales (In Thousands of Dollars)	Total Sales (In Thousands of Dollars)
1961	11,268	732	12,000
1962	12,250	950	13,200
1963	15,827	1,173	17,000
1964	20,130	1,870	22,000
1965	25,284	2,716	28,000
1966	31,045	3,955	35,000
1967	29,002	4,998	34,000
1968	30,492	5,508	36,000
1969	34,080	5,920	40,000
1970	37,395	7,605	45,000
1971	38,125	7,975	46,100
1972	35,399	10,101	45,500
1973	35,664	10,536	46,200
1974	35,176	12,424	47,600
1975	38,425	14,575	53,000
1976	39,875	16,525	56,400
1977	41,311	17,789	59,100

Chapter 3

Index Numbers

3.1 SITUATION I: WHY?

College textbook publishing is a time-consuming process generally taking 10 months to one year between receipt of the final manuscript and appearance of the textbook on the market. The lag between the time when the publisher contracts with an author and when the final manuscript is received can be as long as two years or more. One positive aspect of this time lag is that it allows for planning well into the future. The editor knows exactly what books will be coming out in the next year.

Because there has been a sharp increase in the number of publishers in the college textbook area, there is a glut of textbooks on the market. As a result, established publishers, such as McGraw-Hill, are not realizing the profits of the past. Suppose you are editor of the college textbook division of McGraw-Hill. You feel that the slump in profit is due only in part to the growth in the number of publishers. Perhaps your pricing policy is also partly at fault. You are concerned that the prices of textbooks you are publishing may not have kept pace with the costs of producing them.

If you are right about the pricing policy, you want to begin to rectify the situation immediately. For instance, suppose a book is priced $1 below what its cost warrants. If 20,000 copies of this book are sold per year, profit is $20,000 below what it should be for any particular year. Even if there are only 10 such books out of all the titles published by McGraw-Hill, the lost profit now totals $200,000.

On the other hand, raising prices may lead to a decline in sales for some texts. The total effect may be negative. It is important, then, to avoid raising prices unless it is clearly warranted.

So the decision to raise the prices of some textbooks must be based upon reliable information. You will need to conduct an extensive market survey to determine the economic impact of raising prices. But such an economic analysis is expensive and clearly should not be undertaken if prices and costs are in line. So you have to answer this question first: Have textbook prices kept pace with the costs of production?

The answer to that question must be based upon information which is both trustworthy and inexpensive. You do not want to undertake the more costly economic analysis simply because of a hunch. Nor do you want to spend a lot of money on this preliminary step. You need a simple mechanism for describing cost changes and price changes. An added benefit of such a mechanism would be your ability to constantly and inexpensively monitor the relationship between costs and prices.

THE PRIMARY PURPOSE OF INDEX NUMBERS IS TO FACILITATE COMPARISONS INVOLVING CHANGES OVER TIME.

Often in business, executives need to make comparisons with the past. Usually, these comparisons must describe in the aggregate all the various aspects of their organizations. Each aspect or facet of the operation impacts on the total performance of the firm, and a change in any one of these aspects results in a change in overall performance. These facets, then, warrant continuous monitoring.

3.2 INDEX NUMBERS: HOW?

An index number can be defined as an indicator of the condition of an entity at a point in time. While the primary purpose of an index number is comparison over time, it can also be used for comparisons between two entities at the same point in time. For instance, a company which is deciding between two states for the location of its new production facility would want to compare cost-of-living indices for the two locations. These indices would allow a comparison of future operating costs, for employees would demand higher wages and salaries in the more expensive living environment.

In choosing to portray a situation or condition by an index number, the executive can select one of the many indices regularly published by existing organizations. Many indices are published periodically by the U.S. government. Other organizations, most notably investment counselors, also produce indices on a regular basis.

As an alternative to selecting one or more of the existing indices, the executive can opt for developing his or her own. For a particular firm, the development of an index number, or the selection of an existing one, depends upon the determination of two items. First, the important variables to be used in the calculation must be identified. Second, the appropriate mathematical expression must be specified. But it should be pointed out that the development of an index is an art.

Identification of the key variables depends upon the purpose of the index. Since profit and sales are important measures of performance, many indices use measures which reflect profit or sales, such as production, consumption, or measures of value. Ordinarily these measures are quantities or prices of items. These would be, for instance, quantities produced, quantities sold, or quantities required for some specified purpose. Of course, other measures or numbers can be used.

Many index numbers are ratios. As such they can be interpreted as percentages. (The ratio is usually multiplied by 100 to emphasize this interpretation.) For instance, a cost-of-living index value of 123.0 would indicate a 23 percent increase in the cost of living. When index numbers are ratios, the comparison being made is usually for two different points in time. For instance, if a building contractor were interested in the cost of building materials relative to three years ago, the key variables might be the cost of lumber, the cost of steel, and the cost of cement. Values of these variables are measured at each of these points in time.

One point in time is called the *given period*. It is the given period of which the index number is a description. Usually, the given period is the present or most recent past. (Most recent past is defined as the latest point in time for which data on the key variables are available.) The second point in time is called the *base period*. It is that point in time to which the given period is compared. In most cases, the base period remains fixed while the given period changes as the decision maker's point of interest changes or as more data on the important variables become available.

3.2a Unweighted Index Numbers

Index numbers are descriptive statistics and as such are meant to be easy to understand and appreciate. Consequently, index-number formulas generally utilize relatively simple mathematical expressions. The most common expressions employed are the sum and the arithmetic mean. When we have an index number involving the ratio of two mathematical expressions, such expressions utilize the same formula—the only difference being the values of the important variables.

As the name suggests, unweighted index numbers incorporate the important variables in the formula without any weighting structure. For example, if the index used the sum of price variables at the point in time labeled t, then the unweighted index would have the form

$$\sum_i p_{t,i}$$

where $p_{t,i}$ is the price of item i at time t. For an index number utilizing the ratio form, we would have two sets of prices, namely, the prices of the items for the given period, labeled $p_{t,i}$, and the prices of the items for the base period, labeled $p_{o,i}$. The index-number formula would take the form

$$\frac{\sum p_{t,i}}{\sum p_{o,i}} \times 100$$

Of course, the expressions illustrated above are linked to price variables. If the executive needed an unweighted quantity index, the forms would be, respectively,

$$\sum_i q_{t,i} \qquad \text{and} \qquad \frac{\sum q_{t,i}}{\sum q_{o,i}} \times 100$$

where q represents quantity rather than price.

3.2b Weighted Index Numbers

Because they incorporate no weighting scheme in their computations, unweighted index numbers treat all variable values as equally important. That is, each variable is treated as if it has the same impact on the entity as each of the other variables. Often, we need to acknowledge that certain variables have a greater impact or in general are more important in our description than others. This feature can be incorporated by the use of a *weighted* index number.

Suppose you were determining raises for each of your employees. If you wanted the raises to be compatible with productivity, you might develop a productivity index for each. Certainly you would use quantity variables. How many of each of the various products did the employee make during the given period? How many of each of the various other tasks did the employee complete during the given period? The answers to these questions are important. But you would also want to weight each of the accomplishments by its value; some accomplishments are likely to be more profitable than others. Your purpose is to better reward those who contribute more, that is, make the firm more profitable. In fact, many common indices use either prices as weights for quantity variables or quantities as weights for price variables.

It seems to us that the easiest way to deal with the various weighted index numbers is to simply enumerate them. We present in that format examples of three kinds of price-weighted quantity indices and three corresponding quantity-weighted price indices. Two of these six index numbers are given equation numbers because we will refer back to them later in the course of this chapter. Other weighted index numbers may use different variables and different weights, but the form will likely be the same as listed below.

Before we begin the listing, we identify three points in time. The first is the base period, labeled o as before. Consequently, prices and quantities for this point in time will be identified by $p_{o,i}$ and $q_{o,i}$, respectively.

The second point in time is the given period, labeled, as before, t. Prices and quantities for the given period will appear as $p_{t,i}$ and $q_{t,i}$, respectively.

The third point in time comes in because some statisticians feel that the weights should not come from either the base period or the given period. We identify this third point in time by the subscript s. The variable values for this point in time will then be identified by the notation $p_{s,i}$, for the prices, and $q_{s,i}$, for the quantities.

1. Price-weighted quantity indices will utilize the same price lists in the numerator and denominator of the formula:

   a. Using base-period prices as weights:

$$\frac{\Sigma\, p_{o,i}\, q_{t,i}}{\Sigma\, p_{o,i}\, q_{o,i}} \times 100$$

   b. Using given-period prices as weights:

$$\frac{\Sigma\, p_{t,i}\, q_{t,i}}{\Sigma\, p_{t,i}\, q_{o,i}} \times 100$$

   c. Using third-period prices as weights:

$$\frac{\Sigma\, p_{s,i}\, q_{t,i}}{\Sigma\, p_{s,i}\, q_{o,i}} \times 100$$

Note in all these cases the same prices are used in both the numerator and denominator of the ratio. Only the quantity values change from time-period calculation to time-period calculation. This makes each of these indices a quantity index and suitable for comparing changes in quantities over time.

2. Quantity-weighted price indices use the same quantity values in both the numerator and the denominator of the ratio:

   a. Using base-period quantities as weights:

$$\frac{\Sigma\, q_{o,i}\, p_{t,i}}{\Sigma\, q_{o,i}\, p_{o,i}} \times 100 \qquad (3\text{-}1)$$

   b. Using given-period quantities as weights:

$$\frac{\Sigma\, q_{t,i}\, p_{t,i}}{\Sigma\, q_{t,i}\, p_{o,i}} \times 100$$

   c. Using third-period quantities as weights:

$$\frac{\Sigma\, q_{s,i}\, p_{t,i}}{\Sigma\, q_{s,i}\, p_{o,i}} \times 100 \qquad (3\text{-}2)$$

Again notice the role played by each type of variable. These last three indices are price indices because only the prices differ in the numerator and denominator. As such they measure changes in price over time.

3.3 SITUATION I REVISITED

We now return to the McGraw-Hill problem, in which you want to determine if the prices of McGraw-Hill textbooks have kept pace with the costs. You decide to use a price index and a cost index. Because you want to compare changes, you choose indices involving ratios.

Since 1973 was the first year of the recent slump, you choose 1972 (the last stable year) as the base period for your indices. The given period will be the current year, that is, the most recent year for which values of the key variables are available.

Because of the comparisons you desire to make, the key variables will be prices and costs. However, it is impractical to use the prices and costs of all the hundreds of textbooks McGraw-Hill publishes. First, you restrict the possible texts to only those which were on the market both in the base year, 1972, and in the current year. From that group, you select eight representative titles.

Surely you want quantity-weighted indices for prices and costs. You settle on the expression of Equation (3-1) for each index number.

The data have been gathered and appear in Table 3-1.

The price index, PI, is computed as

$$PI = \frac{21{,}500(14.95) + 13{,}300(15.95) + \cdots + 20{,}900(11.95)}{21{,}500(11.95) + 13{,}300(12.95) + \cdots + 20{,}900(\ 9.95)} \times 100$$

$$= \frac{1{,}964{,}930}{1{,}584{,}615} \times 100 = 124.0$$

The cost index, CI, is calculated to be

$$CI = \frac{21{,}500(11.42) + 13{,}300(10.76) + \cdots + 20{,}900(8.22)}{21{,}500(\ 7.56) + 13{,}300(\ 8.48) + \cdots + 20{,}900(7.02)} \times 100$$

$$= \frac{1{,}353{,}180}{1{,}000{,}183} \times 100 = 135.3$$

Table 3-1 Quantities, Prices, and Costs for Selected Texts

	Quantity	Prices		Costs	
Text	1972	1972	Current	1972	Current
1	21,500	11.95	14.95	7.56	11.42
2	13,300	12.95	15.95	8.48	10.76
3	44,100	7.50	8.95	3.91	5.74
4	19,600	12.95	15.95	8.31	10.95
5	23,400	2.95	4.50	1.37	2.83
6	8,000	13.95	17.50	11.13	12.44
7	17,200	9.95	12.50	6.56	8.68
8	20,900	9.95	11.95	7.02	8.22

Now you see that prices have risen 24 percent since 1972, but costs have increased by 35.3 percent. So, just as you suspected, the prices of McGraw-Hill's textbooks have not kept pace with the costs of producing those textbooks. Now you can turn your attention to the more extensive and complex analysis of determining the economic impact of raising the prices of certain textbooks.

3.4 IMPORTANT GENERAL BUSINESS INDICES

We discuss briefly two groups of indices which are in the news regularly and are of considerable value to businesses as well as to individuals. The first group consists of the two inflation/deflation indices, the consumer price index and the wholesale price index. Not only are businesses consumers, but many of them contract with labor unions which represent consumers. Inflation means that businesses will pay more for their supplies. The wholesale price index describes this effect. Inflation also means that employees must earn more to remain "even." As it describes this need, the consumer price index is surely an important consideration in contract negotiations with labor unions.

The second group, the stock market indices, has measures which describe the general business environment. The Dow-Jones Industrial Average and the Standard & Poor's 425 provide descriptions of investor confidence in the strength of the industrial organizations listed on the New York Stock Exchange. Since issuing stock is a major source of funds for business firms, and since many organizations have substantial sums invested in other companies, these indices are quite valuable. These indices are also used by economists and politicians in gauging the health of the business sector.

3.4a Inflation/Deflation Indices

The consumer price index (CPI) is a measure of the average change in prices for goods and services purchased by urban wage earners and clerical workers, who constitute roughly 40 percent of the total population. Because the CPI is calculated for that specific group within the population, it may not accurately describe what other individuals experience in their purchases.

The CPI is published monthly in the *Monthly Labor Review* by the Bureau of Labor Statistics. A more complete listing, with analysis, appears monthly in the Bureau's *CPI Detailed Report*. The CPI is a percentage-type index with the base period currently identified as 1967. The base period is changed regularly to reflect changes in technology and in life-styles. When a change is made in the base period, the index values reported for periods using the old base period are revised to reflect the new base period. This allows continued long-term comparisons.

In calculating the CPI, the Bureau of Labor Statistics uses the concept of a "market basket," that is, all the goods and services typically purchased for day-to-day living. About 400 items are selected for the market basket from the following general areas: (1) food, (2) housing, (3) clothing, (4) transportation, (5) health and recreation, and (6) other goods and services. These items are believed representative of all goods and services purchased by the group in question. Revisions are made, periodically, in the market basket as well. However, when left unchanged, the 400 items are so strictly defined that any change in the index can only be attributed to price variations.

The key variables are the prices for the 400 items, and their values are received from approximately 40,000 tenants and 18,000 retail establishments at the present time. Data are collected from 56 cities across the country. The prices include purchase prices, maintenance costs, and all taxes directly tied to buying and using the market-basket items.

The formula used for the calculation is similar to the fixed-weight price index of Equation (3-2). The quantity weights used are the result of extensive surveys taken periodically in which the market basket is observed as well. Studies have shown that, under normal conditions, the percentages expended on the various general areas change only slightly from year to year. Consequently, the weights need adjustment infrequently.

Beginning in February 1978, the Bureau of Labor Statistics computed two new indices which reflect consumer price changes. Both of the new indices use a new base year, 1977, and a new market basket which consists of several thousand items including the 400 used previously. Also, more retail establishments (21,000) are being surveyed. One of the new indices incorporates only these changes. The second expands the group of individuals from urban wage earners and clerical workers to all urban consumers. This new group constitutes roughly 80 percent of the total population.

The Bureau of Labor Statistics also calculates the wholesale price index (WPI), which is also published monthly in the *Monthly Labor Review*. The Bureau's *Wholesale Prices and Price Indexes*, also a monthly publication, presents a more detailed report and analysis of the markets described by the WPI.

Like the consumer price index, the WPI is produced by a formula similar to Equation (3-2). The current base period is 1967 as in the CPI calculation. The weights used are revised more frequently than are the weights in the CPI calculation. Beginning January 1976, the weighting structure was revised according to 1972 values.

The key variables utilized in the WPI are prices of over 2000 products and commodities in certain, specified transactions. For goods which proceed from crude to manufactured or crude to processed, the prices at each level of processing are included. Otherwise, only the price from the first commercial transaction is utilized. Prices generally do not include transportation charges or excise taxes. Prices which are assigned to interplant transfers, military production, or goods sold directly to household consumers by the manufacturers or the processors are likewise excluded from the index.

Based solely on the names of the indices, you might expect the CPI to follow, to some extent, the WPI. That is, you might expect to be able to find a particular length of time for which the WPI at the beginning of the period and the CPI at the end of the period would always be nearly equal. However, as we can see from the above discussion the WPI does not really involve the values we generally think of as "wholesale" prices. For that reason, you may notice from time to time that the CPI for a particular month varies considerably from the WPI for the comparable period.

3.4b Stock Market Indices

The Dow-Jones Industrial Average (DJIA) is not a ratio index. It has as its main purpose the description of the multifaceted entity called the New York Stock Exchange (NYSE). The description is in terms of a single "price" which attempts to reflect the present condition of all the issues, taken collectively, which are listed on the NYSE.

The DJIA is computed daily from the closing prices for the stocks of 30 preselected industrial firms. So the price variables are the closing prices of these 30 stocks. The mathematical expression used involves the sum of these 30 prices. The DJIA began as a true average of the 30 prices, that is, the sum of the 30 prices was divided by 30 to get the index number. Then it became necessary to deal with stock dividends and mergers.

A stock dividend is declared when the firm announces that it is paying dividends by increasing the number of shares held by each investor. A 20 percent stock dividend increases the number of shares of each investor by 20 percent. A 100 percent stock dividend is called a 2-for-1 stock split and doubles the number of shares held by each investor. Stock dividends do not increase the total value of the investor's holding. If an issue which is valued at \$50 per share splits 2 for 1, then each share is now worth

$$\frac{\$50}{2} = \$25$$

So the investor now has twice as many shares—each worth half as much.

Since stock dividends do change the price of the particular stock involved, they would tend to change the DJIA. But no real value has been added, and therefore stock dividends should not change the index. To cope with this problem the DJIA uses the "divisor-adjusted" method. When a stock dividend occurs, the denominator of the formula is changed so that the index remains the same. For instance, suppose we have only two stocks used for the index. With selling prices of \$20 and \$60, we have, as our index value,

$$\frac{20 + 60}{2} = 40$$

If the first stock splits 2 for 1, its new price is \$10 per share. But the index value must remain the same. So the denominator of the index is changed from 2 to 1.75, because

$$\frac{10 + 60}{1.75} = 40$$

Stock dividends in which the number of shares increase by less than 10 percent—for instance, a 5 percent stock dividend—are ignored, and thus do not change the divisor in the index formula.

There are two comments which need to be made about the validity of the DJIA as a market index. First, the divisor-adjusted method gives less weight to the stocks of firms which give stock dividends. This creates distortion in the meaning of the average every time the divisor is adjusted. Second, the 30 stocks used are "blue chip" stocks. The small number and high quality of the issues make the group of stocks unrepresentative of the market as a whole.

An alternative index for describing the market at a point in time is the Standard & Poor's 425 (S & P 425). This index is based upon prices and quantities for the stocks of 425 industrial firms. The key variables are the prices per share and the number of shares outstanding for the 425 stocks.

The mathematical expression takes each stock price $p_{t,i}$ and multiplies by the number of shares outstanding $q_{t,i}$ and adds these products together. This sum

$$\Sigma\, q_{t,i}\, p_{t,i}$$

is divided by a similar sum for the base period 1941–1943. The sum in the denominator is

$$\Sigma\, q_{o,i}\, p_{o,i}$$

where $q_{o,i}$ is the average number of shares outstanding during the years 1941–1943, and $p_{o,i}$ is the average price per share during the years 1941–1943. This quotient is multiplied by 10. So,

$$\text{S \& P 425} = \frac{\Sigma\, q_{t,i}\, p_{t,i}}{\Sigma\, q_{o,i}\, p_{o,i}} \times 10$$

Note that this is *not* a quantity-weighted or a price-weighted index. Both prices and quantities in the numerator differ from those in the denominator.

Let's look at the S & P 425 in relation to the two difficulties we had with the DJIA. First, there are 425 stocks involved, not just 30, and not all are blue chip stocks. Second, and more importantly, there is no adjustment needed for stock dividends, so no distortion can result. Suppose one of the 425 stocks has 1 million shares outstanding, each valued at \$50 per share. In the numerator of the formula for the S & P 425, we have

$$qp = (1{,}000{,}000)(50) = 50{,}000{,}000$$

If the stock splits 2 for 1, the price drops to \$25 per share, but the shares outstanding increase to 2 million. In the numerator, we now have

$$(2{,}000{,}000)(25) = 50{,}000{,}000$$

which is the same as we had before the split. For those two reasons, the S & P 425 is preferred to the DJIA as a market indicator.

3.5 EXAMPLES

(1) Armco is entering contract negotiations with a union that represents some 3800 Armco production workers. The workers are paid within a narrow range of rates and the average weekly earnings for each of the past nine years is shown below. While earnings have increased steadily, the union is expected to request a substantial increase to offset the impact of inflation. It is your feeling that wages have more than kept up with the inflation rate. You gather the following data:

Year	Average Weekly Earnings	Consumer Price Index (1967 = 100)
1968	103.20	104.2
1969	110.90	109.8
1970	117.40	116.3
1971	125.10	121.3
1972	129.80	125.3
1973	135.50	133.1
1974	147.90	147.7
1975	164.46	161.2
1976	179.69	171.9

Using the consumer price index as a measure of the inflation rate, you can adjust the average weekly earnings figures to reflect 1967 dollars. Dividing average weekly earnings for any year by CPI/100 for that year, we obtain what is called *real earnings*. The results are the following:

Year	Adjusted Average Weekly Earnings
1968	99.04
1969	101.00
1970	100.95
1971	103.13
1972	103.59
1973	101.80
1974	100.14
1975	102.02
1976	104.53

Although your employees have experienced ups and downs in their earnings, two things are apparent. First, over the nine years, Armco employees have experienced an increase in earnings that has more than offset inflation. Second, over the last two years, there has been a steady increase in real earnings.

(2) A possible rebuttal by union negotiators is that the productivity of Armco employees has increased substantially. Suppose that the average dollar output per week per Armco production employee had increased as shown below. A price index for Armco's products (base 1962–1963) is also shown.

Year	Weekly Output per Employee	Armco Price Index
1968	436.00	136.4
1969	472.00	145.2
1970	514.00	152.3
1971	559.00	163.4
1972	598.00	169.1
1973	663.00	179.1
1974	715.00	192.3
1975	792.00	197.8
1976	864.00	206.4

It is essential to adjust these output figures to determine "real" productivity. We use a formula similar to the one used in Example 1, that is, Weekly Output per Employee is divided by API/100. The values for real productivity are given below.

Year	Adjusted Weekly Output
1968	319.65
1969	325.07
1970	337.49
1971	342.11
1972	353.64
1973	370.18
1974	371.81
1975	400.40
1976	418.60

It is now clear that real productivity has increased steadily over the past eight years. For the most part, real productivity has increased faster than real earnings.

(3) Costs of attending a public four-year college have increased rapidly over the past decade. A current debate is whether or not the increase since 1970 has been relatively more than, less than, or the same as increases in other consumer prices. One estimate of the costs of attending a four-year public school is presented below for each year since 1970.

Year	Estimate
1970	7,800
1971	8,050
1972	8,300
1973	8,610
1974	9,020
1975	9,510
1976	10,200

Using the CPI to adjust these estimates for inflation, we get the following:

Year	Adjusting Estimate
1970	6706.79
1971	6636.44
1972	6624.10
1973	6468.82
1974	6106.97
1975	5899.50
1976	5933.68

After the adjustment, we can easily see that the costs of attending a public four-year college have not risen as fast as the observed inflation rate as measured by the consumer price index.

3.6 IDENTIFICATION OF APPLICATIONS: WHEN?

An index number is a descriptive statistic which facilitates comparisons between (1) conditions of an entity or process at two points in time (thus measuring changes over time), and/or (2) conditions of two entities or processes at a single point in time. It is usually employed to describe complex, multifaceted entities or processes.

For descriptions of changes over time, the index number is a ratio. In the numerator, we use values from the given period; in the denominator, we incorporate values from the base period. These index numbers are used to adjust sets of data for differences in time. For example, a price index would be used to adjust the 1977 sales figure for comparison with 1972 sales.

As in Situation I, two values for an index can be examined to effect the comparison of two processes or entities.

A very common application of index numbers is the periodic monitoring of an entity or process. This is the purpose of each of the four well-known indices discussed in the chapter. The inflation/deflation indices allow monitoring of the economy; the stock market indices provide a monitor for the strength of the business sector.

The application of index numbers would be of value when one needs to compare one entity over time or two entities at a point in time.

PROBLEMS

1. Identify an application of index numbers in your major field of study. Point out the conditions present in your situation that would make this application both appropriate and of value.

2. As a financial analyst for a major trust company, you have been assigned the task of evaluating the profit performance over the past 10 years of a large number of industrial firms. To maintain the confidence and continued patronage of your clients, it is

imperative that you be able to isolate the firm or firms that have shown both rapid and consistent growth. The net-profit performances over the relevant time period for three of the most promising firms are presented below.

a. Develop a net-profit index for each company with 1968 as base.

b. Evaluate the three firms.

Year	Standard Corp.	Atlantic Inc.	Dubow Inc.
1968	567.3	116.5	718.5
1969	590.2	122.7	790.3
1970	671.8	480.3	850.4
1971	750.3	482.1	933.2
1972	818.4	475.3	1020.6
1973	767.6	486.4	1130.2
1974	822.0	463.2	1250.3
1975	840.1	475.1	1306.9
1976	890.2	482.3	1445.1
1977	903.6	491.6	1607.3

3. A municipal bus company has consistently operated at a loss since 1974. The operating loss has been made up by subsidies from city taxes. The city feels justified in this expenditure since reducing the usage of automobiles in the city seems a goal worthy of tax support. The bus line has not raised its fares since 1972 and thus the required subsidy has grown rapidly. The city council has decided that ticket prices must be increased according to some standard plan. One group of council members feels that prices should be adjusted as key operating costs change. Others feel that, since the line is operated as a public service, prices should change in line with other prices facing the consumer. But no plan will work if it cannot be implemented. How might index numbers be appropriate for implementing each of these strategies?

4. As an accountant for Standard Fasteners, you have been working on an accounting system designed to control for the effects of inflation. When uncontrolled, these effects can substantially distort the data and thereby negatively influence decisions. Inflation on the cost side of your firm had been relatively stable through 1968. Therefore you decide to use 1968 as a base year. You want to compare the materials cost involved in producing 10,000 three-inch decorative cabinet hangers in 1968 with the 1977 cost. Since the production process has been changed since 1968, different quantities of each raw material are used. Given the prices and quantities shown below, construct a quantity-weighted price index.

	1968		**1977**	
Raw Material	Price	Quantity	Price	Quantity
Brass	$20 (per lb)	20 lb	$35.00 (per lb)	18 lb
Screws	$ 1 (per case)	100 cases	$ 2.50 (per case)	100 cases
Shellac	$ 2 (per quart)	50 quarts	$ 3.00 (per quart)	70 quarts

5. Sales of Home Appliance Corporation's electric exhaust fans increased by more than 25 percent ($880,000 to $1,114,000) between 1974 and 1977. However, Jane Ellison,

product manager for exhaust fans, is not sure exactly how to interpret the results. The increase in sales does not necessarily indicate better performance. Therefore, she has gathered the information presented below and asked you to prepare a quantity-weighted price index with 1974 as base in order to evaluate the meaning of the sales increase.

Sales of Home Appliance Exhaust Fans

	1974		1977	
Model	Price	Quantity (In Thousands)	Price	Quantity (In Thousands)
Economy	$30	12	$32	15
Standard	$40	4	$46	4
Delux	$45	8	$50	9

6. Valley Furniture handles a complete line of quality home furnishings, appliances, and televisions. For the past eight years, its sales have rather consistently been composed of 65 percent furniture, 20 percent appliances, and 15 percent televisions. Total sales volume for this period is shown below. Norm O'Dell, the firm's general manager, is concerned that sales increases reflect inflationary pressures rather than actual increases in business activities. He has found the retail indices for furniture, appliances, and televisions but is unsure how to utilize them. Make a recommendation to Mr. O'Dell, using an index which is a weighted average of the indices shown below.

Year	Valley Furniture Sales (In Thousands)	Retail Furniture Price Index (1967 = 100)	Retail Appliance Price Index (1967 = 100)	Television Price Index (1967 = 100)
1970	$1218	111.6	105.3	102.1
1971	$1614	117.2	108.5	101.2
1972	$1675	124.2	109.8	103.5
1973	$1810	128.3	114.1	107.8
1974	$2350	136.1	117.6	110.5
1975	$2475	139.8	122.4	119.6
1976	$2630	145.7	128.3	118.7
1977	$2730	156.2	131.2	121.2

7. A person who purchased a home several years ago and sold it on today's market would be likely to realize a substantial profit. Unless another home is purchased within 18 months, this profit is subject to a capital gains tax. Assume that you have been asked to advise a citizens' committee for tax reform. The committee feels that the rule is unfair. In particular, many elderly people sell their homes and rent apartments in retirement communities. Their rents reflect increased construction costs and other inflationary pressures just as do the costs of houses. Yet, they must pay the tax on the sale of their home while individuals buying another home do not.

   Develop a tax reform measure to alleviate this problem, utilizing an appropriate index (indices).

8. Market share (the percent of industry sales held by one brand) is a commonly used measure of the effectiveness of a firm's marketing program. Many firms would like to make more use of this concept but cannot since they do not know total industry sales. Miram Chemical Corporation was in this position with respect to GuardAll, a special gasoline additive. Annual sales of GuardAll had shown a fairly steady growth for the past 10 years (see below). While studying a government report on taxation policies in the chemical industry, Gina Flemming, a tax accountant for Miram Chemical, noticed a statement that an index of tax revenues received from gasoline additives had increased from 136 in 1970 to 193 in 1977. She sent this information and a note that the tax rate (on a per-gallon basis) had not changed in the past 10 years to the product manager of GuardAll. How, if at all, can the product manager use this information to evaluate the performance of GuardAll relative to the rest of the industry?

Sales of GuardAll

Year	Cases (In Thousands)
1965	31.5
1966	33.2
1967	37.1
1968	38.8
1969	39.6
1970	40.3
1971	41.2
1972	43.4
1973	44.2
1974	46.1
1975	47.9
1976	48.5
1977	49.7

Chapter 4

Probability

4.1 SITUATION I: WHY?

Suppose J. Arthur Patterson is Southeast District manager of the Industrial Credit Company. The company has several field offices in the Southeast region which make loans primarily to individuals and small businesses. Organizations like Industrial Credit generally attract loan applications from individuals and businesses whose applications have been denied by local banks. Consequently, the risk of default is greater on these loans. However, Industrial Credit makes these loans at higher interest rates than banks. While the default rate is higher than banks experience on similar loans, the increased interest charged has been sufficient to offset these losses.

A second implication of Patterson's clientele is of major concern to him at this time. As might be expected, risk is a vague concept. It is most certainly related to the frequency of default. But it is customary to include in the measurement of risk the amount of money involved. During certain economic periods, it is very difficult to get a bank loan regardless of one's credit rating. Funds simply are not available. During these times, companies like Industrial Credit thrive. The rate of default experienced is considerably reduced, since many of the clients have excellent credit ratings and have been turned down by other institutions simply because money is "tight." This has been the situation during the three years that Patterson has headed the Southeast District.

Unfortunately, Patterson and Industrial Credit now find themselves on the other side of that coin. With the recent swing in the economy, forecasts indicate that banks will have ample funds to make such loans. Given economic conditions, Industrial Credit can look forward to applications from only those who are least able to meet the loan obligations, a clientele that heretofore Patterson has had the luxury of rejecting.

The Southeast District offices generally make loans totaling approximately $325 million in a given year. Because of certain long-term commitments to organizations, the economic conditions will not affect about $200 million of these loans. However, the remaining $125 million will now be loaned to what Industrial Credit calls the "highest risk" group.

Patterson's deliberations focus on determining the effect of making loans to this highest risk group. As an alternative, it may be advantageous to divert these funds or a substantial portion of them to other long-term investments. But to reach a satisfactory determination, he must come to grips with the uncertainty involved in making loans to the highest risk group.

4.2 SITUATION II: WHY?

Like most life insurance companies, New York Life Insurance Company (NYLIC) has been offering health insurance policies for several years. Suppose you were a vice president with NYLIC and were reviewing the performance of a recently introduced low-cost family health insurance policy.

Insurance companies are major users of statistics and New York Life is no exception. In particular, this new policy was the result of a rather expensive study using 15 years' worth of data on health insurance policies. The study produced what is called an actuarial model, *which was used for predicting future claims. The first few years the policy was offered had produced claims only slightly higher than predicted. Thus, while the policy was not as profitable as originally projected, it was a leading contributor to overall profits. Company executives were pleased with the plan and, one year ago, agreed to increase the marketing effort. At that point, you became the primary officer in charge of the policy.*

However, during the last year, the policy has resulted in substantially higher claims being submitted and a material loss to NYLIC. Your responsibility at this time is to review the policy in terms of the eligible subscribers and the premium. Your analysis focuses on the significance of the past year. Does the amount of claims during the past year indicate a major change in the rate of claims originally projected for the plan? Or is the last year's performance simply a chance fluctuation from the projected rate?

If the marketing effort launched a year ago has attracted a different clientele, then the amount of claims expected in future years may be substantially different from the projected figures. In that case, a change in the policy is prescribed. New York Life can modify the coverage of the policy or increase the premium. There are substantial costs involved in making changes in an insurance policy, such as the costs of filing these changes with state insurance commissioners and the like. Such changes will certainly have an effect on the other insurance policies offered by NYLIC. Studies have shown that many policyholders of the major insurance companies often have several insurance policies with the same company. Thus, major changes in this policy may mean that fewer new life insurance policies will be written by NYLIC agents next year. Because of the various ramifications, you would like to avoid making a major change in the policy, if possible.

On the other hand, continuing to sell an unprofitable policy is costly. While insurance companies profit more from investments than from policy sales, it is certainly true that policies which continually produce substantial losses cannot be tolerated. So, you direct your attention to the question of the nature of the performance of the last year.

You conclude that your decision can be based upon the validity of the actuarial model: If this model is still an accurate portrayal of the market, then you will suggest that the policy remain unchanged. If not, you will recommend a policy change. You feel that the validity of the actuarial model can be judged by the degree to which last year's performance is consistent with that model. If the results of last year are extremely unlikely given the "old" market, it may be more reasonable to conclude that a new market has been developed, and the actuarial study no longer applies. The question is: How likely was last year's performance?

4.3 PROBABILITY: HOW?

Before we go on, perhaps we should briefly discuss what decisions are all about. What components of the above situations are common to most decisions or even all decisions? How typical are Situations I and II as decision problems? In Chapter 9 we deal with what a decision is in greater detail, but getting a perspective at this point in our study seems desirable. Decisions involve choices among alternatives. A prudent choice requires some determination of the consequences of each of the possible alternatives. The decision maker may completely control the outcome by his or her choice. For instance, buying a $25 Series E U.S. Savings Bond costs *exactly* $18.75, and, in seven years, results in *exactly* $25.16. Furthermore, there is an exact schedule of value for each half-year interval between the issue date and the date of maturity. There is no doubt about the consequence of investing in this U.S. Savings Bond. If all alternatives lead to known consequences with no doubt, we call that decision a *decision under certainty*. It is difficult to construct realistic scenarios for such decisions, and, in business, such decisions are nonexistent. Ordinarily there are ranges of consequences for an alternative; that is, an alternative seldom leads to only one outcome.

An alternative may be pictured as in Figure 4-1. Which of these outcomes will actually occur is unknown at the time the decision is made. Business

Figure 4-1

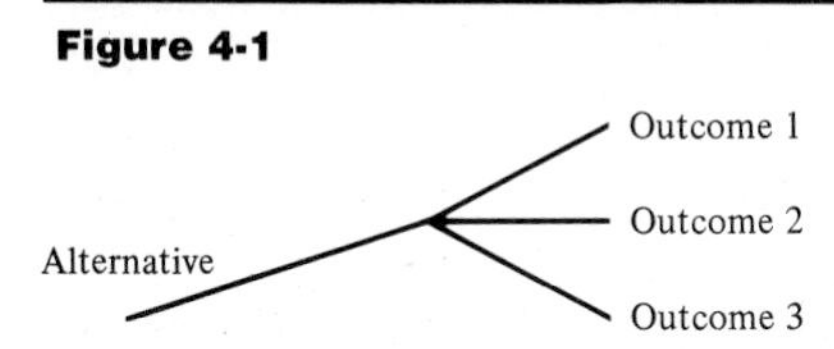

decisions, then, could be called *decisions under uncertainty*. Since uncertainty is present and is an important aspect in determining the consequence of an alternative, it seems advantageous to try to get a handle on it, to get an appreciation for it, to draw a mathematical picture of it.

Each of the decision makers in the above situations is dealing with uncertainty, and each surely must make a prudent choice. In Situation I, Patterson is faced with the uncertain consequences of making the loans to the highest risk group. In Situation II, you struggle with the uncertain applicability of the actuarial model. Each must, therefore, analyze the relationships between the alternatives and their potential consequences. But uncertainty means that these relationships are nebulous, and this vagueness hinders analysis. However, numbers are concrete. As in many other instances, the presence of meaningful numbers here would facilitate understanding and allow analysis. This would be advantageous.

PROBABILITY IS A NUMERICAL MEASURE OF UNCERTAINTY.

There are other advantages in having a numerical measure for uncertainty. Such a number aids in communication between executives. Verbally a manager at a staff meeting might indicate that he or she is "fairly sure" about the success of a particular project. This phrase might mean something quite different to the other executives at the meeting. "Fairly sure" might mean that success will occur 9 times out of 10 to one decision maker while the same phrase might indicate success 7 times out of 10 to another. Numbers eliminate such confusion.

Perhaps the most important advantage of a numerical measure is the ability to make use of mathematics for analysis. Suppose a particular process involves two components. About one of these the manager is "fairly sure" of success, but about the other he or she is "very sure" of success. What phrase does the manager use in identifying this uncertainty about the success of the process which requires both components? Or suppose the executive were choosing between two advertising alternatives—one which was "fairly sure" to produce \$100,000 in sales and another which was "very sure" to generate sales of \$80,000. Which advertising plan should be selected? The numerical measure, and the mathematics associated with it, alleviates these difficulties in communication.

4.3a The Sample Space

To develop the numerical measure of uncertainty, we should identify the fundamental characteristics of an uncertain situation. We can verbally describe uncertainty as doubt about which of several possibilities will occur or will be found to have occurred. The key element is the existence of several possibilities. Anytime an alternative has more than one possible consequence, we have uncertainty. For each customer, Patterson and Industrial Credit realize one of two possibilities, default or payment. During any year, New York Life receives claims which could range from a low of $0 to a high of the total insurance in force (at least theoretically, anyway). So, the first step in the development of a numerical measure of uncertainty must be the listing of all the possible consequences or outcomes of a particular uncertain situation. The set of possibilities is called the *sample space*. Each element of the sample space is called an *outcome*. The outcomes can be thought of as the possible values of a variable. The uncertain situation is called an *experiment* or a *process*.

AN EXPERIMENT OR A PROCESS GENERATES OUTCOMES. THE SET OF ALL POSSIBLE OUTCOMES IS CALLED THE SAMPLE SPACE FOR THE EXPERIMENT OR PROCESS.

Sample spaces take different forms in different situations. The form depends primarily upon the type of variable involved. If the variable is qualitative, the sample space will consist of a list of qualities. If you were hiring a cabinet maker of uncertain abilities, you might find that person's work to be poor, fair, good, or outstanding, for instance. The sample space for this uncertainty would be

{poor, fair, good, outstanding}

In Situation I, Patterson might identify the following sample space:

{payment, default}

If the variable is quantitative, the sample space is a set of numbers. A continuous quantitative variable has a sample space which is an interval of numbers. For instance, the inside diameter of a bearing might assume values between 4.97 and 5.02 inches. In symbolic notation, we would write the sample space for x as

$\{x: 4.97 \leq x \leq 5.02\}$

A discrete quantitative variable has a sample space which is a sequence of numbers. The manager of a clothing store, in describing the uncertainty about the number of blue suits the store will sell this season, may identify that demand for the suits may be as small as 0 or as large as 100. But only a whole number of suits can be sold. The sample space would consist of the whole numbers from 0 to 100 inclusive, or

$\{0, 1, 2, \ldots, 100\}$

The sample space may take different forms depending upon the decision to be made. The descriptions of the outcomes are often dictated by the decision. We already mentioned that Patterson of Industrial Credit might describe his sample space as

{payment, default}

This would be sufficient if his only concern were the rate of default. But, if his primary concern is in terms of profit, this sample space lacks sufficient detail. To provide an adequate description of the possible outcomes, Patterson must recognize that "default" may mean quite different consequences. A loan for $1000 made for a period of one year at 14 percent interest is repaid by 12 equal installments of $95 each. If the customer defaults before the first payment, Industrial Credit has lost the full $1000. If, at the other extreme, the customer defaults on the last payment, Industrial Credit has received $1045, which is actually a profit. So, in terms of profit and loss, the two default outcomes are quite different.

The enumeration of the outcomes in the sample space is critical not only for the above reasons, but also because the determination of probabilities is greatly affected by it. Since the sample space lists all the possible outcomes, it identifies all the values of the variable which should have probabilities associated with them. If the sample space includes outcomes which cannot occur, or does not include outcomes which can, attractive opportunities may be underestimated or overlooked entirely. There have been cases in which undesirable alternatives have been selected because a most undesirable outcome was not accounted for. One of the most spectacular failures in the United States auto industry resulted from a failure to adequately identify the sample space. The impact of that failure was felt for some ten years after the decision was made.

4.3b Discrete Variables

To be useful as a measure of uncertainty, probability must provide some kind of description. To be an aid in decision making, this description should focus on the actual outcome which is observed, since it is this outcome which impacts on the organization. This suggests that probability should be a measure of how likely a particular outcome is to occur—the more likely the outcome, the higher the probability.

Let's isolate a particular outcome b in a sample space S for an experiment. Let's suppose that the process was observed to produce an outcome N times. Let's further suppose that outcome b occurred n times. To have any meaning, the probability assigned to the outcome b, written $Pr(b)$, should be close to the value n/N. If N is a large number, n/N and $Pr(b)$ should be *very* close. Some properties of probabilities can be derived from these observations. For any possible outcome b, $Pr(b) \geq 0$, and at the same time, $Pr(b) \leq 1$, since n/N must be greater than or equal to 0 and less than or equal to 1. If $Pr(b) = 0$, then $n/N = 0$, $n = 0$, so outcome b *never* occurs. If $Pr(b) = 1$, then $n/N = 1$ and $n = N$. In that case, outcome b *always* occurs. Since the sample space S *always* occurs, $Pr(S) = 1$. Finally, if we add all the probabilities assigned to the possible outcomes in a sample space, we must get 1, since adding all the times, n, that each of them occurs will give N. For example, if

$$S = \{\text{payment, default}\}$$

in Situation I, let's examine the loans made to $N = 1000$ customers. Then S will occur 1000 times. Suppose payment occurs $n = 726$ times; then default occurs $m = 274$ times. We have

$$n + m = N$$

since

$$726 + 274 = 1000$$

Let's recap the basic properties of probabilities which we have observed:

1. For any outcome b, $0 \leq Pr(b) \leq 1$.
2. $\Sigma Pr(b) = 1$, where the summation is over all b.
3. $Pr(S) = 1$.

These properties must be kept in mind when the decision maker assigns probabilities.

To discuss the methods for assigning probabilities to the various outcomes in a sample space, let's take

$$S = \{a, b, c\}$$

There are two procedures in common use. One method has the decision maker use his or her experience and insights to assign probabilities. For instance, the decision maker may feel that a will occur 20 percent of the time, b will happen 40 percent of the time, and c will be the outcome 40 percent of the time. The decision maker then assigns probabilities as:

$$Pr(a) = .20 \qquad Pr(b) = .40 \qquad Pr(c) = .40$$

This procedure produces what are called *subjective probabilities*.

The second method has the decision maker express his or her knowledge or assumptions about the process. These translate into relationships between the likelihoods of the various outcomes. From the relationships, the decision maker mathematically determines the probabilities of the possible outcomes. Returning to the above sample space S, suppose that b can occur in twice as many ways as a, and that b and c can occur in the same number of ways. If the decision maker knows or assumes that each of the ways that one outcome can occur is as likely as any other, then he or she can establish that b and c are equally likely and each is twice as likely as a. Solving the implied equations for the desired probabilities gives

$Pr(a) = .2 \qquad Pr(b) = .4 \qquad Pr(c) = .4$

Probabilities developed in this way are called *axiomatic probabilities* (or *objective probabilities* to differentiate them from subjective probabilities). Which ever procedure is used, the association of each possible outcome to its probability is called a *probability function*.

At this point, we want to draw attention to a similarity between the concepts introduced here and the concept of frequency discussed in Chapter 2. In constructing a histogram for data, we computed frequencies for each observation or class of observations. Each of those tallies would be a value of n. In the probability discussion, we find it more desirable to talk about n/N since this ratio also makes use of N, an important piece of information. The value n/N is called the *relative frequency*, and the histogram which plots relative frequencies rather than frequencies is called a *relative frequency histogram*. Given a collection of data, the frequency histogram and the relative frequency histogram will have the same shape. The only difference will be the scaling of the vertical axis. In any application with N observations, the frequency histogram can have rectangles as tall as N units, with the sum of the heights being N. The relative frequency histogram for the same problem can have rectangles no taller than 1 unit, with the sum of the heights equaling 1.

As a simple illustration of this, suppose that you were conducting a market research study on brand preference in which individuals ranked three brands of household cleanser. Suppose brand A received the ranks: 1, 2, 2, 1, 3, 1, 1, 3. You compile the following:

Frequency and Relative Frequency Distributions for Ranking of Brand A

Rank	Frequency	Relative Frequency
1	4	.5
2	2	.25
3	2	.25

Figure 4-2 shows both the frequency histogram and the relative frequency histogram. Graph *a* pictures the frequency histogram; graph *b* shows the relative frequency histogram on a separate pair of axes in which the vertical axis is magnified by a factor of 8.

The same structure of the relative frequency histogram can be applied to provide a graphic representation of probability. For each outcome *b*, draw a rectangle with height equal to the probability of *b*. The resulting diagram is a graph of the probability function and can be used to provide quick communication and appreciation of the probabilities assigned to the possible outcomes.

Figure 4-2

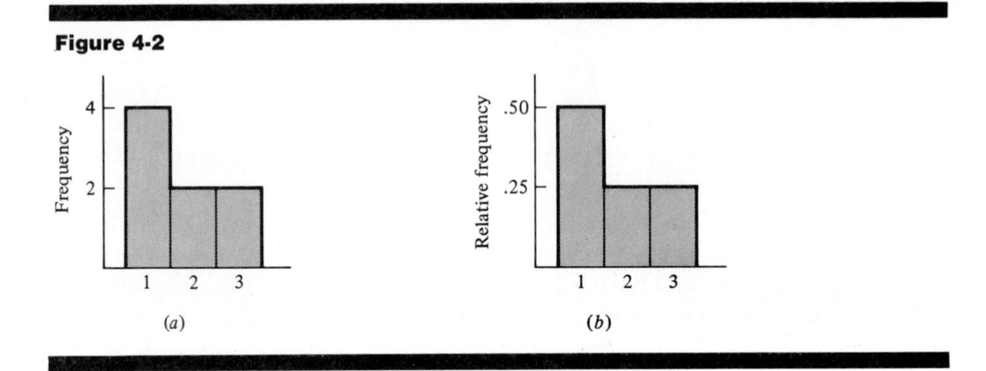

4.3c Continuous Variables

As mentioned above, the sample space S for continuous variables is an interval of values. In particular, S contains an infinite number of possible outcomes. It's unreasonable to expect to be able to assign a probability to each possible outcome. What we can do, however, is to divide the interval S into some collection of subintervals of equal widths. The subintervals should not overlap, and they should make up the entire sample space S. A probability can be assigned to each subinterval. So we get a graph which resembles Figure 4-3. This assignment is an approximation, since the collection of subintervals is an approximation to S. Figure 4-4 is a better approximation still, since the larger number of intervals provides a better approximation to S. If we continue to subdivide, we would get successively better approximations to the assignment

Figure 4-3

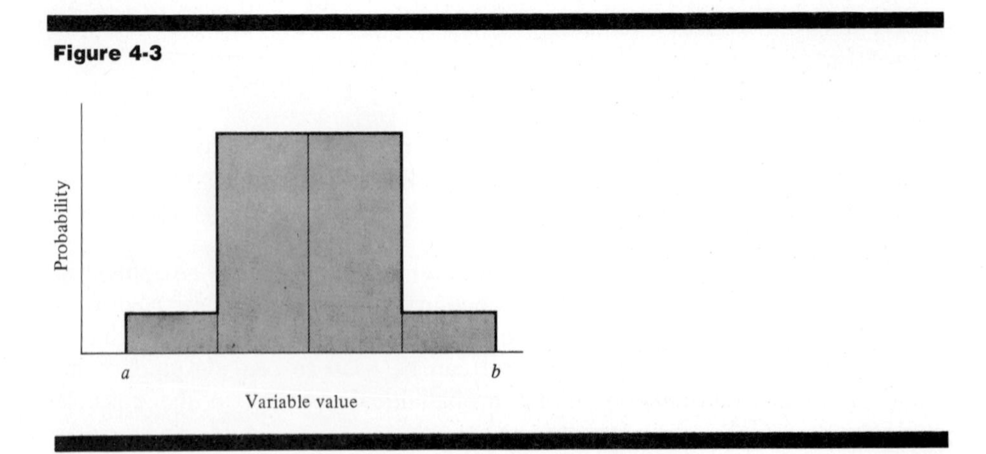

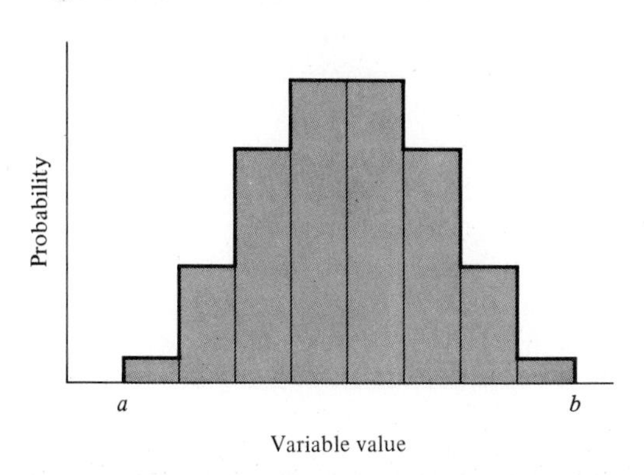

of probability to the sample space S. Eventually, the graph would be very similar to the one pictured in Figure 4-5.

Because the sum of all the heights of the rectangles in Figures 4-3 and 4-4, and all subsequent approximations, is equal to 1, it's reasonable to expect that the area between the curve and the horizontal axis in Figure 4-5 is 1 also. This area is commonly referred to as the area "under the curve." The result of the process illustrated by Figures 4-3, 4-4, and 4-5 is the graph of a mathematical function which describes the way probability is assigned to (or distributed over) the interval sample space S. Such a function is called a *probability density function* (abbreviated as pdf). The probability density function is to a continuous variable what a probability function is to a discrete variable. We would like

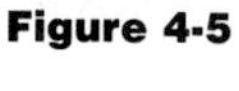

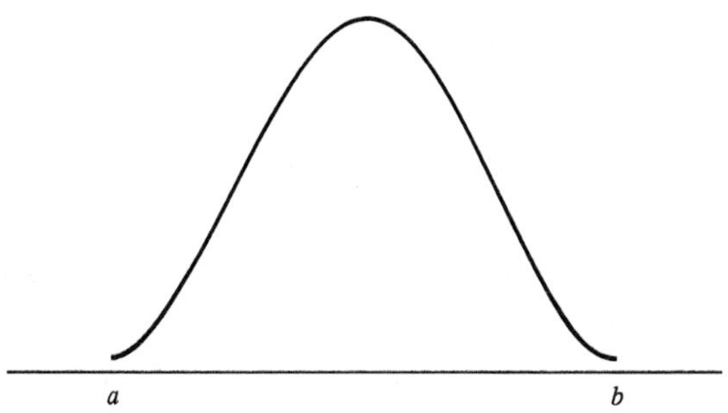

to avoid differentiating between a probability density function and a probability function, since they both describe how probability is assigned. However, the use of the word "density" has a special meaning in mathematics, and it does notify the statistician that a continuous variable is involved.

The probability function for a discrete variable is easily seen to describe the assignment of probabilities for that variable. The probability density function provides the same description in a less obvious way. Its graph gives an excellent visual presentation of this probability assignment to the interval sample space. It can be shown mathematically that the probability assigned to a particular subinterval I can be calculated as an area between the graph of the pdf and the horizontal axis. This relationship can be pictured as shown below.

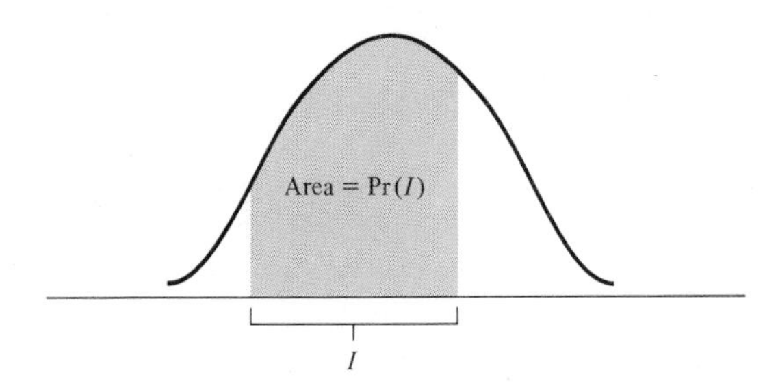

We can think of the assignment of probability to the items in a sample space as a distribution of probability over the sample space. For that reason, it is customary to call the sample space together with its probability function (or probability density function) a *probability distribution*. When the probability distribution is associated with a quantitative variable, we call that variable a *random variable*.

4.4 EVENTS AND THEIR PROBABILITIES

Often in business, management's interest lies in a group of outcomes rather than just one outcome. If, for instance, executives of United Airlines were preparing a schedule change request for the Civil Aeronautics Board, they might formulate a sample space on quarterly profit, measured to the nearest thousand dollars, such as

$S = \{-\$100{,}000, -\$99{,}000, \ldots, \$100{,}000\}$

Of primary interest may be all those possible outcomes which represent realizing a profit. This is expressed as

$A = \{\$1000, \$2000, \ldots, \$100{,}000\}$

Collections of possible outcomes are called *events*. Once a sample space is identified and a probability is assigned to each possible outcome, the probability of an event can be computed. The probability of any event is simply the sum of the probabilities assigned to each possible outcome contained in the event. Thus, for the United Airlines example,

$$Pr(A) = Pr(\$1000) + Pr(\$2000) + \cdots + Pr(\$100{,}000)$$

This process can be related to the graph of the probability function. For discrete variables, the probability of an event is the sum of the heights of all the rectangles associated with outcomes in the event. Events are also of interest when continuous variables are involved. If General Motors were to realize more than a 55 percent share of the market, it might face governmental action dividing it into two or more corporations. Executives of General Motors, then, would be very concerned about all the possible outcomes which represent a threat of governmental action. Symbolically,

$$C = \{\text{market share greater than } 55\%\}$$

For continuous variables, the probability of an event is the area under the portion of the probability density function graph which corresponds to the outcomes in the event. For the General Motors example, we may have a probability density function as described in Figure 4-6, in which case the probability of the event C is identified by the area of the shaded region.

Once we have calculated the probability of an event B, we have readily available the probability of another important event, called its *complement* and written $\overline{B}$. In logical terms, $\overline{B}$ is the negation of B. $\overline{B}$ occurs whenever B does not, and when B occurs, $\overline{B}$ does not. As this suggests, the complement of an event B is the collection of all the possible outcomes that are not in B. This relationship can be pictured as in Figure 4-7, where $\overline{B}$ is the shaded region.

So the sample space has been divided into two parts: B and $\overline{B}$. In any situation, one and only one of the two events will occur. United Airlines will make a profit in the coming quarter, A, or it will not, $\overline{A}$. In any model year, General Motors will get more than 55 percent of the market, C, or it will not, $\overline{C}$.

Figure 4-6

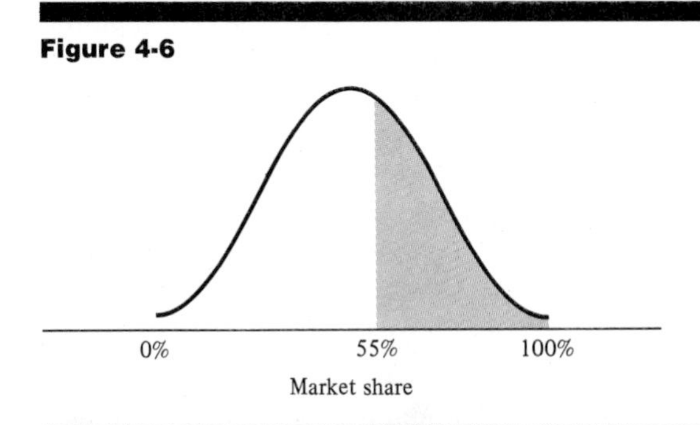

Figure 4-7

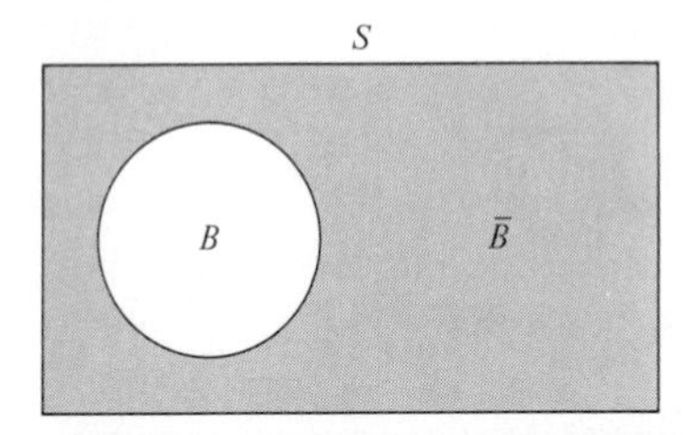

Determining the probability of the complement of an event is easily accomplished. Since the sum of all the probabilities assigned to the possible outcomes must be 1, the total probability assigned to outcomes in $\overline{B}$ must equal whatever is left after the probability of B is determined. So,

$$Pr(\overline{B}) = 1 - Pr(B)$$

Equivalent forms of this equation are:

$$Pr(B) + Pr(\overline{B}) = 1 \qquad \text{and} \qquad Pr(B) = 1 - Pr(\overline{B})$$

The latter equation helps us appreciate that the complement of $\overline{B}$ is B.

The importance of the complement lies in the computation of probabilities. Frequently it will be easier to determine the probability of the complement of an event rather than the probability of the event itself. In later sections, you will make such frequent use of this relationship that you will use it almost instinctively.

Other combinations of events are important also. Suppose you are the director of the research and development group at a large plastics firm. You have submitted requests to the president of the company for additional funds to support two different research projects. To facilitate the discussion, call the event that the first project is funded A, and let B represent the event that the second project is funded. You are interested in the probability that at least one of the projects will be granted additional funding, that is, that at least one of the events will occur. Note this is an important combination of events. Such a combination of events is called the *union* and is denoted by $A \cup B$.

Let's, for the moment, consider two events A and B which have no outcomes in common. Such events are called *mutually exclusive*. If A occurs, B cannot occur, and vice versa. (See Figure 4-8.)

The event $A \cup B$ is made up of the outcomes in A and the outcomes in B. Thus, $Pr(A \cup B)$ can be calculated by summing the probabilities for the outcomes in A and the outcomes in B. Symbolically, this is

$Pr(A \cup B) = Pr(A) + Pr(B)$ when A and B are mutually exclusive (4-1)

Figure 4-8

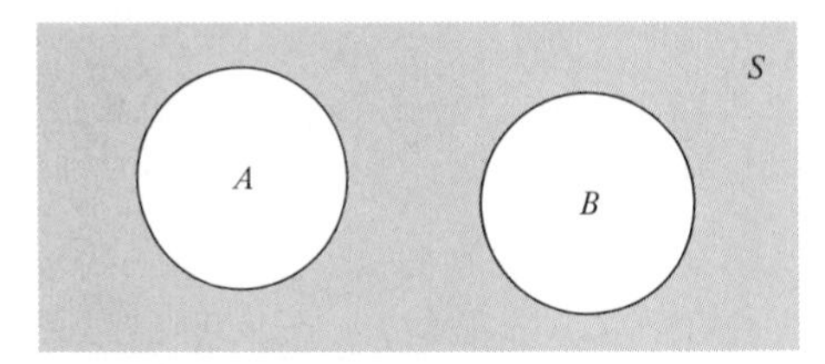

Equation (4-1) is called the *special addition rule*. The word "special" is used because this addition rule works only when there are no outcomes which are both in A and in B.

Now let's look at the case when the events are not mutually exclusive. The event C contains some outcomes which are in D and others which are not. This can be diagrammed as in Figure 4-9. The region labeled "I" represents the outcomes that both of the events, C and D, have in common. This region is called the *intersection* of the events C and D and is symbolized by CD.

To determine the probability of $C \cup D$ we might add $Pr(C)$ and $Pr(D)$. However, when we do that, the probabilities assigned to the outcomes in CD have been included exactly twice—once as outcomes in C and once as outcomes in D. Consequently, $Pr(C) + Pr(D)$ is larger than $Pr(C \cup D)$ by exactly $Pr(CD)$. So

$$Pr(C \cup D) = Pr(C) + Pr(D) - Pr(CD) \tag{4-2}$$

Equation (4-2) is referred to as the *general addition rule*.

We should point out that Equation (4-1) is a special case of Equation (4-2). When two events C and D are mutually exclusive, CD contains no outcomes. Consequently,

Figure 4-9

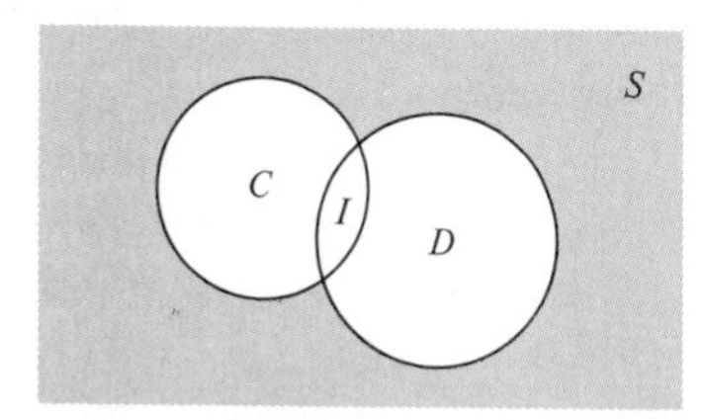

$Pr(CD) = 0$

which makes Equation (4-2) the same as Equation (4-1) in this special case.

The intersection of two events became important in the discussion above when we were considering the probability of the union of two events. However, there are decision situations in which the decision maker is interested in the intersection of two events without any reference to the union. Suppose that General Motors sold 48 percent of all new cars purchased in the United States during the last model year, and that, as in the above discussion, a market share in excess of 55 percent would present a threat of governmental intervention. Executives at General Motors would like to achieve a market share between 48 percent and 55 percent for the current year. This would represent increased market share without the potential for antitrust action. The event

{market share between 48% and 55%}

can be written as the intersection AB, where

A = {more than 48% of the market}

and

B = {less than 55% of the market}

Each of the events A and B individually represents outcomes of interest to General Motors' executives.

We would like to proceed now with our discussion of intersection and develop simple formulas for calculating the probability of the event AB. But we can't. We must first develop the concept of conditional probability.

4.4a Conditional Probability

Conditional probability signifies a probability assignment in the face of new information. Recall that probability is a numerical measure of uncertainty. But uncertainty can change in the face of new data.

Often, for instance, estimates and projections are revised in the light of new information. Chrysler's projections of Plymouth sales for a model year may be quite different in January, when sales figures are known for three months, than projections made in the preceding September. General Electric's estimate of its nationwide inventory of refrigerators may change dramatically when the number of orders received becomes known. Just as estimates of sales and inventory can change in the face of new information, a decision maker should be willing and able to change the assignment of probability. Conditional probability specifies this revision of probabilities. In most decision-making problems and virtually all business decision problems, information is either present or can be gathered. The concept and specification of conditional probability is of value because of the availability of such information.

Conditional probability provides an expression for probabilities of events which may be reassessed after new information becomes known. For events A

and B, the conditional probability of A, given B, written $Pr(A|B)$, is the probability assigned to the event A once the event B has been known to occur. In the Chrysler example above, A may be model year sales in excess of 500,000 units. If B is the event that 200,000 Plymouths are sold in the first three months of the model year, then $Pr(A|B)$ is the probability of selling more than 500,000 cars for the year under the condition that 200,000 are sold in the first three months. The knowledge of B's occurrence revised $Pr(A)$ to $Pr(A|B)$. The knowledge of B's occurrence also means that if A is to occur, it must occur jointly with B. That is, A and B must both occur—the outcome must be in common to both A and B.

Figure 4-10 presents a picture of a sample space S and events A and B. To facilitate the discussion, let's suppose that the probabilities of events are proportional to the areas of the figures representing those events. So, the probability of an event can be represented by the ratio of the area of the figure for that event to the area of the rectangle which represents S.

Suppose we learn that event B has taken place. We now know that only outcomes in B are possible outcomes. Our uncertainty has been reduced to some extent. Knowing that B has occurred, we can now act as if B were the sample space. The conditional probability of event A can be represented by the ratio of the area of AB to the area of B. In mathematical terms, this statement is

$$\Pr(A|B) = \frac{Pr(AB)}{Pr(B)} \qquad \text{if } Pr(B) > 0$$

The stipulation that $Pr(B) > 0$ is included for mathematical precision.

This definition now gives us a means of calculating $Pr(AB)$ as

$$Pr(AB) = Pr(A|B)\,Pr(B) \tag{4-3}$$

Since

$$Pr(B|A) = \frac{Pr(AB)}{Pr(A)}$$

we also have

$$Pr(AB) = Pr(B|A)\,Pr(A)$$

Figure 4-10

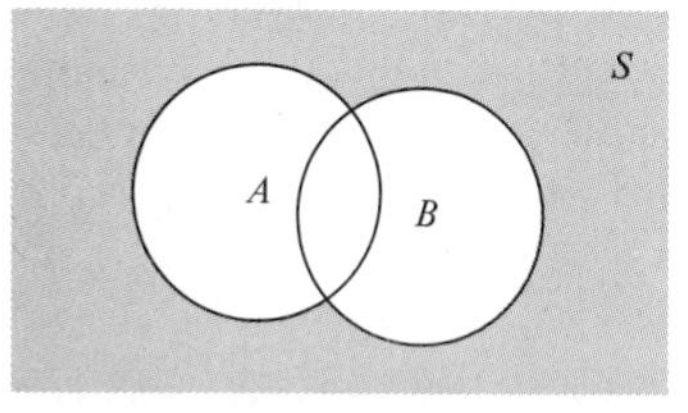

We finally arrive at the *general multiplication rule*:

$$Pr(AB) = Pr(A|B)Pr(B) = Pr(B|A)Pr(A)$$

To solidify these concepts, it's helpful to analyze the following situation. Suppose a bank in a small Midwestern community is considering a merger with a bank in a neighboring town. As part of the information gathering, a bank officer intends to contact some of the other bank's customers to ask their opinions concerning bank characteristics such as courtesy, cooperation, and willingness to be of assistance. Because the expressed opinions may depend upon the amount loaned and outcome (payment versus default); the following information on signature loans over the past six months is gathered and tabulated:

	Amount of Loan			
	<$1000	$1000–$2000	>$2000	Total
Payment	236	118	55	409
Default	48	19	6	73
Total	284	137	61	482

We leave the discussion of this decision problem here to analyze the table. Suppose that each of the 482 accounts has an equal chance of being selected for contact. We can compute probabilities by dividing by 482. So, if

A = account selected was a default

B = selected account involved more than $2000

then

$$Pr(A) = \frac{73}{482} = .151 \qquad Pr(A|B) = \frac{Pr(AB)}{Pr(B)} = \frac{6/482}{61/482} = \frac{6}{61} = .098$$

$$Pr(B) = \frac{61}{482} = .127 \qquad Pr(B|A) = \frac{Pr(AB)}{Pr(A)} = \frac{6/482}{73/482} = \frac{6}{73} = .082$$

$$Pr(AB) = \frac{6}{482} = .012$$

Let's look closely at the calculation of $Pr(A|B)$. The reduction of the compound fraction $\frac{6/482}{61/482}$ to the simple fraction 6/61 is a typical occurrence. But this simple fraction suggests a new table, namely,

	>$2000
Payment	55
Default	6
Total	61

Similar remarks apply to $Pr(B|A)$ leading to the table

	<$1000	$1000–$2000	>$2000	Total
Default	48	19	6	73

Each of these indicates a change in the sample space once an event is known to have occurred. If B is known to have occurred, the sample space which contained 482 possible outcomes is replaced by a sample space with only 61 possible outcomes.

Let's consider two other events.

C = account chosen was paid in full

D = amount loaned was between $1000 and $2000

What information do we get about C from knowing that D has occurred, that is, from knowing that the account selected involved a loan amount between $1000 and $2000? That could be measured by comparing $Pr(C|D)$ and $Pr(C)$.

$$Pr(C) = \frac{409}{482} = .849$$

$$Pr(C|D) = \frac{118}{137} = .861$$

Knowing that D has occurred changes our uncertainty about C only slightly. Also,

$$Pr(D) = \frac{137}{482} = .284$$

$$Pr(D|C) = \frac{118}{409} = .289$$

which is not much change, either. When there is *no change whatsoever* in the probability of an event even though we learn that another event has taken place, we call the two events *independent*. Mathematically, whenever

$$Pr(A|B) = Pr(A)$$

A and B are independent. Incidentally, whenever

$$Pr(A|B) = Pr(A)$$

it is also true that

$$Pr(B|A) = Pr(B)$$

So, two events are independent if knowledge of the occurrence of one does not alter our uncertainty about the other.

If we return to Equation (4-3),

$$Pr(AB) = Pr(A|B)Pr(B) \tag{4-3}$$

and we assume that A and B are independent, the formula becomes

$$Pr(AB) = Pr(A)Pr(B) \qquad \text{when } A \text{ and } B \text{ are independent} \tag{4-4}$$

by substituting $Pr(A)$ for $Pr(A|B)$. Equation (4-4) is called the *special multiplication rule*. "Special" is used here because Equation (4-4) requires the special condition that A and B are independent.

We close our discussion of conditional probability with Bayes' rule. This rule is employed in decision theory (Chapter 9) whenever the decision maker has sample information available. With the data, we revise probabilities. We begin with a probability for an event B and an event A which indicates the outcome of some experiment. The word "experiment" here refers to an information-gathering procedure and might include market surveys, product testing and evaluation, or the like. Suppose you were evaluating your company's new-formula glue. Let B be the event that the new glue is more durable than the old glue. The event A would be the event that a particular test showed the new glue to be more durable. (Recognize that one test does not necessarily provide a perfect indication of the glue's durability.) Your testing may be diagrammed as in Figure 4-11. The equations on the right in Figure 4-11 come from the application of Equation (4-3). The desired conditional probability is

$$Pr(B|A) = \frac{Pr(AB)}{Pr(A)}$$

which expressed the probability that the new glue is more durable given that your test indicated it to be.

We use Equation (4-3) to replace $Pr(AB)$ by $Pr(A|B)Pr(B)$ and get

$$Pr(B|A) = \frac{Pr(A|B)Pr(B)}{Pr(A)} \tag{4-5}$$

From the diagram, we can see that, if the observed outcome is A, then the path resulting from the experiment was either BA or $\bar{B}A$. That is, the new formula is

Figure 4-11

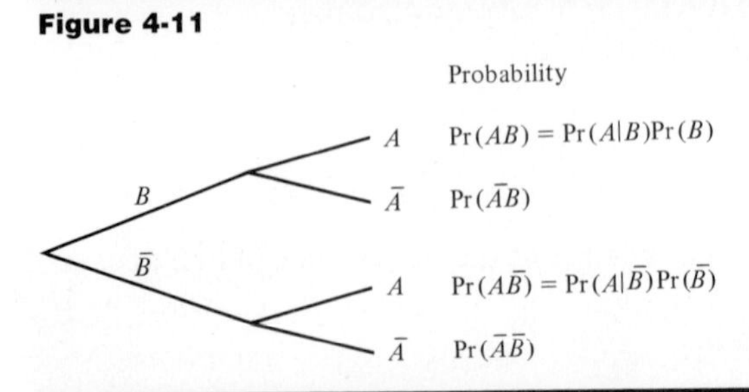

Probability

$\Pr(AB) = \Pr(A|B)\Pr(B)$

$\Pr(\bar{A}B)$

$\Pr(A\bar{B}) = \Pr(A|\bar{B})\Pr(\bar{B})$

$\Pr(\bar{A}\bar{B})$

Figure 4-12

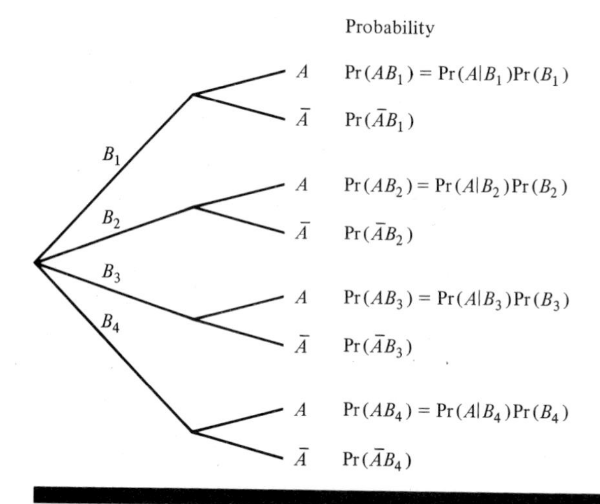

more durable, and it appeared more durable (BA). Or the new glue is not more durable, but it appeared to be more durable in your test ($\overline{B}A$). In that case, the probability that the new glue appeared to be more durable in the test can be calculated using the special addition rule, Equation (4-1).

$$\begin{aligned} Pr(A) &= Pr(BA) + Pr(\overline{B}A) \\ &= Pr(A|B)Pr(B) + Pr(A|\overline{B})Pr(\overline{B}) \end{aligned} \tag{4-6}$$

Substituting Equation (4-6) into Equation (4-5), we get one form of Bayes' rule:

$$Pr(B|A) = \frac{Pr(A|B)Pr(B)}{Pr(A|B)Pr(B) + Pr(A|\overline{B})Pr(\overline{B})}$$

This rationale can be applied to the more complex experimental situation depicted in Figure 4-12. Again, the equations on the right come from Equation (4-3). In this case, suppose you have identified four different levels of durability, and the B's describe these. You would get

$$Pr(B_1|A) = \frac{Pr(A|B_1)Pr(B_1)}{Pr(A|B_1)Pr(B_1) + Pr(A|B_2)Pr(B_2) + Pr(A|B_3)Pr(B_3) + Pr(A|B_4)Pr(B_4)}$$

as the probability that the new glue has durability rating 1 given the test showed it to be more durable than the old glue. As in the previous discussion, the numerator is an expression for $Pr(AB_1)$, and the denominator is the probability of the event A.

Finally, suppose there are n possible events $B_1, B_2, \ldots, B_n$ such that:

1. No two B's have outcomes in common.
2. Taken collectively they include *all* the possibilities.

Suppose, further, that A represents the result of some experiment. Then, for any one of the B's, say B_i, we get

$$Pr(B_i|A) = \frac{Pr(A|B_i)Pr(B_i)}{Pr(A|B_1)Pr(B_1) + Pr(A|B_2)Pr(B_2) + \cdots + Pr(A|B_n)Pr(B_n)} \quad (4\text{-}7)$$

Equation (4-7) is the most general form of Bayes' rule. Again, the numerator is just $Pr(AB_i)$, and the denominator is just $Pr(A)$.

4.4b Distribution Functions and Probability Tables

When we are dealing with a random variable, the concept of cumulative probability plays an important role in applications. The *cumulative distribution function* is defined as

$$F(x) = Pr(X \leq x)$$

As we'll see often, later in the book, nearly all the probabilities used for business decisions are calculated from tables which are readily available. These tables usually give the values of the cumulative distribution function or some modification of these values. Use of these tables is mandatory in most applications of statistics because the probability models of greatest value to the decision maker are not tractable mathematically. The models which are manageable are time-consuming to manipulate. Consequently, the tables are widely used not only by business decision makers but by mathematical statisticians and researchers as well.

4.5 SUMMARY

We have talked about probability as a measure of uncertainty, and a probability model as a model of uncertainty. It may be unclear, at this point, just what we've accomplished in adding probability to our vocabulary. We have not eliminated uncertainty or unpredictability, so the decision problem itself has not changed. What benefits we have realized come from the modeling process in general. Models allow the decision maker to better understand the situation, and we've seen that to be the case here, also. Models usually provide a surrogate for the problem which can be manipulated and analyzed. As you'll see later, the probability distribution does this for us. It is true that you have yet to discover much of the value of probability and probability distributions. Still, as you reflect on Situations I and II, you may already see that some useful information has been generated.

There are two ways in which we are uncertain. We all recognize the uncertainty in the future: Will the union strike against our company this year? Will Congress institute tax incentives for certain investments our firm would like to undertake? This kind of uncertainty is particularly apparent when the future occurrences are beyond our control.

There is also uncertainty which is really lack of knowledge: How many new sales were generated by the advertising campaign we ran last winter? How much did sick leave cost us last month? How many orders for new Pintos did Ford get across the country today? In some cases, this knowledge is easy or inexpensive to obtain, in some cases it is not. Whenever the information is difficult or expensive to collect, or potentially inaccurate, the decision maker faces this uncertainty of incomplete knowledge.

Decision makers, then, face uncertainty whenever future events, particularly those beyond their control, affect profit or whatever measure of performance they feel is important. Decision makers face uncertainty whenever unknown elements of the present have an effect on those measures. It is not difficult to appreciate that *every* major decision problem has one or more uncertain aspects. The decision maker can choose to ignore the uncertainty in the problem or can choose to cope with it via probability. The answer to the question, "When can an executive use probability in making business decisions?" is "Always."

PROBLEMS

1. Identify an application of probability in your major field of study. Point out the conditions present in your situation that would make this application both appropriate and of value.

2. Your company is considering the replacement of its machine for filling aerosol cans. The cost performance of such a machine is evaluated on the basis of its probability of overfill or underfill. A machine having the smallest probability will be the most economical. At your company's current volume, a decrease of .001 in the probability of overfill or underfill would result in weekly savings of nearly $1000. The performance report on the new model indicates that it overfills the cans 5 times out of a 1000 and underfills them 11 times per 1000. What is the probability that a can will be rejected because it is either over- or underfilled?

3. As a consultant you are asked by a client whether he should invest money in an oil venture. Past experience has convinced you that such investments are profitable if the probability of hitting a well is at least .5. Initial geographic surveys of a region indicate that there is a .6 probability that it contains oil. Seismographic surveys were then conducted, which also indicated the presence of oil. These surveys are correct 70 percent of the time. What should be your advice to the client? (Hint: First compute the probability that the area contains oil given the seismographic survey.)

4. Suppose that, because of recent trends in the economy, the management of Procter & Gamble directs its personnel department that from this year on the recruitment of new employees has to be done within the spending limits provided in the budget. This year the company wants to hire 450 management trainees for next year. In the past it has found that 20 percent of those it interviews on college campuses meet its requirements. Of those that meet its requirements, 40 percent accept the offer of employment. How many interviews should Procter & Gamble schedule?

5. Ranox Construction has bid on three major construction projects, all of which are due to begin in the spring. Ranox needs to begin stockpiling certain materials and building up its labor force right away if it receives any of the contracts. Due to the absence of planning in the past, Ranox has lost sizable amounts of time and money. The extent of the required buildup depends on the probabilities of the number of contracts it receives. Gail Williams, president of Ranox, estimates that the probability associated with receiving contract A is .8, contract B is .6, and contract C is .4.

   What is the probability that the firm will receive at least one contract? Any two contracts? All three contracts? (Assume that all events are independent.)

6. Major companies maintain lobbyists in Washington to safeguard their interests. The feedback from lobbyists helps these companies to plan the allocation of resources and funds. On the basis of anticipated congressional actions (as reported by the lobbyist), Tamplin Products estimates the probability of obtaining a major government contract at .8 if the state's senior senator is appointed chairperson of a key committee, and .55 if he is not appointed. The senator's office informs the company that there is a 70 percent chance that the senator will receive the appointment. What is the probability that Tamplin will receive the contract?

7. It is necessary for Emerson Consulting Inc. to plan the efficient use of its labor force and resources depending on the anticipated number of contracts it will receive. The company is negotiating for five separate consulting projects. The negotiators report the following probabilities of obtaining each contract: G.E., .3; Sears, .4; Gulf States, .8; Western International, .1; and American, .6. What is the probability of receiving none of the contracts? All the contracts? Only the G.E. and American contracts? Any three of the contracts?

8. Because of the expense involved in repairing or replacing defective merchandise, management would like to monitor the rate of defectives in the production process. As a quality-control engineer you have been asked to examine a proposed plan for a new production process designed to produce large transformers. In effect, the new process involves two separate processes, each of which produces only part of the transformer. These two parts are then combined into the completed transformer as the last step in the overall process.

   An examination of the machines and operations involved in each half of the production process convinces you that there is a .0014 probability that one part of the process will produce a defective part and a .0025 probability that the other will. What percent of the total output of the process would you estimate to be defective?

9. Management of Esco, Inc., is developing a policy for allocating the advertising budget. Of particular concern is the penetration of the market resulting from ads placed on television. Esco's marketing research department has shown that there is a .7 chance that a member of the target market will watch "Charlie's Angels." The

probabilities associated with "NBC Nightly News," "Six Million Dollar Man," and "Baretta" are .6, .5, and .5, respectively. It appears that the probability of watching any given show is independent of the probabilities for the other shows. If Esco wants to conduct a major campaign by placing ads in each of these four shows, what percent of the target market will be exposed to the ad at least twice in the first week? (That is, what percent of the target market will watch at least two of the four shows?) What percent of the target market will not be exposed to the ad at all?

10. Louise Pickett has just reported the results of her fourth sales call on American Foundries. The purchasing agent has promised to order at least 15 percent of American's annual 100,000-gallon requirement of cleaning solvent from her. He also stated that he would not order more than 55 percent of his firm's requirements from any one supplier. Ms. Pickett believes that he is equally likely to purchase 15, 25, 35, 45, or 55 percent. As production manager, you must begin planning now for the impact of this major purchase on your production requirements. Calculate the probability that the order will be at least 15,000 gallons, 25,000 gallons, 35,000 gallons, 45,000 gallons, and 55,000 gallons. Of what use is knowledge of these probabilities to you?

11. Past experience has shown that the probability of any given typist missing work on a given day is .02. The personnel department has a contingency plan for backup typists to deal with absenteeism on a normal working day. However, during the winter season when flu outbreaks occur, the probability of a typist missing work increases to .15. This results in many complaints from various departments. The personnel department wants to revise its existing plan to account for this seasonal variation. It has learned that local health officials assign a .7 probability to flu outbreak occurring during February. What is the probability that any given typist will miss work on a given day in February? How can this information be incorporated to update the contingency plan?

12. Due to the increasing number of airplane accidents, the Federal Aviation Administration (FAA) has issued strict guidelines regarding the design of aircraft landing gear. To examine the potential compliance with the guidelines, an airplane design engineer in test situations found that, for a certain design, the landing wheels fail to lock into place .1 percent of the time. A manual override system will lock the wheels when the electronic system fails. A warning switch indicates when the wheels are properly locked into place. This switch fails to indicate unlocked wheels .2 percent of the time when the wheels are unlocked and indicates unlocked wheels 1.0 percent of the time when the wheels are, indeed, locked.

    a. What is the probability that the plane will attempt to land with its wheels unlocked and with no warning from the switch?

    b. How would the probability in part *a* (above) change if a second (identical) backup switch were added? (Assume the switches are independent.)

    c. What is the probability that, if the warning has come on, the wheels are not locked in place?

13. Arno Corporation is considering legal action in a patent infringement case. These suits are expensive to undertake but can result in substantial monetary gains if the court ruling is favorable. The chief legal counsel for Arno feels that the company has a probability of .65 of winning the case in court. The chief counsel, however,

indicates that it would be a good idea to try to get a better estimate of this probability. To do this, Arno can utilize a special legal consulting service. This service uses a trained lawyer as a judge, paid volunteers as a jury, and actually conducts a mock trial and reaches a decision. It has agreed with the actual decision 22 of the 25 times that it has been utilized.

a. If the mock trial produces a verdict in Arno's favor, what probability should be assigned to Arno's winning the suit?

b. If the mock trial decision is against Arno, what probability should be assigned to Arno's winning the suit?

Chapter 5

Sampling

5.1 SITUATION I: WHY?

Suppose Rod'n'Reel, Inc., is a small manufacturer of fishing equipment located in Seattle, Washington. All its products are distributed to Pacific Northwest discount stores to be sold as the store's own brand. Sales last year totaled nearly $100,000.

Further suppose that Rod'n'Reel is currently negotiating a long-term contract with Zebco, a nationwide distributor of fishing equipment. The contract would call for Rod'n'Reel to supply a minimum of 200,000 fishing reels per year which would be sold with rods as "beginner's sets." The reel to be supplied has some design features intended to make its operation simple and easily understood by children between six and twelve years old. Zebco's market research department has identified this age group as "potentially hot." Rod'n'Reel currently does not produce a reel identical to the one required by Zebco, although it does have the capability of producing one.

At this stage in the negotiations Rod'n'Reel would receive cost plus 10 percent on all reels supplied up to 300,000 units, and cost plus 15 percent for any units over 300,000. The projected cost of producing the reel is $4 per unit, so the first 300,000 units would be supplied at $4.40 each, and any additional units would be supplied at $4.60 each. Sales of this reel under the contract would be a minimum of $880,000 per year, nearly nine times current sales. Needless to say, Rod'n'Reel is somewhat interested!

The only thing holding Rod'n'Reel back from signing the contract is the clause in the contract relating to defective units. Zebco insists that Rod'n'Reel *absorb all defective units. The contract calls for Rod'n'Reel to replace any and all returned units. Therefore, every defective unit produced would cost Rod'n'Reel $4. This is of no great concern—so long as the proportion of defective units is small. For example, say the defect rate was 5 percent on a day-in, day-out basis. The production cost for a hundred units would be (100)($4) = $400. However, on the average only 95—(.95)(100)—would be good and thus return (95)($4.40) = $418 to Rod'n'Reel. On the other hand, if the defect rate was 10 percent, Rod'n'Reel would receive (90)($4.40) = $396 for each 100 reels they produced—a loss of $4 per hundred units. As the contract calls for a minimum of 200,000 units in the first year, the 10 percent defect rate represents at least an $8000* loss.

The defective rate of the Rod'n'Reel production operation is currently averaging about 6 percent. But the production process called for in the contract involves untried elements and the tremendous increase in volume would require adding a large number of personnel with varying degrees of experience and expertise. Yet, as far as Rod'n'Reel is concerned, the decision hinges on the percent defective produced under the contract, and a decision must be made.

5.2 SITUATION II: WHY?

Suppose Chevrolet Motor Division of General Motors is planning the introduction of a smaller, light-weight "full-size" car in the fall of this year. With the two most recent introductions, the marketing group has had to rely on extensive advertising efforts to approach the projected first-year sales.

Several factors differentiate this introduction from its two predecessors. The marketing group feels that two of these are of utmost importance. First, the new car being developed is considerably larger and more expensive than the last two. Second, while projected first-year sales of the two preceding introductions have been in the neighborhood of 200,000 to 250,000 units, sales projections for the new car call for 375,000 to 400,000 units sold in the first year. Further, selling fewer than 300,000 of the new cars in the first year would be considered somewhat of a setback.

Rather than emphasizing the novel aspects and gimmicks of the new car, marketing executives would rather advertise something of major value to the buyer. In particular, they have approached Frank Matlow, executive vice president, regarding the new car warranty. Current plans are to offer the usual one-year, 12,000-mile warranty on all parts except tires. The marketing group has asked Matlow about the feasibility of a 5-year, 60,000-mile warranty on the power train (engine, transmission, and differential) and a one-year, 12,000-mile warranty on all other parts except tires. Since the cost of major engine work can easily exceed $250, a mere 20,000 cars returned for major engine work can cost Chevrolet Motor Division in excess of $5 million. That does not include transmissions, carburetors, and other costly parts. So, the warranty decision is no small matter.

The economic consequence of changing from the one-year, 12,000-mile warranty to a five-year, 60,000-mile warranty depends upon the trade-off between increased unit sales as a result of the extended warranty and increased repair cost under warranty.

It is currently mid-June and 10,000 cars have already been produced. These can be used in tests to determine the feasibility of the suggested warranty or some desirable alternative. (A three-year, 36,000-mile warranty on the power train, while not being as valuable to the consumer, would be a major selling point.) Each car tested would be driven for 60,000 miles on the test tracks of General Motors, and the extent of repairs needed would be recorded. After the tests, the cars would need some cleanup and repair work (including tires) before they could be sold as used cars. The used cars would sell for about half the base sticker price of $4798. Production costs on the cars average $3078 (including the options).

WAIT A MINUTE! These tests are going to be quite costly. Frank Matlow jots down the following estimates:

Production cost	$3078
Maintenance (tires, oil, lubrication, gasoline, etc.)	1330
Personnel	3000
Clean-up, repair	230
Total	$7638
Selling price	−2399
Testing cost per vehicle	$5239

Matlow certainly cannot test 10,000 cars! The test costs would exceed $52 million. Half of the cars sold could be repaired for that price.

5.3 SITUATION III: WHY?

Wilson is a large manufacturer of sporting goods equipment. The tennis division has become quite important, with the tremendous increase in popularity of the game. This popularity is partly due to the dramatic increase in television coverage of professional tennis tournaments.

At present, Wilson manufactures and markets a full line of wood and steel rackets. Unfortunately for Wilson, the strong market for tennis equipment has attracted some very effective competitors. *As a result of the efforts of some competitors, Wilson sees a trend toward composition rackets—especially for the top of the line.*

Suppose Wilson is not willing to give up this segment of the market and so has initiated some research and development work on a composition aluminum-epoxy racket. Some experimental rackets have been produced and given to a number of tennis players of varying capabilities. The objective of this experimentation is to determine the playing characteristics of the proposed racket. After a specified length of time the rackets are "called back" for examination and the players interviewed for opinions about the rackets. Some of the more accomplished players complain about excessive torque. That is, the players feel a lack of control of the racket head due to a twisting movement. Upon examining the experimental rackets Wilson finds both delamination and small fatigue fractures in the frame. As a result, the research and development unit is instructed to work on alternative constructions. It does, and it has come up with a perfected model. Wilson is happy with the new model and is about to start full production.

Suppose your firm has been hired by Wilson to develop a quality control test for the composition frames. Wilson has informed your firm of the delamination problem under torque and wants to guard against producing and marketing rackets that will delaminate in use. The ill will generated by marketing an inferior product could be devastating and indeed affect the consumer opinion of all Wilson products. Wilson would rather not market the racket than run such a risk.

Your firm has built a special vice which is designed to twist the racket frame and record the number of degrees of twist prior to delamination. A racket frame is twisted until it delaminates and the degrees of twist are recorded. Wilson has set an acceptable standard for the number of degrees a frame should withstand without delamination. Your firm is proposing the use of your vise to monitor production of the new composition frame.

So, where are we? We have seen three different situations and all three have the same need but for different reasons.

1. Rod'n'Reel needs to know what the true proportion of defective reels would be if it manufactured the product for Zebco. If it knew what the true proportion would be, Rod'n'Reel could easily determine whether the contract presents a profit or loss—thus, whether it should sign the contract or not.
2. Frank Matlow needs to know what the mean cost of repairs per car would be if the warranty was for five-years, 60,000 miles. This figure is necessary in order to determine what the cost would be of increasing the warranty.
3. Wilson needs to know if and when its process is producing too many defective frames. "Too many" would be translated by Wilson into a proportion greater than a certain value.

All three need to know the true value for some characteristic. In order to determine true values for characteristics, one must examine or measure *all* the elements in the population. In our three situations, that is unreasonable. Rod'n'Reel cannot examine all the reels it could produce. Theoretically, there are an infinite number and they would never stop. Thus, the company would never quite determine the true proportion of defects. Frank Matlow cannot test all 10,000 existing cars, as the expense would exceed any reasonable limit. Wilson would certainly not agree to have every frame tested, as the test destroys the product. Wilson would finish each day knowing exactly what proportion of their frames were defective—but have nothing but delaminated frames left.

Via the three situations we see three of the most important reasons for *sampling*.

1. Populations are large or perhaps even infinite.
2. Observations or measurements are expensive.
3. Measurement or testing is destructive.

SAMPLING MEANS TESTING, OBSERVING, OR MEASURING *PART* OF THE TOTALITY OR POPULATION.

The simple fact is that in most situations the economic conditions dictate that we must deal with sample information, rather than a census (100 percent sample). Accepting this fact, the real question becomes: "How should a sample be drawn?"

There are, of course, any number of ways we could physically draw or select a sample of elements from a totality or population. However, we will make use of the concepts introduced in Chapter 4. If we know the population and select an element in a very special way (such that each element has an equal chance of being selected), we can determine the probability of every possible sample outcome. It is the fact that each element has an equal or known chance of being selected that allows us to mathematically determine the probabilities associated with the possible outcomes. Without knowing the chance of being drawn for each element, we cannot determine the probabilities associated with an event or outcome. This link is very important, and the process is referred to as *deduction*.

In statistics we reverse this process and conclude or infer about populations based upon sample information. This process of generalizing from the specific sample outcome to the population is known as *induction*. Again, the link that allows us to infer about populations by examining sample results is *the very special way samples are drawn*. Without this link we have no objective, predictable connection between sample observations and populations.

The simplest form of probability sampling is known as *simple random sampling*, which is commonly referred to as "random sampling." Random sampling is actually the prerequisite for a majority of existing statistical techniques, and the only type of probability sampling considered in this text.

5.4 RANDOM SAMPLING

Any procedure of selecting elements where every possible combination of n elements has an equal chance of being drawn would constitute random sampling. Let us restrict our discussion at this point to finite populations. In standard notation, N represents the total number of elements in the finite population, while n is the number of elements in the sample. There are $N!/n!(N - n)!$ possible combinations of the N elements where each combination contains n elements. If n elements are selected in such a way that each of the $N!/n!(N - n)!$ possible combinations is equally likely (with equal probability), the process is known as random sampling and the specific set of n elements selected is called a random sample.

How do we draw a random sample? One way would be to enumerate the $N!/n!(N - n)!$ possible combinations, assign identical poker chips to each, place all chips in a large container, thoroughly stir them, and draw out one chip without looking. Since theoretically each chip has an equal chance of selection, and each chip corresponds to one of the possible combinations, each combination would have an equal chance of selection. The problem with this approach is that N is usually large for most practical applications. For large values of N, $N!/n!(N - n)!$ becomes rather enormous and such a procedure is fairly unrealistic. For example, say we want to draw a random sample of $n = 5$ elements from a small population of only $N = 100$. We have

$$\frac{100!}{5!95!} = 75{,}287{,}520$$

possible combinations of size 5. It would take *a long time* to enumerate the 75,287,520 combinations. Fortunately there is a simple process that gets us to the same point. If we select the first element in our sample such that each element in the population has an equal chance of selection and such that, for each subsequent selection (second through n), the element selected has an equal chance with those elements remaining, then all possible combinations of size n have an equal chance of selection. This can indeed be shown to be true. The advantage of this approach is that we do not need to enumerate each of the $N!/n!(N - n)!$ possible combinations.

Here is how we draw a random sample. First we assign a unique code number to each element in the population, and all code numbers must have the same number of digits. If, say, $N = 342$, we would assign three-digit code numbers to the elements, while if $N = 1428$, our code would have to have four digits for each element to have a unique number. Next we draw code numbers with equal probability. This is accomplished by using a random number table such as the one shown in Table A-2 in the Appendix. Random number tables are generated by drawing single digits 0 through 9 with equal and independent probabilities. Thus, the order of digits in the random number table is "random."

In effect, the random number table represents a random selection or arrangement of code numbers. Therefore, in order to draw a random sample of unique code numbers, we arbitrarily select a starting position (row and column or columns depending on the number of digits in the code numbers) and adopt a sequence of movement. Note we do not look at the table of random numbers, select a code number, and glance at some other location to obtain another number. As we cannot be assured that our eyes will stop on each digit or combinations of digits with equal probability, this procedure would or could alter the probabilities. The specified sequence of movement is followed until we obtain n unique code numbers associated with n elements in our population. If in following the sequence we come to a code number that either has already been selected or is not associated with an element in our population, it is skipped.

As an example, suppose we were involved in the auditing of a small retail store. Specifically we want to draw a random sample of accounts receivable in order to verify the stated figure on the income statement. Assume the ledger shows 149 accounts receivable. We assign code numbers 001 through 149 to the accounts such that we know which account is associated with which code number. Next we select a starting position in our random number table. Say we chose page 1, the first three digits in column 5, line 22, and our sequence was to move down the column. If we wanted a random sample of $n = 10$ accounts, our sample would contain the accounts with the following code numbers:

1. 133
2. 040
3. 023
4. 016
5. 148
6. 105
7. 047
8. 053
9. 058
10. 009

Note that 105 was skipped the second time it was encountered. If we had gone through all 200 lines in our table and still not reached n elements in our population, we would merely have continued to another column until we obtained n elements. Our sample is one of the $149!/10!139!$ possible combinations and our process was such that each possible combination had an equal chance of being drawn. Thus, our selected 10 accounts constitute a random sample.

The above procedure is straightforward and relatively simple to apply for finite populations. If our population were infinite, we would have a problem. We would never stop assigning code numbers. Let us view infinite populations as the output of a process. There is really no way to be assured that every possible combination of size n has an equal probability of selection. However, if the process is stable over time and sequential observations are independent, the sequential output of the process can be viewed as a random sample from the infinite population. So, when we treat the next n items produced by a machine or process as a random sample, we are assuming that the difference between sequential elements is random.

5.5 PARAMETERS VERSUS STATISTICS

Throughout this text we will refer to both parameters and statistics.

A *PARAMETER* IS A CHARACTERISTIC OF A POPULATION. A *STATISTIC* IS A CHARACTERISTIC OF A SAMPLE.

A parameter is any numeric value computed from *all* elements in a population. A statistic is any numeric value computed from a sample of elements. Thus, populations have parameters and samples have statistics. It has become customary to denote parameters by Greek letters.

You may recall in Chapter 2 that all data sets were regarded as populations. Therefore such characteristics as means and variances were denoted by Greek letters.

5.6 SAMPLING DISTRIBUTIONS

HERE COMES THE MOST IMPORTANT CONCEPT IN STATISTICS!

If you truly understand this section, all that follows will be logical and almost obvious. The concept of sampling distributions is indeed the foundation for statistical inference. In all seriousness, if you master this section, this is going to be *one easy course*! On the other hand, if you do not, what follows will be black magic. You will find yourself memorizing steps that appear to have no logic, and you will just be plugging numbers into formulas and regurgitating conclusions.

HERE IT COMES!

Every statistic has a sampling distribution. A sampling distribution gives every possible value the statistic could assume and the corresponding probabilities of each. That is, if we draw a random sample of size n from a finite population, there is a certain set of values the statistic could assume. And, each value or interval of values has a certain probability of occurring or being drawn. The set of values and corresponding probabilities constitute the sampling distribution for the statistic.

Let us generate a sampling distribution. First, we must apologize for the following example, as it certainly does not justify sampling, sampling distributions, or statistics as a useful area of study. The problem is that we need to deal with a very small population in order to show *every possible sample* of size n.

Say we have five coins in a box—two quarters, one dime, one nickel, and one penny—and we choose to draw a random sample of two coins. The two quarters are easily distinguished, as one was minted in 1968 and the other in 1975. There are ten possible samples of size $n = 2$ that can be drawn from these $N = 5$ coins:

1. Q_{68}, Q_{75}
2. Q_{68}, D
3. Q_{75}, D
4. Q_{68}, N
5. Q_{75}, N
6. Q_{68}, P
7. Q_{75}, P
8. D, N
9. D, P
10. N, P

Our random sample of two coins *must* be one of these ten combinations. Say we are interested in the sum or total value of the two coins. Let's denote the sum as T. T is a statistic, as it is computed from sample information. Each possible sample has a value of T shown in Table 5-1.

Since we are drawing a random sample, each of the ten possible samples will have an equal chance of being drawn. Thus, each possible sample has probability 1/10 or .1 of selection, as shown in Table 5-1. For convenience let us rearrange the values for T in ascending order and show their corresponding probabilities of occurrence.

Table 5-2 is the sampling distribution of T. The sampling distribution shows both the possible values the statistic could assume and the corresponding probability for each value. Ordinarily we find it convenient to view a sampling distribution graphically. Thus Figure 5-1 is a graphic portrayal of our sampling distribution of T. Note that Figure 5-1 shows both every possible value for the statistic T (in this case there are seven possible values) and the corresponding probability for each. That constitutes a sampling distribution!

Table 5-1

Sample	T	Probability
1. Q_{68}, Q_{75}	.50	.1
2. Q_{68}, D	.35	.1
3. Q_{75}, D	.35	.1
4. Q_{68}, N	.30	.1
5. Q_{75}, N	.30	.1
6. Q_{68}, P	.26	.1
7. Q_{75}, P	.26	.1
8. D, N	.15	.1
9. D, P	.11	.1
10. N, P	.06	.1
		1.00

Table 5-2 Sampling Distribution of *T*

T	Probability
.06	.1
.11	.1
.15	.1
.26	.2
.30	.2
.35	.2
.50	.1

MAKE SURE YOU UNDERSTAND THIS:

We draw a random sample of size n from a population. We compute a value from the n observations, and it is called a *statistic*. Thus, each sample of size n has a value for the statistic. As every possible sample has an equal chance of selection, the values for the statistic have known probabilities. The probability distribution of the statistic is known as the *sampling distribution* of the statistic. Although in practice we draw only one random sample of n observations and therefore observe only one value for the statistic, the probability distribution (sampling distribution) of the statistic DOES EXIST! We could actually view the sampling process as randomly selecting one value of the statistic from its sampling distribution.

5.7 MATHEMATICAL EXPECTATION

There is another very important concept which we should discuss at this point. It is the concept of *mathematical expectation*. We can define a *random variable* as any quantitative variable with a probability distribution. The mathematical expectation of a random variable is defined as the weighted average of the random variable, where the weights are the probabilities. That is, if x is a discrete random variable with probability function $f(x)$, then

$$E(x) = \Sigma x f(x)$$

where $E(x)$ is read as "the expected value of x" and the products of x and $f(x)$ are summed over all values of x. $E(x)$ is actually the average or mean value of x per trial or observation.

A statistic is a random variable, as its sampling distribution is indeed a probability distribution. Therefore, every statistic has an expected value. For example, let us take our coin-sampling problem. T, which is the sum of the values for two coins selected from five via random sampling, is our statistic and

Figure 5-1 Sampling Distribution of *T*

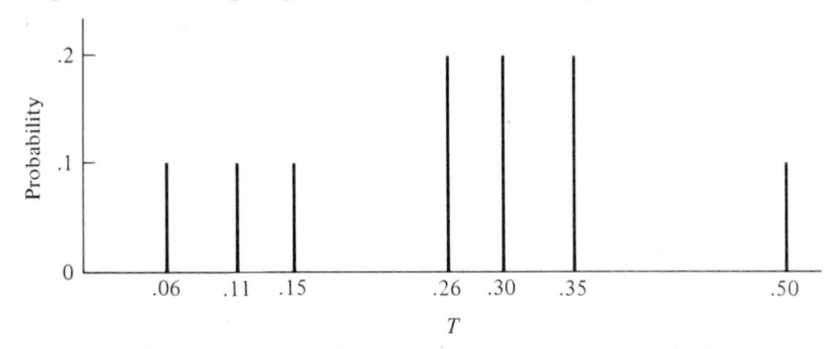

its sampling distribution is given in Table 5-2. The expected value of T is therefore

$$E(T) = (.06)(.1) + (.11)(.1) + (.15)(.1) + (.26)(.2) + (.30)(.2) + (.35)(.2) + (.50)(.1)$$

$$E(T) = .006 + .011 + .015 + .052 + .060 + .070 + .050 = .264$$

Theoretically, if we were to draw a random sample of two coins from the five, compute and record T, replace the two coins and sample again, and again, and again forever, the mean value of T per trial or sample would be exactly .264. That is, we can view the expected value of a statistic as its "long run average."

Clearly, mathematical expectation is the foundation for insurance premium determination and finding payoffs for virtually any form of gambling.

One further note on mathematical expectation. We have discussed the expected value for a discrete random variable. Continuous random variables also have expected values. The only difference is that instead of summing a finite number of products, we must use the concept of the definite integral. Thus, the expected value of a continuous random variable x defined over the interval a to b with probability density function $f(x)$ is by definition

$$E(x) = \int_a^b xf(x)dx$$

Don't worry, we are not going to have you compute the expected value for a continuous random variable. The point is, however, that continuous random variables do have mathematical expectations.

5.8 SUMMARY

We have seen both the reasons for sampling and in detail how one draws a random sample from a finite population. And we have seen what assumptions are necessary for us to regard a sequence of output from a continuing process as a random sample. A very deliberate distinction was made between parameters and statistics. Finally, the most important concept in the area of statistics was introduced—namely, the sampling distribution of a statistic—along with mathematical expectations.

IF YOU DO NOT UNDERSTAND EXACTLY WHAT A SAMPLING DISTRIBUTION OF A STATISTIC IS AND THE MEANING OF MATHEMATICAL EXPECTATION—DO NOT GO ON. GO BACK AND READ SECTIONS 5.6 AND 5.7 AGAIN.

PROBLEMS

1. Identify an application of sampling in your major field of study. Point out the conditions present in your situation that would make this application both appropriate and of value.

2. Omac Industries is in the process of designing a test market for its newly developed lawn mower. The company wishes to test four different price levels, so it needs four cities with similar population and climatic characteristics. The research department has found ten medium-sized cities that meet the requirements. Dianne Radley, product manager for the new product, calls you in and says: "We want to take a random sample of four of these ten cities. A random sample means that every possible combination of four cities must have an equal chance of being selected. Therefore, I want you to identify all the possible combinations of four cities that we could select from these ten."

   Is Ms. Radley's statement concerning a random sample correct? If you were to comply with her request, how many combinations would you be able to isolate?

3. Barbara Dessler, comptroller for Stacy Manufacturing, is concerned about the firm's short-term cash requirements. Firms need short-term cash to meet their day-to-day transaction needs. The sales department has submitted bids on three government contracts and is certain that the firm will be awarded at least one of the contracts. The probability of being awarded just one contract is estimated at .5, two contracts is .3, and all three contracts is .2. Each contract will place a short-term demand on the firm for $200,000. In order to plan future cash requirements, Ms. Dessler needs to have some idea as to the magnitude of the cash demands that will result from the three contract bids. What is the expected cash demand resulting from these bids?

4. Evaluation of two mutually exclusive and independent projects: Bill Wheeling, president of Wheeling Enterprises, is investigating two real estate investment opportunities. The corporation's research department has isolated four potential levels of success and has assigned probabilities to the occurrence of each. These are presented below.

Investment Opportunities with Payoffs

Northwood Center		Western Hills	
Payoff	Probability	Payoff	Probability
−4,000,000	.2	0	.4
+1,000,000	.3	2,000,000	.3
+4,000,000	.3	4,000,000	.2
+8,000,000	.2	6,000,000	.1

Which should Mr. Wheeling choose, assuming that his resources will allow him to make only one of the investments? Why?

5. Ralph Elwood is trying to decide whether or not to appeal an unfavorable judgment against his firm. This judgment has given a bad name to the firm which he wants to clear. At the same time he cannot overlook the financial implications should the appeal fail. His lawyers believe that the appeal has a 40 percent chance of succeeding. If it suceeds, it will be worth $800,000 to the firm. However, if it fails, it will cost the firm an additional $500,000. What should Mr. Elwood do? Why?

6. The objective of many firms is to maximize profits. The Standard General Insurance Company is no exception. It is considering offering a major medical plan to college students. The unique feature of the plan is that it will be a $5000 deductible. That is, the insurance company pays nothing until the total bill passes $5000. Studies of college students reveal that 2 out of every 1000 will be able to file a claim in a given year. The average cost of these claims to the insurance company would be $2000. What is the minimum annual premium the company could charge and still break even? (Assume no administrative cost.)

7. A production process is designed to produce five hundred 17-centimeter perforated copper tubes per hour. A length between 16.94 and 17.06 centimeters is acceptable. About 1 tube out of every 500 is unacceptable. This is usually not discovered until it is returned from a customer. These returns are viewed as quite acceptable by the firm. However, on occasion the machine that determines the length of the tubing loses its adjustment and begins to produce about 50 unacceptable tubes per hour. When these reach the firm's customers, there are serious financial consequences for the firm. What would you suggest the firm do? What assumptions do you find necessary?

8. A local university is considering opening a campus tavern. Before proceeding with a full-scale investigation of the matter, the administration wants to find out what the parents of the 13,000 undergraduate students think of the idea. Each student's name and record, including parents' names and addresses, is kept on file in alphabetical order. The university feels that it can spend only $100 for this part of its investigation. How should it proceed? Explain in detail.

9. Gayle Myers, a staff attorney for the FTC (Federal Trade Commission), was recently assigned a deceptive-advertising case. The firm in question readily admitted that its advertising claim was false. In fact, its defense was that the claim was so blatantly false that no one would be fooled by it. The claim was meant to amuse, not

to deceive. Ms. Myers decided that she needed to determine if consumers were mislead by the claim. How should she proceed? Explain in detail.

10. As an assistant to the purchasing manager of Talbert Construction, you decide to test a random sample of the 2000 iron stress rods that recently arrived. You perform these tests to make sure that not more than the acceptable level of defective rods are present in your purchase. You randomly pick 25 and find that 15 are below standard. Therefore, you reject the entire shipment. The supplier's sales representative is furious. Deleting certain phrases, his basic question is, "How can you justify rejecting 2000 rods when you found only 15 defective? In fact, you only tested a few." How would you respond?

Chapter 6

Probability Distributions

6.1 SITUATION I: WHY?

Consumer research organizations are one of the primary sources of statistical information used by business executives, and they provide a wide range of marketing information. Studies might include brand-preference surveys, the test marketing of a new product, or consumer preferences on particular qualities of, perhaps, laundry detergents or power lawn mowers. Data collection varies from self-administered questionnaires or personal interviews administered by relatively untrained interviewers to consumer panel discussions which require a moderator with extensive training.

In the consumer panel discussions, six to ten individuals are assembled in a relaxed setting. The product to be evaluated is shown, and a brief description or demonstration of its purpose is presented. The moderator then leads the group in a discussion of the advantages and disadvantages of the product. Groups such as these are easily distracted and tend to digress from the issues of import to the manufacturer. It is the interviewer's responsibility to keep the group's discussion on the topics of interest and yet not stifle the conversation or offend any of the participants. The training of such interviewers includes education, usually a college degree in the behavioral sciences, and two to three years of working with groups as an observer. During the training period, the prospective moderator progresses from no involvement in the discussion to, finally, the role of the primary interviewer. The training of such an interviewer is indeed expensive.

After a consumer panel discussion is completed, the moderator presents a report to the manufacturer. The quality of this report provides some indication of the interviewer's effectiveness in the discussion. The director of the consumer research organization is not personally involved in the interaction between the interviewer and the client manufacturer. Consequently, the director cannot judge the report by reading it, for he or she does not have complete understanding of the objectives of the study. The director can, however, contact the manufacturer and ask if the report was deemed satisfactory.

Suppose the director of such an organization is concerned about the performance of one of the organization's experienced interviewers. The employee is in the middle of lengthy and bitter divorce proceedings, and there may be some evidence that the individual's performance is affected.

If the performance of the interviewer in question is negatively affected by the divorce proceedings, the director can either dismiss the employee or initiate counseling. Each of these actions would require that a potential replacement be trained. If the interviewer is not negatively affected, dismissal or the initiation of counseling may have severe negative effects on the morale of all the employees. So the director must proceed with care in judging the interviewer's effectiveness.

Based on historical records, the director has concluded that an experienced interviewer will produce reports rated as unsatisfactory by clients about 15 percent of the time. It should be recognized, however, that the majority of the unsatisfactory reports are rated so because the product was unfavorably received by the consumer panel. That is, if the interviewer has a representative mix of products which were favorably and unfavorably received, the probability of a report's being rated unsatisfactory by the client is .15. The director has contacted ten recent clients serviced by the interviewer in question. Four of them rated the reports they received as unsatisfactory.

The question is: How does the director view this information? Is the observation of four unsatisfactory reports out of ten extreme enough to indicate that the interviewer is adversely affected by the divorce proceedings? The director needs to determine how likely it is that the interviewer would produce, under normal conditions, four or more unsatisfactory reports out of ten reports submitted.

6.2 THE BINOMIAL DISTRIBUTION

In Chapter 4, we mentioned two methods of assigning probabilities to the items in a sample space. Of particular interest here is the axiomatic approach. This procedure is most often employed by recognizing the elements of the situation which are common to some well-known phenomenon and developing measures of the unique characteristics of the situation at hand. As might be expected, one must have the familiarity with many phenomena to allow this kind of probability assignment. The recognition of well-known processes gives a form to the probability distribution. What remains is the measurement of the unique characteristics which identify this particular application. These unique characteristics are called *parameters*.

In business decisions, it is often the executive who is familiar with the particular situation at hand. It is the executive who is in a position to recognize commonalities between a particular problem and well-known phenomena. So, it is essential that the business executive have some appreciation for certain well-known processes. In this chapter, we examine the essential characteristics of two of these processes. These two are the most important for dealing with business problems, as they are found most often in applications.

We return to Situation I. Let's extract from that problem some key elements. First, we have a process, the rating of ten reports, which is comprised of several very simple processes called *trials*. Each of these trials results in one of two possible outcomes: satisfactory or unsatisfactory. In general, we can talk about the outcomes in terms of success and failure. As the ratings are from different individuals, they are independent. Thus we have a probability of failure which remains constant for each trial. Our primary interest lies in the total number of failures, and we seek a probability distribution on the total number of failures. The binomial distribution applies under these circumstances.

The binomial distribution can be used for experiments with these key elements:

1. The process is the joint occurrence of n trials with identical, dichotomous sample spaces.
2. The trials are independent.
3. The probabilities assigned to the possible outcomes are the same for each trial.

The binomial distribution assigns probabilities to the total number of failures or successes obtained on n trials.

Before we analyze Situation I, let's take a close look at an example with fewer trials. Suppose your employees report late for work 10 percent of the time. You need to assign a crew of three to begin a particular project first thing in the morning. Since you keep no records on the promptness of an individual employee, you cannot make use of the promptness criterion when selecting the three. You can only assign a probability of .10 that any given employee will report late. Still you would like to know the probabilities associated with the possible values for the number reporting late. You select three to begin on the project in the morning. This gives you a process with $n = 3$ trials. Since the three employees live in different areas of the city, it seems reasonable for you to assume that the three employees report for work independently. So the trials are independent. Each employee will either be prompt, P, or late, L, for work tomorrow. The three trials then have identical, dichotomous sample spaces. The three employees can be arranged in order (alphabetically, perhaps), and you can identify an outcome such as PPL, the first two employees were on

Figure 6-1

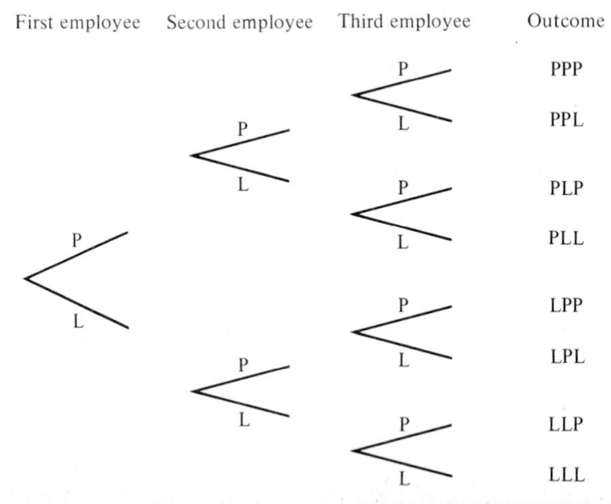

time but the third was late. The diagram in Figure 6-1 illustrates the sample space.

Since you are assuming the employees arrive independently of each other, the probability of each outcome is just the product of probabilities. Table 6-1 shows the calculations of these probabilities.

The number reporting late was included in Table 6-1 to illustrate the relationship between probability and the number reporting late. Note that each of the outcomes with one employee reporting late has a probability of .081. If we calculate the probabilities associated with a particular number reporting late, we get Table 6-2.

This last table identifies the binomial probability distribution with three trials and a probability of failure (reporting late) of .1. The sample space is

$$S = \{0, 1, 2, 3\}$$

and the probability function is

$$f(0) = .729, \quad f(1) = .243, \quad f(2) = .027, \quad f(3) = .001$$

This distribution can be depicted as

x	0	1	2	3
$f(x)$	.729	.243	.027	.001

where x is the number reporting late. This probability distribution can be pictured as in Figure 6-2.

Looking carefully at Table 6-2, we notice two distinct parts of each probability. The first is the part which is a product of probabilities. In numbers, these are .729, .081, .009, and .001, the probabilities computed in Table 6-1. The

Table 6-1

Outcome	Probability	Number Reporting Late
PPP	$(.9)(.9)(.9) = (.9)^3 = .729$	0
PPL	$(.9)(.9)(.1) = (.9)^2(.1) = .081$	1
PLP	$(.9)(.1)(.9) = (.9)^2(.1) = .081$	1
PLL	$(.9)(.1)(.1) = (.9)(.1)^2 = .009$	2
LPP	$(.1)(.9)(.9) = (.9)^2(.1) = .081$	1
LPL	$(.1)(.9)(.1) = (.9)(.1)^2 = .009$	2
LLP	$(.1)(.1)(.9) = (.9)(.1)^2 = .009$	2
LLL	$(.1)(.1)(.1) = (.1)^3 = .001$	3

second part is the count, that is, the number of outcomes which share the same number of failures. There is one outcome in Table 6-1 with no late arrivals (*PPP*). There are three outcomes with one late arrival (*PPL*, *PLP*, *LPP*), three outcomes with two late arrivals (*PLL*, *LPL*, *LLP*), and there is one outcome corresponding to three late arrivals (*LLL*).

We can generalize our discussion of the binomial distribution by focusing on the parts. If we have a random variable which is described by a binomial distribution with parameters n and π, then n is the number of trials and π is the probability of failure of any one trial. The product-of-probabilities part is the probability of any specific arrangement of x failures out of n trials, which is

$$\pi^x(1 - \pi)^{n-x} \qquad \text{for } x = 0, 1, 2, \ldots, n$$

Since we have x failures each occurring with probability π and $n - x$ successes occurring with probability $1 - \pi$, the product of probabilities will have x factors of π and $n - x$ factors of $1 - \pi$. Thus, each arrangement with x failures will have the same probability of occurrence.

The count part of the binomial probability is simply the number of ways to select or arrange x failures out of n outcomes. We could derive the mathematical formula for determining that number, but that derivation seems to be of limited value. The number of ways to select or arrange x failures out of n outcomes is written $\binom{n}{x}$ and is given by

Table 6-2

Number Reporting Late	Associated Outcomes	Probability
0	*PPP*	$.729 = 1(.729) = .729$
1	*PPL, PLP, LPP*	$.081 + .081 + .081 = 3(.081) = .243$
2	*PLL, LPL, LLP*	$.009 + .009 + .009 = 3(.009) = .027$
3	*LLL*	$.001 = 1(.001) = .001$

Figure 6-2

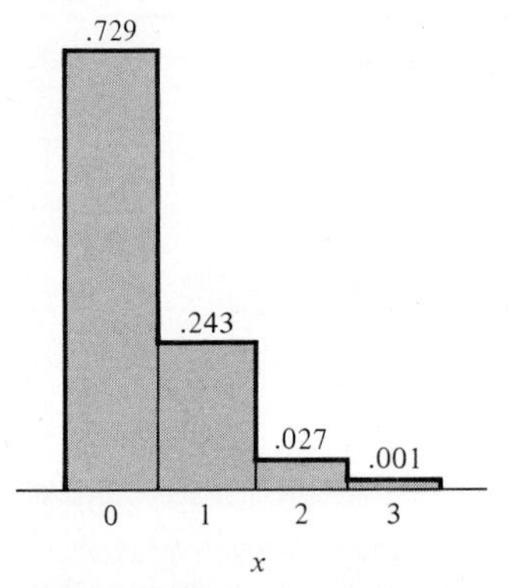

$$\binom{n}{x} = \frac{n!}{x!(n-x)!} \qquad \text{for } x = 0, 1, 2, \ldots, n$$

where $n!$ indicates "n factorial" (see Chapter 1).

The binomial probability distribution with parameters n and π is identified by the probability function

$$f(x) = \overbrace{\binom{n}{x}}^{\text{count part}} \underbrace{\pi^x(1-\pi)^{n-x}}_{\text{probability part}} \qquad \text{for } x = 0, 1, 2, \ldots, n \tag{6-1}$$

The mean μ and variance σ^2 for this probability distribution are expressible in terms of the parameters as

$$\mu = n\pi \tag{6-2}$$

and

$$\sigma^2 = n\pi(1-\pi) \tag{6-3}$$

Now let us return to Situation I. The director of the consumer research organization makes the determination that the binomial distribution models a process which is quite similar to the situation under consideration. There are ten independent trials, each of which produces an unsatisfactory rating with probability .15, if normal conditions prevail. The director's interest lies in determining the probability that four or more of the ten are rated as unsatisfactory. That is, the director is modeling the total number of failures. All the conditions of the binomial distribution being satisfied, the director calls the number of unsatisfactory reports x and computes

$$
\begin{aligned}
Pr(x \geq 4) &= 1 - Pr(x \leq 3) \\
&= 1 - Pr(x = 0) - Pr(x = 1) - Pr(x = 2) - Pr(x = 3) \\
&= 1 - \binom{10}{0}(.15)^0(.85)^{10} - \binom{10}{1}(.15)^1(.85)^9 - \\
&\quad \binom{10}{2}(.15)^2(.85)^8 - \binom{10}{3}(.15)^3(.85)^7 \\
&= 1 - 1(.196874) - 10(.0347425) - 45(.0061310) - 120(.0010819) \\
&= .05
\end{aligned}
$$

This is a rather small probability. On the average, a trained interviewer working under normal circumstances will be rated this poorly only five times out of 100. The director might feel that a probability of .05 is *so small* as to warrant the following conclusion: The interviewer is adversely affected by the divorce proceedings, and the director should take action.

The director in Situation I was interested in the number of unsatisfactory reports, the number of failures. When we deal with dichotomous sample spaces, we usually identify the two possible outcomes as success and failure. Do not read more into the terms "success" and "failure" than is intended. Just what constitutes a "success" in any problem is *totally* at the discretion of the decision maker. Whether the decision maker is interested in the number of failures or the number of successes is immaterial to the applicability of the binomial distribution.

Let's consider another simple example. Suppose that 60 percent of those who shop at A&P supermarkets return to A&P the next time they need groceries. If we were to select a shopper at an A&P store, we could assign a probability of .6 to the outcome that that shopper will return to A&P for his or her next grocery purchase. Each shopper is one trial with his or her next visit to a grocery store being considered a success (to A&P) or failure (not to A&P). If we were to analyze the shopping behavior of four shoppers in an A&P supermarket, we would use a binomial distribution with $n = 4$ and $\pi = .6$. Using x to identify the number of shoppers who return to A&P, we calculate

$$f(4) = Pr(x = 4) = \binom{4}{4}(.6)^4(.4)^0 = \frac{4!}{4!(4-4)!}(.6)^4(.4)^0 = 1(.1296) = .1296$$

$$f(3) = Pr(x = 3) = \binom{4}{3}(.6)^3(.4)^1 = .3456$$

$$f(2) = Pr(x = 2) = \binom{4}{2}(.6)^2(.4)^2 = .3456$$

$$f(1) = Pr(x = 1) = \binom{4}{1}(.6)^1(.4)^3 = .1536$$

$$f(0) = Pr(x = 0) = \binom{4}{0}(.6)^0(.4)^4 = .0256$$

This probability distribution can give us some insight into the prevalence of store loyalty in this situation. For instance, under the assumptions made to develop this probability distribution, we can calculate the probability that at least half of the four shoppers return to A&P to be

$$Pr(x \geq 2) = Pr(x = 2) + Pr(x = 3) + Pr(x = 4) = .8208$$

To reiterate, the binomial probability distribution can be used when an experiment can be viewed as a number of trials (four shoppers). Each trial:

1. Has the same dichotomous sample space as the others (return to A&P or do not return).
2. Is independent of the others (whether one shopper returns has no bearing on the likelihood that any other shopper will return).
3. Has the same probability distribution associated with its sample space as the others (in each of the four cases, the probability of the shopper's returning to A&P is .6).

Using a calculator, we can compute the probabilities for the binomial distribution. However, the calculations get more tedious as n gets larger. Since the binomial is such a useful probability distribution, tables are available for use in determining the appropriate probabilities. Table A-3 in the Appendix is a tabulation of the cumulative distribution function, $F(a) = Pr(x \leq a)$, of the binomial probability distribution.

Table A-3 is really several tables—one for each value of n. Each column in these tables is associated with a different binomial distribution, which is identified by the value of π, the probability of success (or failure) in any subprocess. The number in the row labeled "a" is $Pr(x \leq a)$. These values can be employed to compute $Pr(x = a)$, $Pr(x \geq a)$, $Pr(x > a)$, and $Pr(x < a)$. The appropriate formulas are:

$$Pr(x = a) = Pr(x \leq a) - Pr(x \leq a - 1)$$
$$Pr(x \geq a) = 1 - Pr(x \leq a - 1)$$
$$Pr(x > a) = 1 - Pr(x \leq a)$$
$$Pr(x < a) = Pr(x \leq a - 1)$$

To illustrate, let $n = 15$ and $\pi = .2$. This distribution is pictured in Figure 6-3. Then, from Table A-3,

$$Pr(x = 4) = Pr(x \leq 4) - Pr(x \leq 3) = .8358 - .6482 = .1876$$
$$Pr(x \geq 5) = 1 - Pr(x \leq 4) = 1 - .8358 = .1642$$
$$Pr(x > 5) = 1 - Pr(x \leq 5) = 1 - .9389 = .0611$$
$$Pr(x < 3) = Pr(x \leq 2) = .3980$$

Figure 6-3

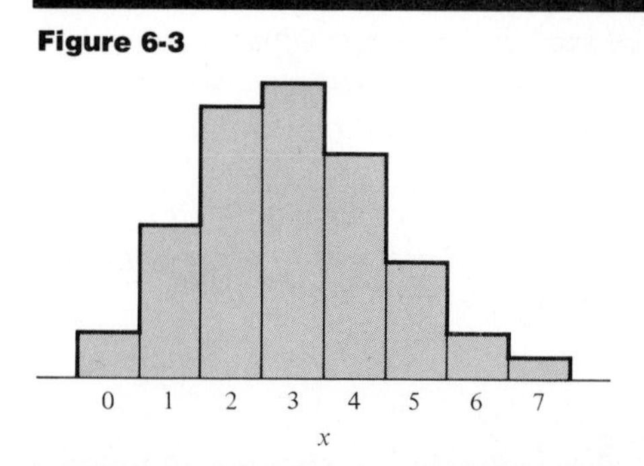

6.3 SITUATION II: WHY?

Probability models and probabilistic concepts have been used in investment analysis for many years. The primary application has been in the modeling of stock price behavior, and the emphasis has been on the mean and variance of the probability distributions which model or describe price movements.

Suppose you have just taken a position in retail sales with Merrill, Lynch, Pierce, Fenner and Smith, Inc. As a leading brokerage firm, Merrill Lynch is very advanced in its use of statistical techniques. Still, the emphasis is on the mean and variance of price movements. Having learned statistics well, you know that, if you limit your analysis to the parameters mean and variance, you may overlook potentially valuable information about stock price behavior. Some changes in stock price behavior do not alter the mean and variance a great deal. Changes such as these may well represent opportunities for investors to realize substantial gains or avoid substantial losses. Since Merrill Lynch and its competitors earn their reputations by informing clients of the appropriate opportunities, the ability to recognize changes in the character of an investment opportunity is quite valuable. Having probability distributions of price movements may lead to detection of otherwise unforeseen changes, allowing Merrill Lynch to better serve clients. Consequently, Merrill Lynch may increase the number of transactions it completes. Since revenues for brokerage firms are generated by completing transactions, this capability takes on increased importance.

On several occasions, you have talked to the manager of your office about this potential information loss. Finally, you propose a research project to develop probability distributions for several stocks. You would like to take the data you collect to a mathematical statistician on the faculty at a nearby university. After due deliberation, the manager is convinced that the potential gains far outweigh the costs and approves your proposal.

You collect data on three stocks, one each from three very different industries. You take these to the mathematical statistician, who will develop probability models for each of the three while you continue to gather more data. Two days later the statistician calls you, and you schedule an appointment to discuss the findings.

During the meeting, your consultant presents you with the three relative frequency histograms pictured in Figure 6-4 and suggests that the appropriate probabilistic model for these histograms is the normal distribution.

6.4 THE NORMAL DISTRIBUTION

If the behavior of the random variable x can be described by the normal distribution, x is said to be *normally distributed*. The specification of the pdf (probability density function) for this random variable is completely in terms of the parameters mean and standard deviation or variance. If x is a normally distributed random variable with mean μ and standard deviation σ, the pdf for x is

Figure 6-4

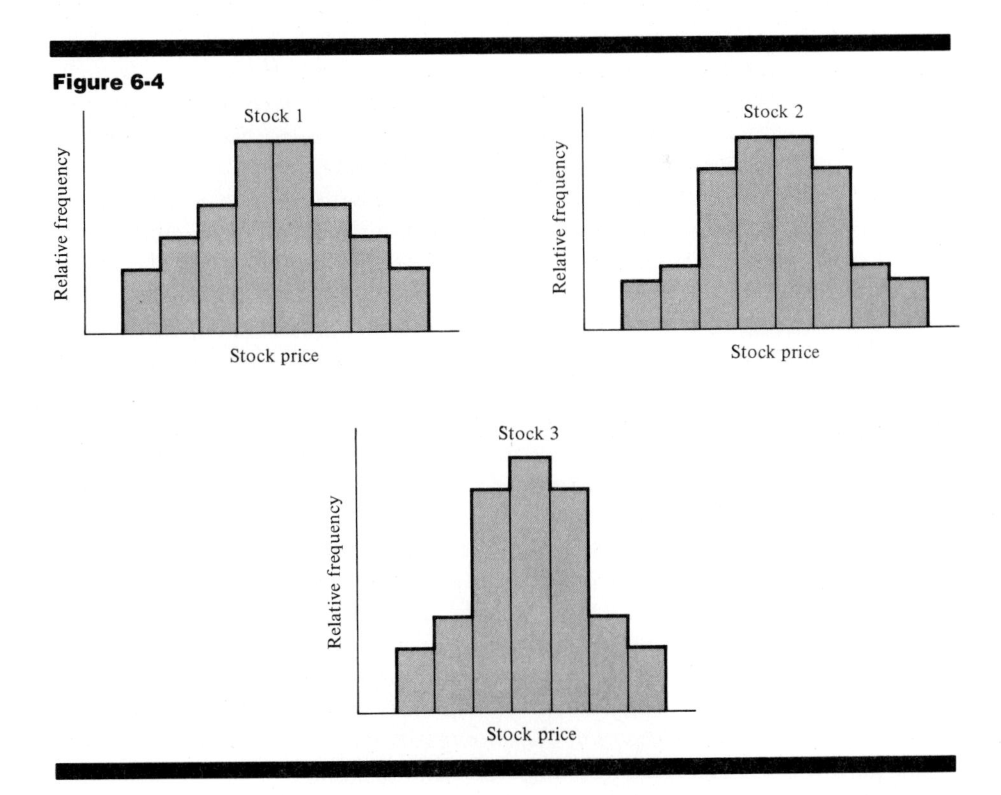

Figure 6-5

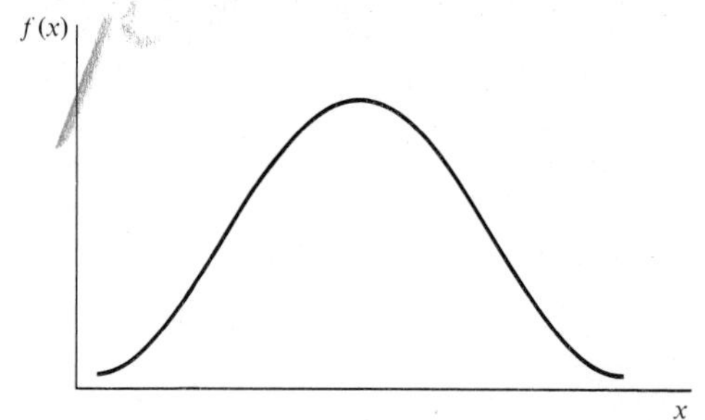

$$f(x) = \frac{1}{\sigma\sqrt{2\pi}} e^{-\frac{1}{2}\left(\frac{x-\mu}{\sigma}\right)^2} \qquad -\infty < x < \infty \qquad (6\text{-}4)$$

The symbols π and e represent specific numbers. Approximate values for these are 3.1416 for π and 2.71828 for e.

The graph of the function identified by Equation (6-4) always resembles the form illustrated in Figure 6-5. This form is often referred to as "bell-shaped" or sometimes "mound-shaped." An important characteristic of this shape is its symmetry. From the peak of the graph, the curve's extensions are mirror images of each other. Since Figure 6-5 is the graph of a pdf, the area of the region between the curve and the horizontal axis is 1. The symmetry of the curve tells us that half of the area lies to the left of the peak and half to the right. So, half the probability is associated with the values to the left of the peak and half associated with values to the right. Because of this symmetry, the positive and negative deviations from the x value where the peak is located are weighted equally and therefore offset each other. This helps us appreciate that the peak occurs at $x = \mu$. Figure 6-6 shows three normal pdf's, all with the same variance but with different means.

Figure 6-6 also allows us to appreciate that the larger the mean, the more probability is associated with the larger values of x. So, if the normally distributed random variable x represents a desirable quantity, such as profit or rate of return, then we would prefer to have a larger mean, for this increases our chances for the more desirable outcomes. Of course, if the normally distributed variable represents an undesirable quantity (cost or the rate of interest we pay), we prefer the smaller mean.

The shape of the graph of the pdf is determined by the standard deviation or variance. We know that variance measures dispersion; with the normal distribution, that is particularly evident. Two graphs of normal pdf's have been superimposed in Figure 6-7. The two pdf's have the same mean but different

Figure 6-6

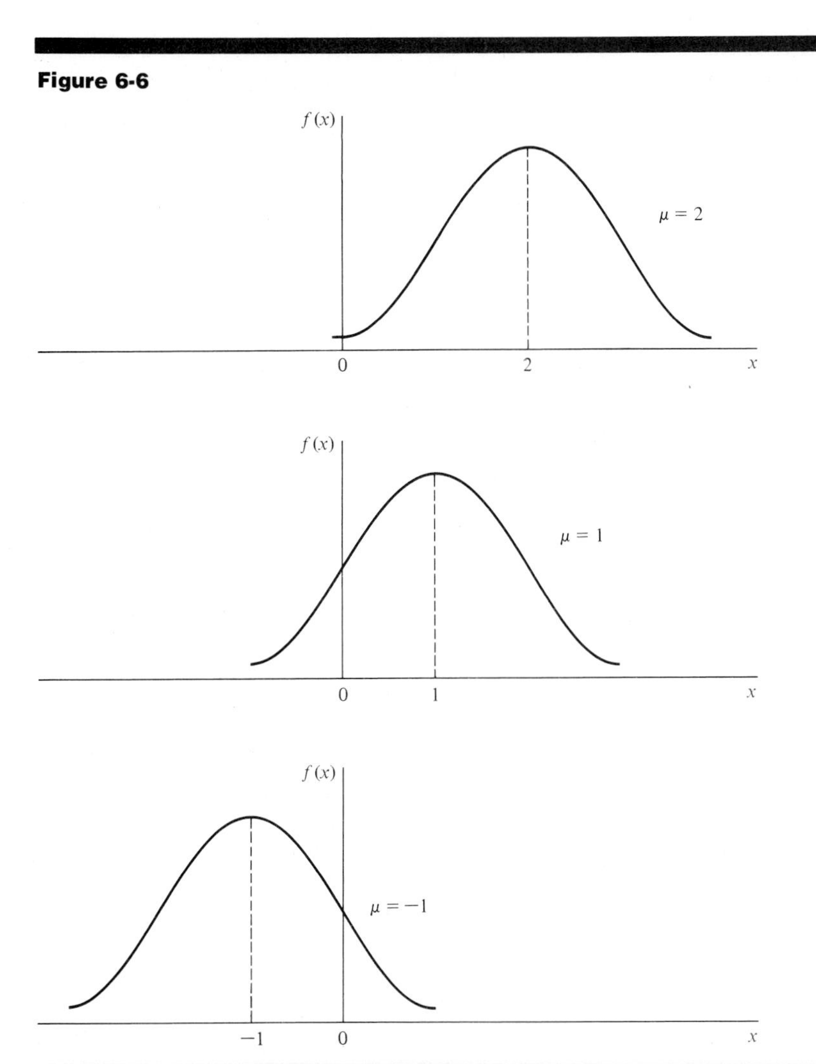

variances. Increasing the variance has the effect of stretching the graph of the pdf.

Suppose x is a normally distributed random variable whose value represents a desirable quantity. It is a little more difficult to appreciate the desirability of a small variance than it was the desirability of a large mean. The larger variance means two things. From Figure 6-7 we see that the larger variance means more probability associated with larger values of x. However, a larger variance also means that more probability is associated with the smaller values of x as well. Generally, decision makers are more willing to forego the greater probability of higher payoffs in order to avoid the greater probability of smaller

payoffs. For that reason, it is generally accepted that the smaller variance is preferred when the random variable x represents a desirable outcome.

The same argument can be used when x represents cost or another such variable. The larger the variance, the more probability is associated with the less desirable as well as the more desirable values of x. So, generally speaking, the smaller variance is preferred regardless of the interpretation of x values.

6.4a The Standard Normal Distribution

As the normal distribution is determined by the mean and variance, there are an infinite number of normal distributions. This would create problems in applications by complicating the calculations of probabilities. Fortunately, all normal distributions can be transformed into what is called the *standard normal distribution*. The random variable for this distribution is given a special letter, z. The mean of z is zero, and its standard deviation is 1. This identifies its pdf as

$$f(z) = \frac{1}{\sqrt{2\pi}} e^{-(1/2)z^2} \tag{6-5}$$

which is the simplest form possible for Equation (6-4). The graph of the expression in Equation (6-5) is shown in Figure 6-8.

The need for a standard form for the normal distribution arises because the evaluation of probabilities for a normal distribution is difficult. Mathematical techniques are ineffective in producing accurate probabilities. As a consequence, computer approximations must be used. These approximations are complex and costly, so we would like to compute these probabilities as infrequently as possible. In particular, we would like to compute only one normal probability table—and the simplest one at that. But a standard form would be useless if we could not derive all the probabilities for other normally distributed random variables from it. Mathematics supports us here. If x is a normally

Figure 6-7

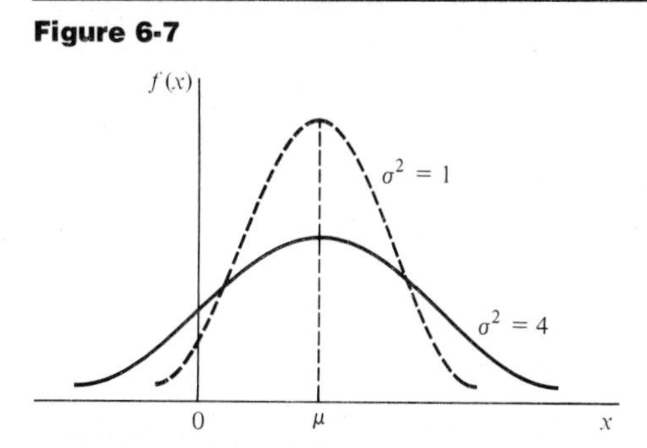

Figure 6-8

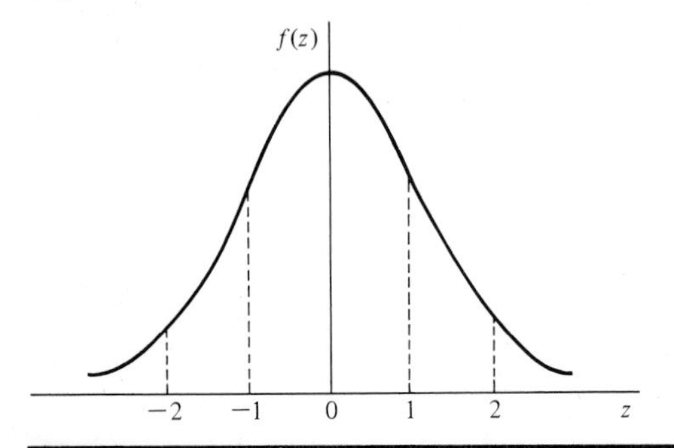

distributed random variable with mean μ and variance σ^2, then the variable

$$z = \frac{x - \mu}{\sigma} \tag{6-6}$$

can be shown to be a standard normal random variable. With any normally distributed random variable, then, if we subtract its mean and divide this difference by its standard deviation, we have a standard normal random variable.

This relationship allows us to determine probabilities for any normally distributed random variable. Let x be normally distributed with mean 50 and variance 100, and z be the standard normal random variable. Then, to determine

$$Pr(x \leq 58)$$

we note that $x \leq 58$ is the same as $x - 50 \leq 58 - 50$. This is equivalent to

$$\frac{x - 50}{10} \leq \frac{58 - 50}{10}$$

or

$$z \leq .8$$

since

$$z = \frac{x - 50}{10}$$

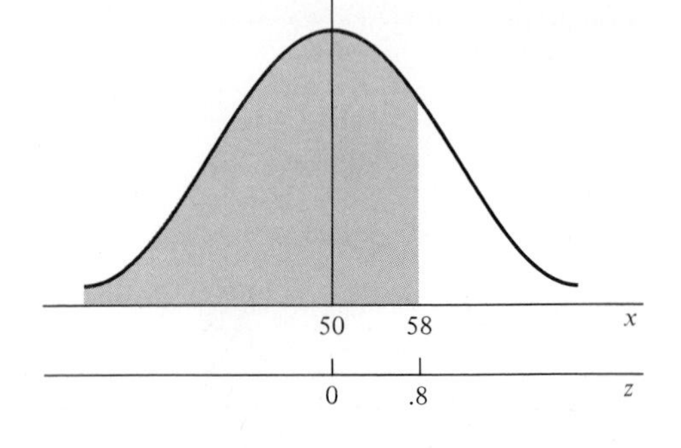

Consequently,

$$Pr(x \leq 58) = Pr(z \leq .8)$$

This tells us that the distribution function for any normally distributed random variable can be determined if we have a distribution function for the standard normal random variable. Table A-4 identifies the distribution function for z. It was, of course, developed using a computer approximation to the mathematical expression which we said earlier cannot be calculated. Table A-4 lists the values $Pr(0 \leq z \leq a)$. This probability is pictured in Figure 6-9. Again, because of the symmetry of this graph about the point $z = 0$, we know that half the probability is associated with z's being negative [$Pr(z \leq 0) = .5$]. So the distribution function values, $Pr(z \leq a)$, can be determined easily by adding .5 to the number given in Table A-4. For instance,

$$\begin{aligned} Pr(z \leq 1.96) &= .5 + Pr(0 \leq z \leq 1.96) \\ &= .5 + .4750 = .9750 \end{aligned}$$

This makes use of the symmetry in Figure 6-9.

We also make use of this symmetry when we calculate probabilities associated with negative values. When a is positive, $-a$ is negative. Now

$$Pr(z \leq -a) = Pr(z \geq a)$$

and

$$Pr(-a \leq z \leq 0) = Pr(0 \leq z \leq a)$$

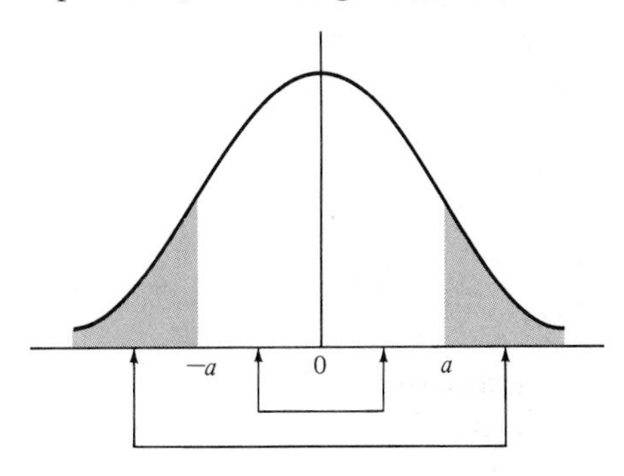

by the symmetry. Consequently,

$$\begin{aligned} Pr(z \leq -a) &= .5 - Pr(0 \leq z \leq a) \\ Pr(z \geq -a) &= .5 + Pr(0 \leq z \leq a) \end{aligned} \tag{6-7}$$

Figure 6-9

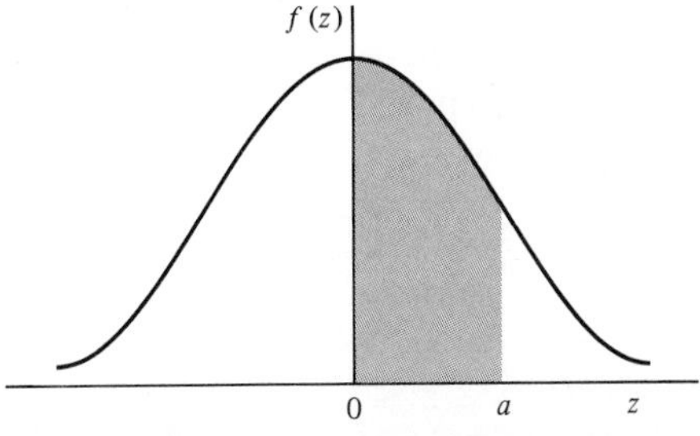

The probabilities on the right-hand side of Equations (6-7) are given in Table A-4.

Another feature of Table A-4 is that it expresses a values to two decimal places. Undoubtedly, there will be times when more accuracy is needed. For instance, suppose we want to calculate $Pr(z \leq .3473)$. Using Table A-4 gives $Pr(z \leq .34) = .6331$ and $Pr(z \leq .35) = .6368$. We recognize that .3473 is 73 percent of the way from .34 to .35. We estimate that $Pr(z \leq .3473)$ is 73 percent of the way from $Pr(z \leq .34)$ to $Pr(z \leq .35)$. Mathematically, because

$$.3473 = .34 + .73(.35 - .34)$$

we approximate

$$\begin{aligned} Pr(z \leq .3473) &= Pr(z \leq .34) + .73(Pr(z \leq .35) - Pr(z \leq .34)) \\ &= .6331 + .73(.6368 - .6331) \\ &= .6358 \end{aligned}$$

This process is called *linear interpolation* and can be employed in the reverse procedure as well, that is, finding a number a which gives $Pr(z \leq a)$ a value which has already been specified. Suppose we want to find the a which gives $Pr(z \leq a) = .98$. From Table A-4,

$$Pr(z \leq 2.05) = .9798 \qquad \text{and} \qquad Pr(z \leq 2.06) = .9803$$

Because

$$.98 = .9798 + .4(.9803 - .9798)$$

we approximate a to be

$$\begin{aligned} a &= 2.05 + .4(2.06 - 2.05) \\ &= 2.054 \end{aligned}$$

Making use of Equations (6-7) when a is negative, and using the other relationships with distribution functions discussed in Chapter 4, we can determine any probability of interest involving the standard normal random variable, z. Because of the relationship between z and any other normally distributed random variable x, exemplified by Equation (6-6), Table A-4 can be used for calculating probabilities involving any normally distributed random variable.

At this point, it seems as though it would be helpful to perform the calculations for some probability problems. Events of the form $a \leq z \leq b$ are particularly useful in statistical analysis, so we compute two probabilities for events of this type. The diagram illustrates the area of interest.

$$\begin{aligned} Pr(.44 \leq z \leq 2.01) &= Pr(0 \leq z \leq 2.01) - Pr(0 \leq z \leq .44) \\ &= .4778 - .1700 = .3078 \end{aligned}$$

$$\begin{aligned} Pr(-1.17 \leq z \leq .78) &= Pr(-1.17 \leq z \leq 0) + Pr(0 \leq z \leq .78) \\ &= Pr(0 \leq z \leq 1.17) + Pr(0 \leq z \leq .78) \\ &= .3790 + .2823 = .6613 \end{aligned}$$

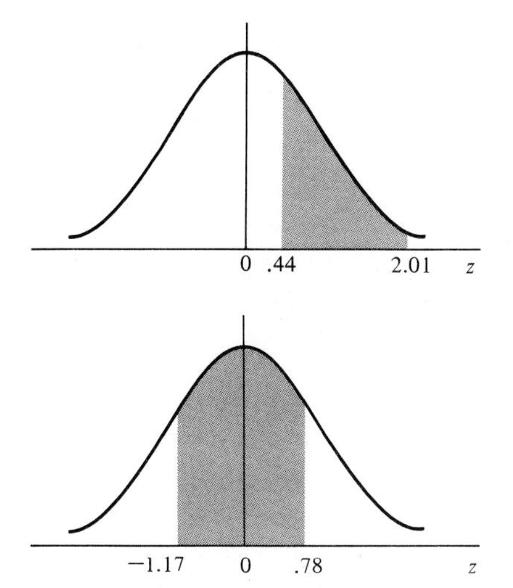

When we are dealing with a normally distributed random variable x with a nonzero mean and/or a variance different from 1, we must first use Equation (6-6). If x has mean 3 and standard deviation 2, then $Pr(4.10 \leq x \leq 5.06)$ is calculated by changing

$$4.10 \leq x \leq 5.06$$

to

$$\frac{4.10 - 3}{2} \leq \frac{x - 3}{2} \leq \frac{5.06 - 3}{2}$$

or

$$.55 \leq z \leq 1.03$$

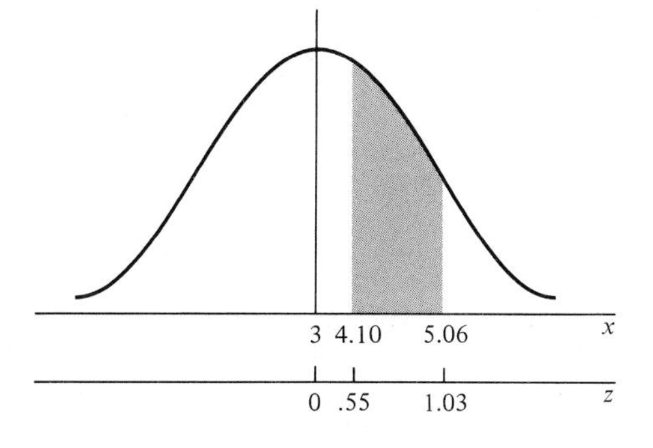

Now

$$\begin{aligned} Pr(4.10 \leq x \leq 5.06) &= Pr(.55 \leq z \leq 1.03) \\ &= .3485 - .2088 = .1397 \end{aligned}$$

6.5 SITUATION III: WHY?

Suppose when you completed the modeling project at Merrill Lynch, you monitored the price movements you had modeled. During the monitoring process, you detected a shift in the rate of return for one of the stocks. You presented a detailed report to the manager. He was very impressed with the work and requested that the research staff at the main office develop probability models for other issues.

It's two months later, now, and the manager faces a serious dilemma. Because of the recent trend in the stock market, activity in the office has been very light. The problem specifically is monthly cash flow. For several months the office has produced negative cash flows, that is, it is spending more than it is receiving each month. If this condition persists much longer, the manager may be seeking employment elsewhere. An alternative would be to dismiss one or more members of the staff. However, he is very impressed with the potential of each and every member and would hate to lose any of them.

The manager realizes that there are a number of aspects which are beyond his control and which contribute to monthly cash flow figures. He has come to you for help. He would like you to approach the cash flow problem as you did the stock returns problem. You will not be able to contact the mathematical statistician, however, since such an expenditure might not be justifiable under these circumstances. The manager would like a probability model, in a mathematical form, to aid in his determination of a need to dismiss one or more staff members.

You accept the challenge and begin to think about what data you will need. You convey your data needs to the manager the next day, and he informs you that only he can gather that information, and it will take him two or three days.

Meanwhile, you have a strange feeling that the stock problem and the cash flow problem are similar. You recognize that the net cash flow figure for a particular month is the result of several positive and negative increments. Each of these increments is a transaction, such as a payment on accounts receivable or a supplies purchase. During the course of one month, there are a large number of transactions. The net cash flow variable is the sum of the cash flows for all these transactions. This random variable, net cash flow, is the sum of a large number of variables.

You focus again on the process of stock price changes. What causes the changes in a stock's price from period to period? Each transaction in the stock market alters supply and demand conditions and, therefore, results in a change in the stock price. So, the price of a stock at the end of the week, for instance, is the sum of the price at the beginning of the week and the changes resulting from the many transactions during the week. Again, this random variable, stock price, is the sum of a large number of variables.

These two random variables aren't exactly the same, but they are certainly similar. Each is the sum of a large number of variables. You feel that the normal distribution may provide a good probability model for monthly cash flow.

The manager brings you the data, and you develop the relative frequency histogram pictured in Figure 6-10. This relative frequency histogram does indeed look similar to the shape of the normal distribution.

6.6 THE CENTRAL-LIMIT THEOREM

Was it a fluke that Situations II and III both led us to normal distributions? Or is the behavior of a sum of variables always describable by a normal distribution?

Figure 6-10 Relative Frequency Histogram

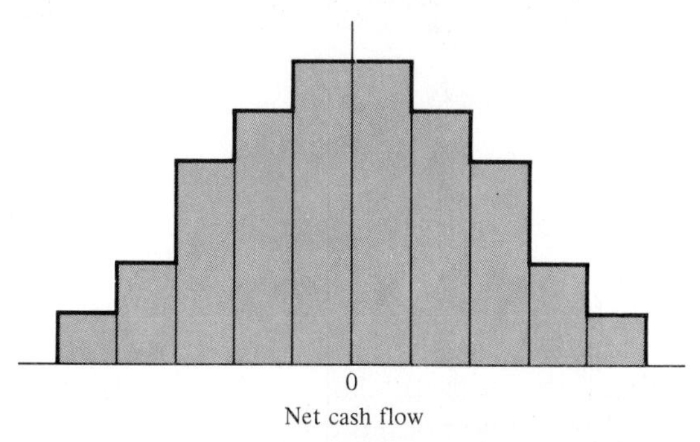

The central-limit theorem provides the answers to each of these questions. As the name implies, this mathematical result deals with the distribution of the center of a random sample. Remember that a random sample consists of independent observations of random variables with the same probability distribution. The center in this case is measured by the mean. If the random sample is written as $x_1, x_2, \ldots, x_n$, then the *mean*, written $\bar{x}$, is the sum of the observations, Σx_i, divided by the number of observations, n. Symbolically,

$$\bar{x} = \frac{1}{n} \Sigma x_i$$

Because the sampling distribution of the mean of a random sample consistently takes the same form as the sample size increases, we talk about a limit. The central-limit theorem, then, identifies the probability distribution for the mean or sum of *any* random sample with a large number of observations.

Central-Limit Theorem

For *any* population, identify the random sample as $x_1, x_2, \ldots, x_n$, the sample mean as $\bar{x}$, and the sum of the random sample as Σx_i. As the sample size n increases, the shapes of the sampling distribution of $\bar{x}$ and the sampling distribution of Σx_i become more and more similar to the shape of the normal probability density function.

The central-limit theorem is such an important result in basic statistical analysis that it would be very difficult for us to overemphasize or overstate the impact of this theorem.

Regardless of the shape of the population from which the random sample is drawn, the shapes of the sampling distributions of $\bar{x}$ and Σx_i tend to or approach normality as the sample size n *increases.*

Because of the central-limit theorem, we can assert the following:

1. For large random samples, the sampling distribution of $\bar{x}$ can be approximated by the normal distribution.
2. For large random samples, the sampling distribution of the sum $\Sigma x_i = n\bar{x}$ can be approximated by a normal distribution.

These statements tell us that almost any time we encounter a process which can be viewed as a sum or an average of independent observations from the same probability distribution, we can use the normal distribution as an approximation to that process. In Chapter 7, we address the question of which normal distribution gives the best approximation. In particular, we will identify the appropriate parameter values to be used in the normal approximations.

Throughout all the statements of the central-limit theorem and its implications, we have used the nebulous phrases "large values of n" and "large random samples." What is a large value for n? Figure 6-11 will help us answer this question as well as demonstrate the central-limit theorem. In that figure we have the graphs of the probability functions for three processes, all of which take one of the three values -1, 0, or 1. Also included are graphs of the sampling distributions for the means of 2, 3, and 10 independent observations from the same distributions. Ten independent observations from one of the distributions constitute a random sample of size 10 from that distribution. Similar statements apply to 2 and 3 independent observations from one of the probability distributions.

As in most usages of the word, the term "large" here is relative. In some cases, $n = 10$ is large; in others, $n = 1200$ is not. The criterion describing a large random sample depends upon several factors. The first major factor is the form of the underlying process. Note in Figure 6-11 that, for the symmetric populations (parts *a* and *b*), the sampling distributions for the means are also symmetric and are quite close in shape to the normal for n as small as 10. For the third population pictured, the sampling distribution is not even close to being symmetric for an n of 10.

The second factor which determines whether a sample size is large enough is the degree of accuracy required in the particular problem. If the probabilities to be approximated need to be accurate to two decimal places, the approximation can be used for relatively small n, perhaps as small as 30 for nearly any distribution. On the other hand, if accuracy to four decimal places is required, the sample size must be considerably larger. For distributions which are not symmetric, sample sizes in excess of 1000 may be necessary before the normal distribution gives approximations which are accurate to four decimal places.

6.7 NORMAL APPROXIMATION TO THE BINOMIAL

It may already have occurred to you that the normal distribution may provide a good approximation for the binomial distribution. The binomial random variable is a sum of random variables which constitute a random sample—each trial

Figure 6-11

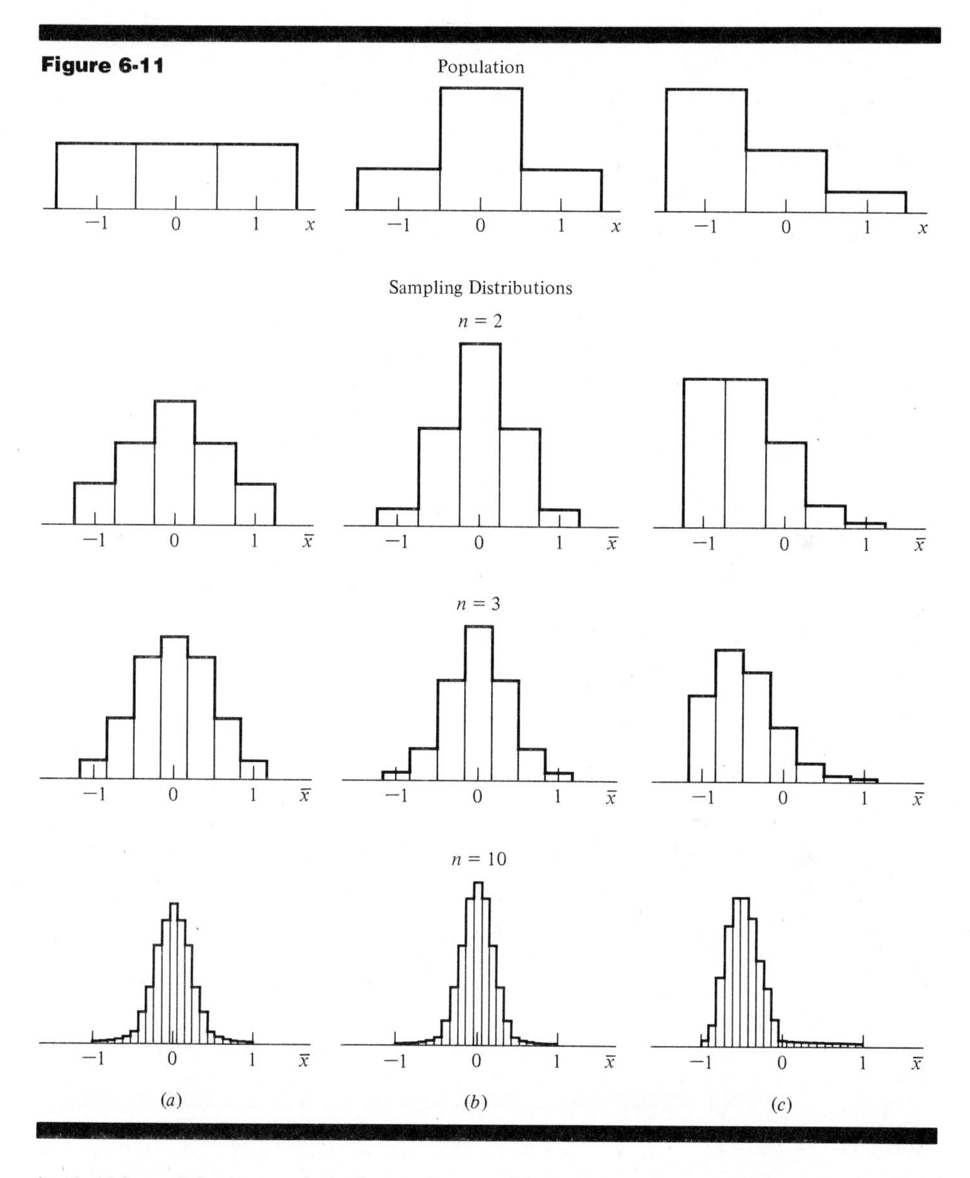

in the binomial process is independent and has the same probability distribution as the other trials. Consequently, for large values of n, the normal distribution will provide a good approximation. Figure 6-12 shows graphs of the probability functions for several binomial distributions which are arranged according to the value of n. Two different probabilities of success are used, namely, $\pi = .1$ and $\pi = .6$. For the different probabilities of success we have different values for n for which the normal distribution becomes a close approximation. This, of course, we've seen before in Figure 6-11 with the other probability distributions.

Figure 6-12

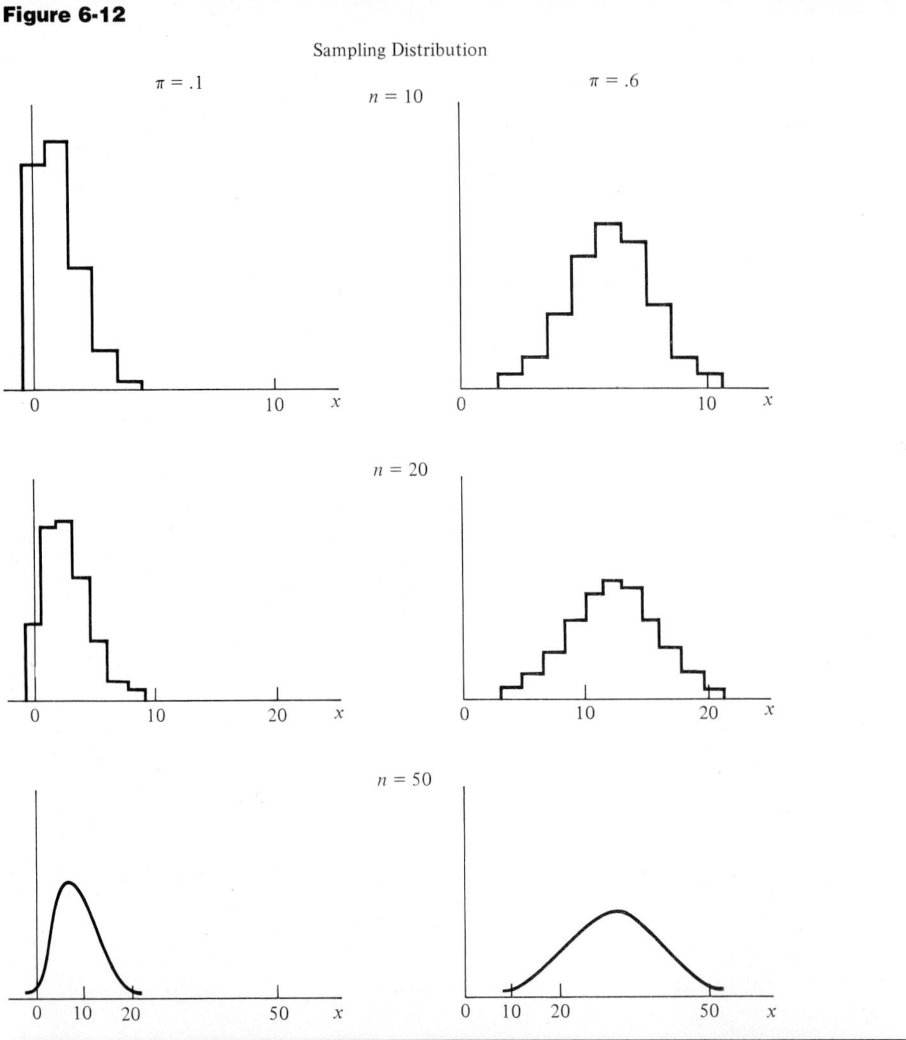

If x is a binomial random variable with parameters n and π, then

$$\mu = n\pi \qquad \text{and} \qquad \sigma^2 = n\pi(1 - \pi)$$

We could define the variable y_i as 0 if the ith observation is a failure and as 1 if it is a success. As the sum of the n y_i values actually equals the number of successes, the binomial random variable x equals Σy_i. Thus, x can be viewed as the sum of n random variables. According to the central-limit theorem, the distribution of a sum of random variables approaches normality as n increases. Thus, the sampling distribution of x is approximately normal for large n. So

$$z = \frac{x - n\pi}{\sqrt{n\pi(1 - \pi)}}$$

is approximately distributed as a standard normal random variable.

Can we make any definitive statements about the minimum value of n necessary to make the normal approximation a good one? We might think that this question can be answered by utilizing the specific knowledge that we have about the binomial distribution. An often-used rule which is easy to apply is to require that

$$n\pi \geq 5 \qquad \textit{and} \qquad n(1 - \pi) \geq 5$$

This rule has been found to give quite satisfactory results in many applications of the normal distribution as an approximation to the binomial distribution.

6.8 SUMMARY

In this chapter, we have presented the two most important probability distributions for business decision making. It is critical that the future decision maker be familiar with these probability models in order that the appropriate applications can be recognized. Recognition is that most important first step in the application of statistical analysis in business decisions.

The binomial distribution models situations which are characterized by:

1. Dichotomous sample spaces (each trial results in one of two possible outcomes).
2. Constant probability of success from trial to trial.
3. Independent trials.
4. The importance in the situation of the total number of successes.

The normal distribution was seen to model experiments in which the outcome was the sum or the average of a random sample. The actual value of the normal probability model, because of the central-limit theorem, will be appreciated more fully in the next two chapters. We stressed the importance of the concept of a sampling distribution in the last chapter. The central-limit theorem tells us that the sampling distribution of the sample mean is quite close to the normal distribution if we have a large sample size. So, in every application of statistics involving the sample mean, we are going to try to use the normal distribution. Because the mean is such an important characteristic of a probability distribution, we would not be exaggerating if we called the normal probability model the single most important probability distribution in statistics.

It would probably not hurt to mention that there are many other well-known probability models. Since these occur less often in business situations or are considerably more complex, we have not introduced them to you. In any event, if you grasp the basic concepts presented here, you can effectively interpret the output of a statistical analysis which uses one of these other models.

PROBLEMS

1. Identify an application of the binomial distribution in your major field of study. Point out the characteristics of this application which make it appropriate and of value.

2. Identify an application of the normal distribution in your major field. Point out the features of this application which make it appropriate and valuable.

3. Most readers pay little attention to the ads in the magazines they read. That is why Trico's advertising director before starting a major consumer advertising campaign estimated that the probability of a subscriber to *Playboy* magazine reading an ad in *Playboy* is .6. The director feels that a minimum of two exposures are necessary for the ad to have the desired impact. *Playboy* has approximately 6.9 million subscribers. If the firm places four ads in *Playboy* over a four-month period, how many subscribers will read at least two of them? (You can assume that reading each ad is an independent trial.)

4. "No-shows" are a critical problem for airlines and have resulted in the highly publicized and frequently criticized practice of "overbooking." A no-show is a person with a reservation who does not show up for a scheduled flight. Overbooking is the practice of making more reservations than there are seats on a plane. Every empty seat on an airplane costs the airline money. However, the inconvenience and anger of customers who are denied space for which they have reservations costs the airline in other ways.

   Suppose that you have been asked to deal with this problem for a small airline. Your first task is to develop an overbooking guideline for one of the small commuter runs. The plane holds 16 passengers. Past experience has shown that 80 percent of those with reservations will actually make their flight. How many tickets should you sell if you want to sell as many as possible for each flight and yet not have one or more passengers with reservations denied space on more than 10 percent of the flights?

5. Any profit-maximizing firm would like a larger share of the market. The advertising campaign in different communications media is the most effective method to reach the consumers. Chuck Atwood, Optico Corporation's sales manager, recently concluded a six-month advertising campaign which featured a series of ads in major business periodicals. The firm's advertising agency assured Optico management that 60 percent of the purchasing agents in their target industries would be aware of the campaign. At a recent sales convention, only 4 of the 10 purchasing agents whom Chuck Atwood questioned could remember seeing any of the ads. Should Optico undertake a more detailed (and expensive) study of the advertising agency's claim? Why?

6. Grady Canning's three-year-old "lunch only" restaurant has been a tremendous success. Although the restaurant seats 100 people, it is frequently necessary to turn customers away. Recently he was given the option of leasing a substantial amount of space across the street from his current establishment. His current concern is how much seating to place in his new establishment. Canning feels that the current number of people showing up at his establishment will not change over the next few years. He has observed that the number of people arriving at his restaurant on any given day is approximately normally distributed with a mean of 92 and a standard

deviation of 15. Canning wants his new establishment to have a sufficient seating to handle all arrivals 90 percent of the time. How much seating should he have?

7. As an innovative executive of Chrysler Corporation, you have always wanted to make use of current techniques of personnel selection. Recently you read an article on a new aptitude test for selecting management trainees. This instrument has been tested by other corporations on several thousand college seniors and the scores are normally distributed with a mean of 100 and a standard deviation of 10. For your purpose, you decide to set a cutoff score so high that only the top 20 percent of those who took the test would be successful. What cutoff score should you use?

8. In consumer surveys it is common practice to "verify" a certain percentage, say 10 percent, of each interviewer's interviews. Verification is needed to insure that the respondents were contacted by the interviewer and that the interviewer asked all the questions. Thus a physical contact between the verifier and the selected respondent is needed for verification, which is both time-consuming and expensive. It is also common that it will not be possible to contact the respondent or that the respondent will not cooperate in 15 percent of these verification calls. A recent verification procedure involved recontacting 20 respondents for each of five interviewers. For each interviewer, the following numbers of respondents produced a negative verification (could not or would not recall the interview): Marshall, 5; Ornales, 2; Greer, 3; Washington, 1; and Jones, 4. On the basis of the above, would you recommend that any of the interviewers be subject to complete (100 percent) verification? Why?

9. Being a broker is like walking on a double-edged sword. It takes just two or three adverse market reactions (e.g., you advised stock would go up and it went down) for your clients as well as your boss to stop going to lunch with you. As a broker you recently recommended 14 stocks that were listed on the New York Stock Exchange to a client looking to pick up a 4 percent gain during a six-month period. Eight of the stocks increased by 4 percent or more during the six-month time period. During this same time period, 55 percent of all the stocks listed on the Exchange increased in value by 4 percent or more. Do you think that your client will recommend your services to his friends as well? If yes, why?

10. In an economy move, Great Basin Corporation formed a typing pool consisting of 15 typists. All the typists used electric typewriters which occasionally would malfunction. The office manager, Ms. Streema, noticed that when a machine malfunctioned, the typist had nothing to do until the machine was repaired, which generally required at least four hours. Thus, the company was losing as much in wages for typists who were unable to type as it was in idle time before the economy move. Therefore, Ms. Streema decided to acquire enough reserve typewriters to ensure that the probability of a typist being unoccupied was 3 percent or less. An examination of the repair records revealed that the probability of any given typewriter malfunctioning was .05. How many reserve machines should Great Basin acquire? Under what conditions would the acquisition of reserve typewriters not be an economical move?

11. As a resident director of a newly opened supermarket in a large city, you are facing an uphill task to attract customers. Your advertising manager has advised you to start a promotional campaign in which randomly selected customers will not have to pay for their purchases. The customers will be selected at the checkout counter and

so cannot increase their purchases once selected. Since you have always wanted to play the game safely, you ask your lawyer to investigate any legal or tax implications of this campaign. Your lawyer learns that you would have to submit a form to the Internal Revenue Service for each prize in excess of $50 awarded. This appears to represent a substantial problem, since you or your representative will have to personally complete the forms and have the customers sign them. Before deciding whether or not to hold the contest, you would like to estimate the percent of the prizes that will be greater than $50. You know that purchases at the store are approximately normally distributed with a mean of $32.50 and a standard deviation of $10.70. What should be your estimate?

12. Pat Lowe, general manager of Chicago Foundries, is facing a difficult situation. Her plant is currently operating at 60 percent of optimal capacity and is losing money. She has submitted bids on two government contracts, either of which would bring the plant up to capacity. However, if she were to receive both, she would have to subcontract one of them and would be no better off than she currently is. Lowe recently received an invitation to bid on another government contract of the same nature and magnitude as the two she currently has bids on. She is unsure whether she should bid. She would prefer to operate at 60 percent of capacity than to receive all three bids. Past experience has convinced her that she has received contracts of this nature 20 percent of the time. Advise Ms. Lowe.

13. Suppose you are the manager of Stoller Inc., which provides maintenance service to other business organizations. It is in the nature of this business that all contracts are secured by competitive bids. Since the costs are always estimated, not all the bids wind up being profitable. In fact, the gross profit per contract as a percent of the contract value appears to be normally distributed with a mean of 12 percent and a standard deviation of 6 percent. You have just begun work on a major new contract. If the gross profit from this contract is less than 10 percent, you will need to secure a short-term loan to assist in the firm's cash flow situation. Interest rates are expected to rise substantially in the next few days. Would you advise the firm to secure a loan now? Why?

14. Suppose a major competitor of General Electric recently extended the guarantee on its small appliances from one year to 18 months. If you, as a manager of the customer relations division of G.E., do not match the guarantee of the competitor, your products may lose a considerable share of the market. Therefore, you also want to increase the guarantee on your small appliances from one year to 18 months. From past experience you know that such appliances have an average life of 30 months with a standard deviation of 9 months. If the guarantee is extended, how much will the percent of appliances replaced under the guarantee increase?

15. Alliance Corporation, a major oil drilling contractor in the North Sea, is concerned about the seals used in drilling rigs. These seals are generally used in a triple sequence to allow a margin of safety, since 20 percent of them fail before their 100-hour replacement time. The firm recently purchased 5000 seals from a new distributor and tested 225 of them. Fifty-two of the seals failed. Can Alliance conclude that the new seals are inferior to the ones that they have been using? Justify your answer.

16. Blithe Electronics produces and sells "do-it-yourself" electronic kits. Recently, a member of the staff of a hobby magazine purchased a Blithe kit for testing and found that it was missing one important printed circuit. Rather than contact the company, the staff member purchased a second kit, only to find that it contained a defective switch. The magazine then published the story of the staff member's experience along with some negative comments about Blithe's quality control. Naturally, the president of Blithe Electronics was quite upset by the article. Ed Manning, the quality control manager, received a sharp memorandum asking him to prepare a report "explaining how something like this could happen." Manning was not sure what to do. Although the kit in question contained over 200 items, his control system was designed to allow a kit to be sold with a defective or missing part only 4 percent of the time. Advise Mr. Manning.

Chapter 7

Estimation

7.1 SITUATION I: WHY?

Suppose you are a purchasing agent for Anheuser-Busch Inc., brewers of Budweiser beer. Anheuser-Busch has a number of purchasing agents, each with a specified list of products for which they are responsible. You are responsible for all purchases regarding product distribution. One of your major areas of concern is the large fleet of trucks used to transport your product from breweries to distributors spread throughout the nation. Anheuser-Busch has nine breweries, and a number of trucks are assigned to each.

Suppose further that Bob Peterson is a sales representative for a tire manufacturer, and he has been trying to get you to purchase his brand of tire. So far, he has not succeeded. Bob comes into your office and says, "OK, I know you purchase some five to six thousand tires per year for your fleet of trucks. That many tires represents around $900,000 a year, and I want those sales. I have asked my company to help me obtain your business by offering a special discount price. They in turn asked me what that price would have to be. So, now I'm here to ask you: What would our price have to be for you to switch to our tire?"

Since you don't know how well his brand of tire would perform, you really can't answer his question. If you could buy his tires at a price which would make them more economical than the tires you are now purchasing, you would like to switch. After all, it is your job to get the most for your money. Since you can't answer Bob's question just now, you tell him that you will figure out what kind of information you will need and call him tomorrow.

Let's see where we are. The size and type of tire that you purchase is 11 × 24.5 tubeless biased ply. The cost is $157.07 per tire plus $12.18 federal excise *tax. Your experience is that this tire lasts an average of 67,700 miles before the first recap. Thus, your cost per mile is*

$169.25/67,700 = $0.0025

Or, the brand of tire you are now purchasing costs you a quarter of a cent per mile per tire. Thus, if you could buy the other brand (Bob's brand) at such a price that the cost per mile was less than $0.0025, you would want to switch. At what price would Bob's brand yield a cost per mile that is less than $0.0025? The answer hinges on one thing: the average miles to the first recap for Bob's brand. If you knew what this average would be, you could answer his question.

There is no way that you would be able to determine the exact value for the average miles to the first recap. This is a parameter for a theoretically infinite population. Your only hope is to somehow estimate this parameter based upon sample information. This describes a process called statistical inference.

STATISTICAL INFERENCE REFERS TO METHODS AND/OR PROCESSES OF DRAWING CONCLUSIONS ABOUT POPULATIONS BASED UPON SAMPLE INFORMATION.

Our job is clear at this point. We need to estimate the parameter μ *(the true mean or average miles to first recap for Bob Peterson's brand of tire) based upon sample information.*

Your day has come and gone, so you call Bob and tell him that you will need to test his brand of tire before you can determine what the price must be for you to switch brands. You agree to purchase, say, 50 tires and plan to randomly assign one to each of 50 different trucks in your fleet. After these tires have been run for some time you will get back to Bob with your answer as to the "magic" price.

After the 50 test tires have run for, say, one month, you have your maintenance people measure tire wear and they are able to determine miles to first recap for each of the 50 tires as shown in Table 7-1.

Now we are ready to estimate the mean based upon our sample information.

7.2 ESTIMATE A POPULATION MEAN: HOW?

How can we estimate the population mean μ based upon sample information? Recall in Chapter 6 that $\bar{x}$ was introduced as the *sample mean*, and was defined as

$$\bar{x} = \frac{1}{n}\sum_{i=1}^{n} x_i \qquad (7\text{-}1)$$

Table 7-1 Miles to First Recap for Test Tires
(Miles in Thousands)

Tire	Miles	Tire	Miles	Tire	Miles	Tire	Mile	Tire	Mile
1	80.1	11	73.4	21	58.8	31	69.9	41	75.5
2	62.4	12	60.1	22	69.4	32	57.4	42	62.1
3	79.3	13	58.3	23	71.2	33	66.6	43	74.6
4	77.2	14	73.7	24	64.4	34	71.1	44	82.2
5	54.1	15	79.2	25	65.7	35	55.3	45	64.1
6	68.2	16	60.8	26	65.6	36	58.2	46	58.3
7	70.0	17	62.2	27	59.8	37	62.1	47	65.1
8	49.8	18	58.1	28	64.7	38	59.4	48	70.0
9	52.7	19	73.0	29	78.2	39	70.5	49	63.2
10	68.7	20	62.4	30	71.1	40	68.7	50	58.4

Given that the n sample observations are the result of random sampling, we can indeed prove that the expected value of $\bar{x}$ is μ regardless of the size or type of population. That is,

$$E(\bar{x}) = \mu$$

We therefore say that $\bar{x}$, as defined in Equation (7-1), is an *unbiased estimator* of μ. If the expected value of a statistic is the parameter of interest, the statistic is referred to as being unbiased. Otherwise, the statistic or estimator is *biased*. Thus, $\bar{x}$ provides us with an unbiased estimate of μ. There is no question but that the property of our estimate being unbiased is desirable. It means that the "long run" average of the estimates is the true value of the parameter.

If we compute $\bar{x}$ for our random sample of n observations, we obtain a numeric value, called a *statistic*, and it would be our *point estimate* of the parameter μ. There, we can estimate the population mean based upon sample information. If we return to the data in Table 7-1, we find

$$\sum_{i=1}^{50} x_i = 3305.3$$

Thus, our point estimate of the true mean for this brand of tire run on our trucks is

$$\begin{aligned} \bar{x} &= 3305.3/50 \\ &= 66.106 \text{ thousand miles} \end{aligned}$$

Are we finished? We now have an estimate of μ, so we could determine what the price must be to "obtain" a cost per mile equal to that for the brand of tire you now purchase. But how good is our point estimate? We know that 66.106 thousand miles is an unbiased estimate of μ. However, $\bar{x}$ is a statistic and has a sampling distribution. We would not be surprised to find that $\bar{x}$ is not equal to μ. How close do you think $\bar{x}$ is to μ? Do you think μ could really be as small as 65,000 miles? Or, could μ be as large as 70,000 miles? We don't know either—yet.

The problem is that a point estimate does not contain information about how close one might expect it to be to the true parameter value. We have probably all heard the advertisement claiming that four out of five dentists favor a particular mouthwash. What does this statement mean to you? It is probably an estimate rather than the parameter, as we doubt that all dentists were polled. In this case we have an estimate of π, the true proportion of all dentists who favor the particular mouthwash. Do you think π could be .7, or .5, or maybe .25? Based upon the point estimate, we just don't know.

What we need is some way of indicating how good our point estimate is. We need something to accompany our point estimate that contains information on how good or close we might expect our point estimate to be to the true parameter.

Actually, "how good" our estimate is depends primarily upon two things: (1) the sample size and (2) the variance of the population from which we sample. Your interpretation of the point estimate, four out of five ($4/5 = .8$) dentists, would no doubt change if you were told the sample size was 50,000 rather than say 5. If indeed n was 50,000, you might expect π to be quite close to .8. The other important factor is the variability of the population. If all values are similar in magnitude (small variance), a small sample would yield good estimates in terms of the likelihood of being close to the parameter.

Statisticians have developed an *interval estimator* that incorporates *both* sample size *and* population variance. This interval estimator is called a *confidence interval*.

Let's see how to obtain a confidence interval estimate and its interpretation. In order to develop the logic and theory *exactly*, we must make certain assumptions. The required assumptions are never true, however; we will be able to relax them—soon. Assume we draw a random sample of size n from an infinite population. We could prove that

$$E(\bar{x}) = \mu$$

and that the variance of the sampling distribution of $\bar{x}$ is σ^2/n, where σ^2 is the variance of the infinite population of x values. Further, if the population of x values is normally distributed, the sampling distribution of $\bar{x}$ will also be normal. So, given our assumptions about random sampling and the normality of the population, the sampling distribution of $\bar{x}$ is as shown in Figure 7-1. Note this distribution has mean μ and is normally distributed with variance σ^2/n. As standard deviation is defined as the square root of variance, we see that the standard deviation of the sampling distribution of $\bar{x}$ is $\sigma/\sqrt{n}$. Given that this distribution is normal, we could use Table A-4 to determine areas and therefore the probabilities associated with certain intervals.

Recall in Chapter 6 we were able to transform any normal into the standard normal via a particular transformation. The formula given for the transformation was

Figure 7-1 The Sampling Distribution of $\bar{x}$

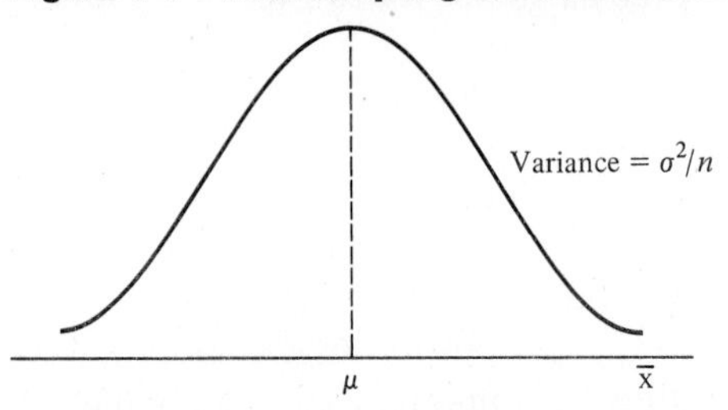

$$z = \frac{x - \mu}{\sigma} \tag{6-6}$$

This transformation can be generalized; x is a normally distributed random variable with expected value μ and standard deviation σ. So, what we really have in this transformation is a normally distributed random variable minus its expected value all divided by its standard deviation. In our present case, $\bar{x}$ is a normally distributed random variable with expectation μ and standard deviation $\sigma/\sqrt{n}$. Thus, the sampling distribution of $\bar{x}$ can be transformed into the standard normal by

$$z = \frac{\bar{x} - \mu}{\sigma/\sqrt{n}} \tag{7-2}$$

Suppose we pick the point $\mu + 1.96\,\sigma/\sqrt{n}$ in Figure 7-1. You'll see why in a minute. This particular point transforms into $z = 1.96$ as shown below:

Let $\bar{x} = \mu + 1.96\,\sigma/\sqrt{n}$

$$z = \frac{(\mu + 1.96\sigma/\sqrt{n}) - \mu}{\sigma/\sqrt{n}}$$

$$= \frac{\mu - \mu + 1.96\,\sigma/\sqrt{n}}{\sigma/\sqrt{n}}$$

$$= \frac{1.96\,\sigma/\sqrt{n}}{\sigma/\sqrt{n}}$$

$$= 1.96$$

So, you can actually view z as the number of standard deviations a point is from μ. Table A-4 shows that a z value of $+1.96$ cuts off .025 of the area in the upper tail of the standard normal. Since a z value of $+1.96$ corresponds to the point $\mu + 1.96\,\sigma/\sqrt{n}$, z of -1.96 must correspond to $\mu - 1.96\sigma/\sqrt{n}$. Figure 7-2 shows this correspondence.

Clearly, the area of the sampling distribution of $\bar{x}$ in the interval

$$\mu - 1.96\,\sigma/\sqrt{n} \quad \text{to} \quad \mu + 1.96\,\sigma/\sqrt{n}$$

is .95. Therefore, 95 percent of all sample means $\bar{x}$ are contained in the interval

$$\mu - 1.96\ \sigma/\sqrt{n} \qquad \text{to} \qquad \mu + 1.96\ \sigma/\sqrt{n}$$

Thus, the probability of drawing a random sample of size n and obtaining an $\bar{x}$ in this interval is .95.

$$Pr(\mu - 1.96\ \sigma/\sqrt{n} \leq \bar{x} \leq \mu + 1.96\ \sigma/\sqrt{n}) = .95$$

Suppose we adopt the following rule: $\bar{x} \pm 1.96\ \sigma/\sqrt{n}$. This gives us the interval

$$\bar{x} - 1.96\ \sigma/\sqrt{n} \qquad \text{to} \qquad \bar{x} + 1.96\ \sigma/\sqrt{n}$$

This interval is *exactly* as wide, $2(1.96\ \sigma/\sqrt{n})$, as $\mu \pm 1.96\ \sigma/\sqrt{n}$, as shown on Figure 7-2. The only difference is that our rule centers the interval around $\bar{x}$ rather than around μ. Since the probability is .95 of obtaining an $\bar{x}$ value in the interval $\mu \pm 1.96\ \sigma/\sqrt{n}$, it follows that 95 percent of all the intervals generated by our rule would contain μ. Think about this! This interval is what is known as a 95 percent *confidence interval*. The two end points of the interval are given specific names:

$$\text{Lower confidence limit} = \bar{x} - 1.96\ \sigma/\sqrt{n}$$
$$\text{Upper confidence limit} = \bar{x} + 1.96\ \sigma/\sqrt{n}$$

The interpretation of the confidence interval is obvious. Once we have drawn our sample and computed the interval, we say we are 95 percent confident that the mean μ is contained in this specific interval. Note, we do *not* say that the probability is .95 that μ is contained in this specific interval. This statement would imply that μ varies and is in this interval 95 times out of 100. The population mean μ is a parameter, *it is a constant*, and it does not vary. It is the location of the interval which varies, because $\bar{x}$, the center, may differ from sample to sample.

Figure 7-2 The Sampling Distribution of $\bar{x}$

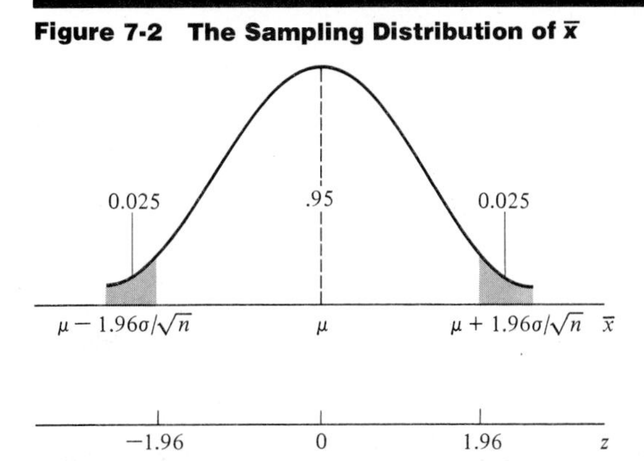

We can see by the way the confidence interval is constructed that the strict interpretation is that 95 percent of all possible intervals generated by our rule would contain μ. Whether the particular interval we obtain by sampling does or does not contain μ will not be known to us. However, we can view this process as drawing one confidence interval at random from a pool of confidence intervals where 95 percent of the intervals in the pool do contain μ. Figure 7-3 graphically depicts the interpretation of our confidence intervals.

Note that for any $\bar{x}$ in the interval $\mu \pm 1.96\,\sigma/\sqrt{n}$, the resulting confidence interval contains μ. And, the probability is .95 that our random sample of size n would yield an $\bar{x}$ in this interval $\mu \pm 1.96\,\sigma/\sqrt{n}$. Make *sure* you understand this interpretation of confidence intervals. We will be interested in generating confidence interval estimates for several different parameters. The procedure will be basically the same as above and the interpretation will be *exactly* the same.

The unique feature in our rule that makes this interval a 95 percent confidence interval is the z value of 1.96. By changing the value of z we change the level of confidence. The probability that z is between -1.645 and 1.645 is .90. Thus, if we choose $z = 1.645$,

$$Pr(\mu - 1.645\,\sigma/\sqrt{n} \leq \bar{x} \leq \mu + 1.645\,\sigma/\sqrt{n}) = .90$$

Consequently, the resulting intervals ($\bar{x} \pm 1.645\,\sigma/\sqrt{n}$) would be 90 percent

Figure 7-3

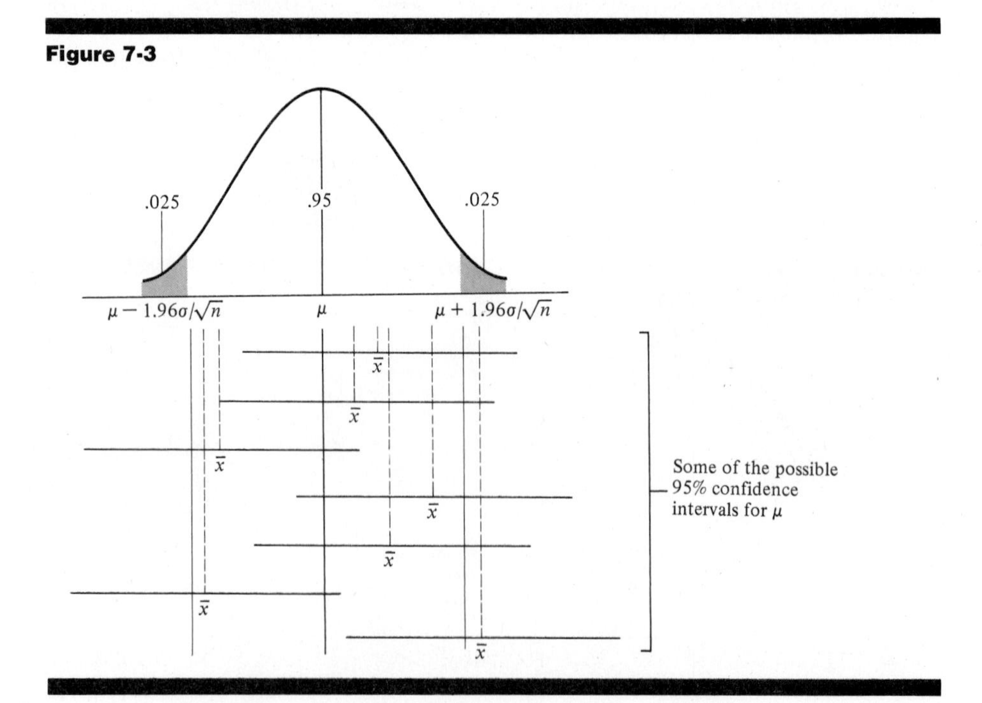

confidence intervals and only 90 percent of these possible intervals would contain μ. What value of z would you use if you wanted a 99 percent confidence interval? ($z = 2.575$)

At this point it should be clear that the information in both n and σ is incorporated in the confidence interval. The narrower the confidence interval, given a specific level of confidence, the closer we might expect our point estimate to be to the parameter. If, for example, we obtained a 95 percent confidence interval for μ of 110.5 to 115.5, we would expect that μ is somewhere between 110.5 and 115.5 and would be very surprised if it were as low as 100 or as high as say 125 or so.

Now that we understand the logic of the confidence interval estimate and know its interpretation, we are ready to relax our assumptions and be more realistic.

7.2a Large Sample Estimation of μ

Our "unrealistic" assumptions were that the population of x was normal and its variance σ^2 was known. In reality, neither of these two would be exactly true. However, we know that as our sample size n increases, the shape of the sampling distribution of $\bar{x}$, via the central limit theorem, will approach normality. And, *it is the normality of the sampling distribution that is needed—not the normality of the population*. So, for relatively large n, the normality assumption will be approximately true.

As for our other assumption (variance known), if we define the *sample variance* as

$$s^2 = \sum_{i=1}^{n} \frac{(x_i - \bar{x})^2}{n - 1} \tag{7-3}$$

we can prove that $E(s^2) = \sigma^2$ given random sampling. That is, s^2 defined by Equation (7-3) is an unbiased estimator of σ^2. Although Equation (7-3) defines sample variance, it is not the easiest form for computation. Algebraically, we could show that

$$\sum_{i=1}^{n} (x_i - \bar{x})^2 = \sum_{i=1}^{n} x_i^2 - \left(\sum_{i=1}^{n} x_i\right)^2 / n$$

Although it does not look like it, the term on the right-hand side is typically easier to compute than $\Sigma(x_i - \bar{x})^2$. Thus, computationally

$$s^2 = \frac{\sum_{i=1}^{n} x_i^2 - \left(\sum_{i=1}^{n} x_i\right)^2 / n}{n - 1} \tag{7-4}$$

As the square root of variance is standard deviation, the *sample standard deviation s* is

$$s = \sqrt{\frac{\Sigma x_i^2 - (\Sigma x_i)^2/n}{n - 1}} \tag{7-5}$$

Yes, we know there are n terms in the numerator, and one might think we should divide this by n rather than $n - 1$. However, there are only $n - 1$ independent terms in the numerator, as $\Sigma(x_i - \bar{x})$ must equal 0. Suffice it to say that Equation (7-3) is the definition of sample variance and it is an unbiased estimator of σ^2. The larger the sample size n, the closer we might expect s^2 to be to σ^2.

Therefore, for large samples, our "unrealistic" assumptions are approximately true. We will assume that n is sufficiently large to justify close approximation of normality of the sampling distribution of $\bar{x}$ and that s^2 is a close approximation to σ^2. With these assumptions we are ready.

Parameter	Point Estimator	Interval Estimator
μ	$\bar{x} = \frac{1}{n}\sum_{i=1}^{n} x_i$	$\bar{x} \pm z_{\alpha/2} s/\sqrt{n}$

The notation $z_{\alpha/2}$ denotes the z value which cuts off $\alpha/2$ in the upper tail of the standard normal distribution.

7.3 SITUATION I REVISITED

Returning to Table 7-1, we find

$$\sum_{i=1}^{50} x_i = 3{,}305.3$$

$$\sum_{i=1}^{50} x_i^2 = 221{,}443.59$$

So our point estimate of the mean number of miles per tire to first recap for Bob Peterson's brand of tire is

$$\begin{aligned} \bar{x} &= 3{,}305.3/50 \\ &= 66.106 \text{ thousand miles} \end{aligned}$$

The sample variance is, using Equation (7-4),

$$\begin{aligned} s^2 &= \frac{\Sigma x_i^2 - (\Sigma x_i)^2/n}{n - 1} \qquad (7\text{-}4) \\ &= \frac{221{,}443.59 - 3{,}305.3^2/50}{49} \\ &= 2{,}943.43/49 \\ &= 60.07 \\ s &= \sqrt{60.07} \\ &= 7.75 \end{aligned}$$

Our 95 percent confidence interval estimate of μ would be

$$\bar{x} \pm 1.96\ s/\sqrt{n}$$

$$66.106 \pm (1.96)(7.75)/\sqrt{50}$$

$$66.106 \pm 2.148$$

Therefore, our point estimate of μ is 66.106 thousand miles and we are 95 percent confident that $63.958 \leq \mu \leq 68.254$ in thousands of miles. This confidence interval certainly helps us evaluate the goodness of our point estimate. We believe μ is somewhere in this interval and we would be very surprised if μ was very far below 63,958 miles or much above 68,254 miles.

Now recall what you wanted to do was to find the price for Bob's brand of tire such that your cost per mile per tire was below \$0.0025, the per-mile cost of your present brand of tire.

To be conservative, we would probably choose our lower confidence limit of 63,958 and find the price per tire that would give a cost per mile equal to \$0.0025.

$$\text{Price}/63{,}958 = \$0.0025$$
$$\text{Price} = \$159.90$$

As this price includes \$12.18 federal excise tax, the tire price would be

$$\$159.90 - \$12.18 = \$147.72$$

Based upon the sample evidence, it looks like you should be willing to pay up to but probably not exceeding \$147.72 per tire. The farther below \$147.72 per tire you can get the price, the more interested in switching brands you would become. Pick your price and give Bob a call.

7.4 SITUATION II: WHY?

Suppose you are the production manager for a medium-sized machine shop. An "inventor" has approached your firm with the drawings for a new disk brake for bicycles. The inventor has asked you to bid on manufacturing the disk brake. Assume your shop has the capability to manufacture the brake as specified on the drawings. However, you haven't been making such an item and really don't know what your labor costs would be for this project. The inventor is convinced that the demand will exceed 50,000 units per year and is quite anxious to get your bid on the manufacturing. If indeed you end up making some 50,000 units per year, you certainly don't want to underestimate your cost per unit for production and base your bid on this underestimate. In order to arrive at your bid price, you would need to have an estimate of the employee-hours per unit required for manufacturing. So, you propose that you go ahead and make, say, 15 units and bill the inventor for your costs. The inventor in turn would have the 15 units and could use these as samples to try to initiate sales. It is agreed. You make the 15 and record the required employee-hours for each unit as shown in Table 7-2.

Table 7-2

Unit	Labor in Employee-Hours
1	1.20
2	0.83
3	0.72
4	0.55
5	0.39
6	0.28
7	0.31
8	0.30
9	0.27
10	0.25
11	0.34
12	0.29
13	0.31
14	0.33
15	0.28

It has been your experience that a certain amount of learning is involved with the first few units for custom jobs. The data in Table 7-2 seem to support this, as it appears that by the fifth unit or so, the required time stabilizes. You reason that the required times for units 6 through 15 are typical or indicative of those you would probably experience in the long run. Thus, your relevant data are the last 10 observations, which are reproduced in Table 7-3.

Your objective is to estimate μ (the true mean number of employee-hours per unit) for a theoretically infinite population. However, a sample of only 10 observations is not large. You can't really buy more observations, as the inventor will not agree to pay for more than the 15 units just to have you arrive at the bid and you certainly don't want to make extra units at your expense! So you will have to get by with the small sample of 10.

Table 7-3

Observation	x_i
1	0.28
2	0.31
3	0.30
4	0.27
5	0.25
6	0.34
7	0.29
8	0.31
9	0.33
10	0.28

7.5 SMALL SAMPLE ESTIMATION OF μ: HOW?

We again must make certain assumptions to develop the exact theory. Any violations in these assumptions will render our conclusions approximate. However, we have found that minor violations do not greatly affect the procedure you are about to see.

Assumptions:

1. The population of x values is normally distributed.
2. The n sample observations constitute a random sample from the infinite population.

It looks as though we could merely apply the same procedure as that for a large sample size. In fact, we almost do. There is only one change and that relates to the use of z, the *standard normal random variable*. Rather than using z in the interval estimator, we will use a similar term which is denoted as t.

Parameter	Point Estimator	Interval Estimator
μ	$\bar{x} = \frac{1}{n}\sum_{i=1}^{n} x_i$	$\bar{x} \pm t(\alpha/2; n-1)\, s/\sqrt{n}$

Note the only difference is that the interval estimator has $t(\alpha/2; n - 1)$ in place of $z_{\alpha/2}$. What is t, and where do we get its value?

The symbol t represents a random variable with a probability distribution which is referred to as the *student t distribution*. A student t distribution looks very much like the standard normal distribution z, as it is a continuous, symmetric probability distribution with mean equal to zero. However, the shape of a particular student t distribution depends upon something called *degrees of freedom*. Figure 7-4 shows several superimposed student t distributions. We can see that as the number of degrees of freedom increases, the student t distribution approaches z, the standard normal. Indeed, the limit of the student t distribution is z.

Figure 7-4

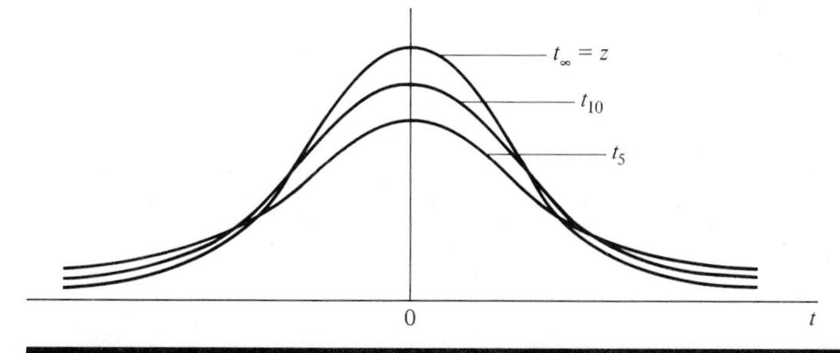

Unfortunately, we cannot transform all student t distributions into some standard as we did with normal distributions. Table A-5 gives selected t values corresponding to various areas for a number of student t distributions. Specifically, Table A-5 gives t values that cut off .1, .05, .025, .01, and .005 in the upper tail of the t distributions for degrees of freedom 1 through 29 plus those for the t with infinite degrees of freedom. Let's look at the student t distribution with, say, 5 degrees of freedom. The t values and corresponding areas given in Table A-5 are shown in Figure 7-5.

One more piece of notation and we will be finished. When we constructed a 95 percent confidence interval for μ (large samples), we chose the value of z that cut off .025 of the area of the standard normal in the upper tail. If we view the confidence level (in this case .95) as being $1 - \alpha$, we chose the z value that cut off $\alpha/2$ of the area in the upper tail. Thus, in our rule or expression for the interval estimator of μ with small sample size, we see the notation $t(\alpha/2; n - 1)$, which means the t value that cuts off $\alpha/2$ in the upper tail of the student t distribution with $n - 1$ degrees of freedom. Now we are ready to return to Situation II.

7.6 SITUATION II REVISITED

From Table 7-3 we find

$$\sum_{i=1}^{10} x_i = 2.96$$

$$\sum_{i=1}^{10} x_i^2 = .883$$

$$\begin{aligned}\bar{x} &= 2.96/10 \\ &= .296\end{aligned}$$

$$\begin{aligned}s^2 &= \frac{\Sigma x^2 - (\Sigma x)^2/n}{n - 1} \\ &= \frac{.883 - 2.96^2/10}{9} \\ &= .00076\end{aligned}$$

$$\begin{aligned}s &= \sqrt{.00076} \\ &= .0276\end{aligned}$$

Our point estimate of μ (the mean number of employee-hours each brake requires in manufacturing) is our sample mean of .296 employee-hours. Suppose we would like a 99 percent confidence interval estimate. Here, $1 - \alpha = .99$, thus $\alpha = .01$ and $\alpha/2 = .005$. As $n = 10$, we need the tabulated value for $t(.005; 9)$. The 99 percent confidence interval for μ is

Figure 7-5 Upper-tail Areas of the Student t Distribution with 5 Degrees of Freedom

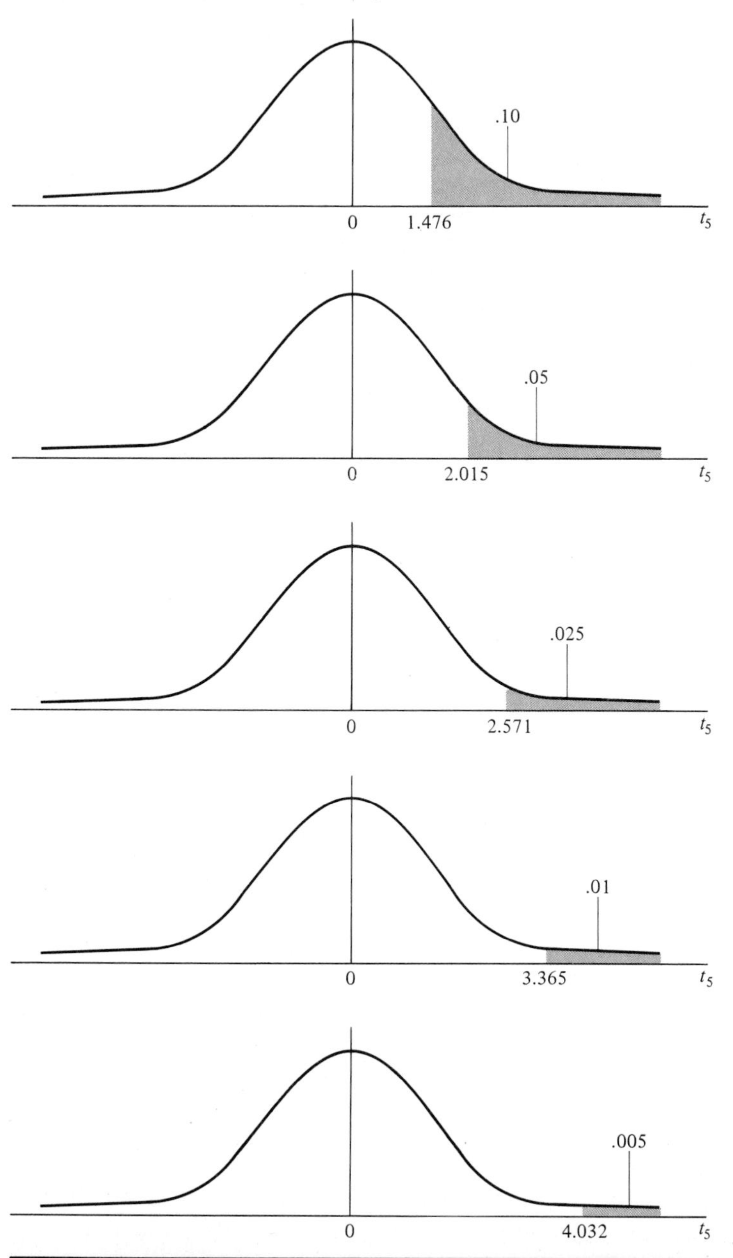

$$\bar{x} \pm t(.005;\, 9)s/\sqrt{n}$$
$$.296 \pm (3.250).0276/\sqrt{10}$$
$$.296 \pm .028$$

We could then say that we are 99 percent confident that $.268 \leq \mu \leq .324$. The interpretation of this interval is *exactly* the same as before. We generated this interval in such a way that 99 percent of all possible intervals generated in this manner (by our rule) will contain μ. We would probably be quite safe if we used .324 as our employee-hours per brake unit to formulate the bid.

Before we go on, there is one rather glaring unanswered question. What constitutes a large (versus small) sample? If we have a large sample, our rule for constructing a confidence interval is $\bar{x} \pm z_{\alpha/2}\, s/\sqrt{n}$. If, on the other hand, n is small, our rule is $\bar{x} \pm t(\alpha/2;\, n-1)\, s/\sqrt{n}$. The difference in rules is whether we use z or t. Actually, unless we know σ^2, we should *always* use t rather than z. But look again at Table A-5. Note that, for any α (which means for any column), as the degrees of freedom increase the tabulated t value decreases and approaches the t value for infinite degrees of freedom. Values in this last row (for infinite degrees of freedom) are really z values. You can see there is relatively little difference between the t values for 29 degrees of freedom and the corresponding z values. Therefore, it is typical to use z as an approximation for t when the degrees of freedom equal or exceed 30. Thus, we will define large sample size as $n \geq 31$.

7.7 SITUATION III: WHY?

Suppose you are involved with the promotion of the bank card VISA. Prior to 1977 your credit card was called BankAmericard. You are aware that a growing number of companies are using direct-mail selling by enclosing advertisements in the customers' regular monthly billings. Large department stores with their own credit cards and national oil companies are the typical firms engaged in this type of selling. It strikes you that VISA may be missing a real opportunity by not participating in the direct-mail type of merchandising.

Suppose you propose the idea to the right people in the VISA organization and they give you the go-ahead for a restricted campaign to see if your credit card holders would sufficiently accept such a project. You select a particular item to offer and check to make sure you can obtain up to 500 units of this product, if you need them. You have 500 small color inserts printed which show a picture of the item and contain an order form.

What you would like to know is what proportion of all VISA credit card holders would purchase this item, if offered the opportunity in their monthly statement. Since you believe the particular item to be "typical," you expect this proportion to be a good estimate of what you might expect for other products.

Say you draw a random sample of 500 VISA card holders and enclose your color inserts with their monthly statements. Of the 500 offered, 63 actually order the item. How do you estimate the proportion of all VISA card holders that would have purchased, if offered? Note VISA has several million card holders.

7.8 ESTIMATING A PROPORTION: HOW?

Let's define the *sample proportion* as

$$p = x/n \tag{7-6}$$

where n is the number of sample observations and x is the number of observations in the sample of n that have a particular characteristic of interest.

Given that the n sample observations are the result of random sampling, we can prove that the expected value of p is π, where π is the true proportion of the population. That is,

$$E(p) = \pi$$

Therefore p, as defined in Equation (7-6), is an unbiased estimator of π. If we compute p for our sample, the resulting value would be our point estimate of π.

As in most situations, we also seek an interval estimate. Note the similarity between the statistic p and our binomial variable x, which was introduced in Chapter 6. Random sampling gives us independence of observations, and if the population is infinite, the probability of drawing a success on any given trial remains constant. Thus, the numerator of p, Equation (7-6), is a binomially distributed random variable. As p equals a binomially distributed random variable divided by a constant n, the only difference in the distributions of p and x is scale: where $0 \le x \le n$, $0 \le p \le 1$. If we are sampling from a finite population and n is small *relative* to N, the probability will not change greatly from trial to trial, so the sampling distribution of p can be approximated using the binomial. Also recall that if n is large the binomial distribution can be approximated by the normal. Thus, for relatively large n from infinite or very large finite populations, the sample proportion p is approximately normally distributed with

$$\mu = \pi$$

$$\sigma^2 = \frac{\pi(1 - \pi)}{n}$$

This leads us to the following rule.

Parameter	Point Estimator	Interval Estimator
π	$p = x/n$	$p \pm z_{\alpha/2}\sqrt{\dfrac{p(1-p)}{n}}$

7.9 SITUATION III REVISITED

Recall that 63 out of the 500 ordered your product. Thus,

$$p = x/n$$
$$= 63/500$$
$$= .126$$

Our 90 percent confidence interval for π would be

$$p \pm 1.645\sqrt{\frac{p(1-p)}{n}}$$

$$.126 \pm 1.645\sqrt{\frac{(.126)(.874)}{500}}$$

$$.126 \pm .024$$

We are 90 percent confident that

$$.102 \leq \pi \leq .150$$

Whether a value of π in this range would make the direct-mail merchandising profitable would depend upon a variety of factors, such as handling costs and profit margins. We will leave your problem at this point, but you are well on your way to a solution.

7.10 *SITUATION IV: WHY?*

Suppose you are a practicing accountant, and one of your clients has come to you for advice. The client has an opportunity to buy an automobile parts and supplies firm and would like you to look at the books (the accounting records for the firm). Specifically your client would like to have your opinion on the profitability of this firm and to know what you think of the asking price of $157,000. After you examine the books you get back to your client with the following: "The books seem to be in order and the enterprise has shown a respectable profit for an investment of $157,000. The only item that concerns me is the book value of inventory. The books show $108,570 for inventory on hand as of the end of last year. Eight months have transpired since then and there are no interim inventory valuations available. If there really is $108,570 or so of inventory on hand, I would think that such a purchase would be a profitable investment. Can you validate the $108,570 figure?" Answer: "No, that's your job!" Accepting the challenge, how would you validate the present level of inventory?

Suppose you can obtain a listing of inventory for the firm in question. This list shows each item carried but does not give the number of units per item on hand. A quick glance at the list reveals that the firm stocks 1015 distinct items. Since it would be rather time-consuming to take a physical inventory for all 1015 items, you resign yourself to sampling. Just what do you need to be able to do? You need to be able to estimate the total for a finite population based upon sample observations.

7.11 ESTIMATION OF THE TOTAL FOR A FINITE POPULATION: HOW?

Our first step would be to draw a random sample of n items from the list of 1015. For each item selected, we would count the number of units on hand and determine the item's dollar value. The result of this process would be the set of sampled values x_i, $i = 1, 2, \ldots, n$. Our objective is to estimate the total of all x_i values, $\sum_{i=1}^{N} x_i$. Let's denote the total x as τ (the Greek letter tau).

$$\tau = \sum_{i=1}^{N} x_i \tag{7-7}$$

Recall from Chapter 2 that the mean for a finite population is defined as

$$\mu = \frac{1}{N}\sum_{i=1}^{N} x_i$$

So

$$N\mu = \sum_{i=1}^{N} x_i$$

and the parameter τ can also be expressed as

$$\tau = N\mu \tag{7-8}$$

Since $\bar{x}$ is an unbiased estimator of μ, we might reason that $N\bar{x}$ should provide us with an estimator of τ. Indeed, given that the n sample observations are the result of random sampling, we can prove that the expected value of $N\bar{x}$ is τ. That is,

$$E(N\bar{x}) = \tau$$

Therefore, our point estimator of τ is

$$N\bar{x} = \frac{N}{n}\sum_{i=1}^{n} x_i \tag{7-9}$$

Of course, we are again interested in obtaining an interval estimator. The statistic $N\bar{x}$ has a sampling distribution that will be approximately normal if n is large but small relative to N. An unbiased estimator for the variance of the sampling distribution of $N\bar{x}$, written var $(N\bar{x})$, is given in Equation (7-10).

$$\text{var}\,(N\bar{x}) = N^2\frac{s^2}{n}\left(\frac{N-n}{N}\right) \tag{7-10}$$

This leads us to the following rule.

Parameter	Point Estimator	Interval Estimator
τ	$N\bar{x}$	$N\bar{x} \pm z_{\alpha/2}\sqrt{N^2\frac{s^2}{n}\left(\frac{N-n}{N}\right)}$

7.12 SITUATION IV REVISITED

Suppose you decided to draw a random sample of 50 items from the inventory list of 1015. You have one of your office assistants actually conduct the sampling and valuation, and you have been presented with the data in Table 7-4. For the data in Table 7-4 we obtain

$$\sum_{i=1}^{50} x_i = 5{,}008.63$$

$$\sum_{i=1}^{50} x^2 = 979{,}477.55$$

$$\bar{x} = 5{,}008.63/50$$
$$= 100.17$$

$$s^2 = \frac{\Sigma x_i^2 - (\Sigma x_i)^2/n}{n-1}$$
$$= \frac{979{,}477.5 - 5{,}008.63^2/50}{49}$$
$$= 9{,}750.00$$

Our point estimate of τ is

$$N\bar{x} = (1{,}015)(100.17)$$
$$= \$101{,}672.55$$

How good is this estimate? Well, let's see. If we wanted a 90 percent confidence interval estimate for τ we would compute the following:

Table 7-4

Inventory Item Number	Inventory Value	Inventory Item Number	Inventory Value
A-107	$138.75	B-111	$110.00
A-110	114.62	B-092	172.92
A-015	55.12	C-702	212.68
A-347	274.10	C-880	32.40
A-342	83.19	A-214	0
A-417	5.26	A-987	0
A-512	101.40	B-073	5.12
A-001	138.20	B-718	108.50
A-905	39.56	C-005	220.15
C-602	62.10	C-719	232.01
B-105	0	A-806	19.05
A-553	214.12	B-929	0
C-222	347.14	A-707	3.19
C-247	73.21	B-711	25.00
C-462	6.19	C-409	321.92
A-915	10.10	B-419	16.50
A-732	14.00	A-782	12.48
A-333	101.32	A-419	118.20
B-101	40.95	C-073	202.56
B-108	25.10	A-181	118.21
B-347	0	B-779	202.95
B-492	215.00	C-081	138.42
B-003	221.93	B-231	7.18
B-667	118.14	B-088	309.95
C-449	5.62	A-299	14.12

$$N\bar{x} \pm 1.645\sqrt{N^2 \frac{s^2}{n}\left(\frac{N-n}{N}\right)}$$

$$\$101{,}672.55 \pm 1.645\sqrt{1{,}015^2 \frac{9{,}750}{50}\left(\frac{1{,}015 - 50}{1{,}015}\right)}$$

$$\$101{,}672.55 \pm 22{,}725.36$$

We would conclude that we are 90 percent confident that

$$\$78{,}947.19 \leq \tau \leq \$124{,}397.91$$

This is a fairly wide interval. What we are saying is that the total value of inventory on hand could be as low as around $79,000 or as high as maybe $124,500. At this stage our best estimate of the total dollar value of inventory is $101,672.55 (our point estimate). However, we realize that it may be as low as $79,000. If you think the asking price of $157,000 for this firm would be reasonable if indeed there was only $79,000 of inventory, you would recommend that the client go ahead with the purchase. Otherwise you might recommend that the client make a counteroffer for less than the $157,000, not make an offer at all, or reduce the width of the confidence interval. How could that be done?

What determines the width of a confidence interval? If we are sampling from a large or infinite population, there are three primary factors that affect the width of a confidence interval. First is the level of confidence selected. As the level of confidence increases, so does the width of the resulting interval. This factor enters the computation through z or t. If we desire a 99 percent rather than a 90 percent confidence interval, we use $z = 2.575$ rather than 1.645 in our rule. Thus the 99 percent confidence interval will be $2.575/1.645 \approx 1.57$ times as wide as the 90 percent confidence interval. The second factor is the variance of the population. This enters our computation through the unbiased estimate of variance s^2. The larger σ^2 is, the larger we expect s^2 to be and thus the wider the resulting confidence interval. The final factor is truly the most important, as it is the one we can effectively control; it is the sample size. As the sample size increases, the standard deviation of the sampling distribution of our statistic decreases. Thus the width of our confidence interval *decreases*. Yes, we could decrease the width of our confidence interval for τ by taking additional sample observations.

7.13 SAMPLE SIZE DETERMINATION

Sample size determination really hinges on the sampling distribution of the statistic used to estimate the parameter. In determining sample size we will assume that the sampling distribution of our statistic is normal. We know that for large samples this is approximately true. Figure 7-6 shows the sampling distribution of $\bar{x}$. We can see that $(1 - \alpha)$ 100 percent of all sample means are contained in the interval

$$\mu - E \quad \text{to} \quad \mu + E$$

That is, $(1 - \alpha)$ 100 percent of all sample means $\bar{x}$ lie within E units of μ. Thus, the probability of drawing a random sample of size n and obtaining an $\bar{x}$ that is within E units of μ is $(1 - \alpha)$.

Figure 7-6 Sampling Distribution of $\bar{x}$

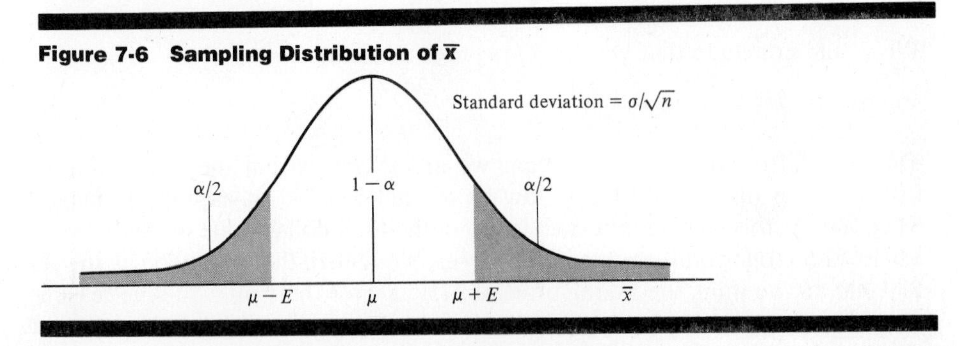

Now, if we were to increase n, the standard deviation of the sampling distribution of $\bar{x}$ would decrease. This means that a larger proportion of sample means will lie within E units of μ. Increasing n has the effect of "pushing in" on both sides of the sampling distribution, thus forcing more of the area closer to the mean μ.

In order to answer the question "How large should the sample size be?" we must specify both *how close* we want our resulting estimate to be to the parameter and *what probability* we want to have that our estimate *is* within the desired number of units. We will define E as the *tolerable error*; it represents how close we want our sample estimate to be to the parameter. So, we specify the value for E and answer the question: What probability do we want that our statistic (estimate) will not differ from the parameter by more than E units? That is,

$$Pr(|\text{statistic} - \text{parameter}| \leq E) = ?$$

Once we answer this question we know what z value would correspond to the point $\mu + E$. That is, what value of z must the point for $\bar{x}$ of $\mu + E$ transform into to have $(1 - \alpha)$ 100 percent of all sample means be within E units of μ?

The only other piece of information we need is the variance of the population from which we will be sampling. If we are estimating μ or τ, we might have an estimate of σ^2 from a prior study or we could take a *small* preliminary sample and use these data to estimate σ^2 via s^2. If we are estimating π, our σ^2 is actually $\pi(1 - \pi)$. If we have no information about π, we choose $\pi = .5$, as $\pi(1 - \pi)$ is maximized when $\pi = .5$. This is the worst case, and the resulting sample size will be *at least* large enough to meet the specifications. Sample size determination formulas are given in Table 7-5.

Table 7-5

Parameter of Interest	Required n
μ	$n = \dfrac{\sigma^2 z^2}{E^2}$
π	$n = \dfrac{z^2\pi(1 - \pi)}{E^2}$
τ	$n = \dfrac{N\sigma^2}{(N - 1)(E^2/z^2N^2) + \sigma^2}$

Let's see how to determine the required sample size by returning to our previous situations.

Example A Suppose in Situation I you wanted to estimate μ for Bob Peterson's brand of tire within 1000 miles and you wanted the probability to be .90 that your resulting $\bar{x}$ would not differ from μ by more than 1000 miles. Suppose further that you knew that your present tire brand had a standard deviation of about 8000 miles and suspected that the standard deviation for Bob's brand would be about the same. Thus we have

$$E = 1000$$

$$\sigma = 8000$$

$$z = 1.645$$

$$n = \frac{z^2\sigma^2}{E^2}$$

$$n = \frac{(1.645^2)(8000^2)}{1000^2}$$

$$= 173$$

Example B Suppose in Situation III you wanted to estimate the proportion of VISA cardholders that would purchase the offered product and you wanted your estimate p to be within .05 of the true π with probability of .99. If you had no idea of what π might be, you would use $\pi = .5$ for calculating n. However, let's suppose you think that $0 \leq \pi \leq .30$. In this case we choose the value for π that is closest to .5. Thus we would use $\pi = .30$.

$$E = .05$$

$$\pi = .30$$

$$z = 2.575$$

$$n = \frac{z^2\pi(1 - \pi)}{E^2}$$

$$= \frac{(2.575^2)(.3)(1 - .3)}{.05^2}$$

$$= 557$$

Example C Suppose in Situation IV you wanted your estimate of the total value of inventory to be within $10,000 with probability of .90. Since you would probably not have a prior estimate of the variance of item values, let's say you have your assistant draw a random sample of 10 items and compute s^2 for the 10. You are told s^2 is 8240.

$$N = 1{,}015$$

$$E = 10{,}000$$

$$\sigma^2 = 8{,}240$$

$$z = 1.645$$

$$n = \frac{N\sigma^2}{(N - 1)(E^2/z^2N^2) + \sigma^2}$$

$$= \frac{(1{,}015)(8{,}240)}{(1{,}014)[10{,}000^2/(1.645^2)(1{,}015^2)] + 8{,}240}$$

$$= 188$$

If we were to go back to Situation IV and increase our random sample of items to 188, the resulting 90 percent confidence interval would be approximately $N\bar{x} \pm 10{,}000$.

7.14 IDENTIFICATION OF APPLICATIONS: WHEN?

We have seen that *statistical inference* refers to methods and/or processes of *drawing conclusions about populations based upon sample information*. We have been concerned in this chapter with *estimating parameters* based upon sample information. Recall in Chapter 4 we defined a decision as the choice among alternatives. Once we select an alternative, we realize the consequences, good or bad. What determines the consequence given a particular alternative is the true state of nature. Often the true state of nature that impacts on the resulting consequence can be described by a single parameter. For example, the true mean μ miles per tire certainly affects the cost per mile of tires. The true proportion π of your VISA credit cardholders that would purchase a given product would affect the profitability of conducting such a merchandising venture. Finally, knowing the true total value of inventory would affect the profitability and/or advisability of purchasing our automobile parts firm.

The simple fact is that for many decision situations the consequences of various alternatives are a function of, or determined by, the value of a certain parameter. Therefore, it is of value to the decision maker to be able to estimate parameters.

What are the conditions under which statistical estimation would be of value? Look back at why we sample rather than take a complete census. If you have a situation where you will be relying on sample information and the outcomes or consequences of the available alternatives are a function of a parameter, you have a situation where statistical estimation would be of value!

PROBLEMS

1. Identify an application of estimation in your major area of study. Point out the features which make this application appropriate and of value.

2. Max Brothers is a chain of 87 men's clothing stores located throughout the western half of the United States. It carries a complete line of high quality, "high fashion" men's clothing. Stewart Max has observed that although the stores carry only men's clothing, approximately 35 percent of its customers are women who are purchasing clothing items for their husbands or as gifts.

   Stewart Max is currently trying to develop company-wide policies concerning the number of salespersons each store should have on the floor. As part of his future plans he wants to know the percentage of customers that want sales help while in the store and the percent that prefer to help themselves. He feels that he will need this information for both males and females, since the customer mix varies at each store. An in-store random sample of 500 customers showed that 300 preferred the assistance of a salesperson on that store visit. Of the 300 males, 165 desired assistance, and 135 of the 200 females desired assistance. In order to be very sure that he does not underestimate the percent of shoppers that want assistance, he needs from you the following:

a. An estimate for the percentage of male customers who want assistance.

b. An estimate for the percentage of female customers who want assistance.

c. An estimate for the percentage of all customers who want assistance.

Give a 95 percent confidence interval estimate for part *a*, a 90 percent confidence interval for part *b*,and a 99 percent confidence interval estimate for part *c*.

3. The management of Eljun Electronics Corporation marketed a new battery for operating the flash unit of 35-mm cameras under the slogan, "Why not use the best? On the average our batteries give 20,000 flashes." Yesterday evening the marketing manager got very upset while watching KMTW-TV local news. The company's claim was criticized by the Consumer Vigil announcer Bibly Kisone. The next day, in order to refute Bibly Kisone's criticism, the manager randomly selected 23 different flash units and tested the battery in them. The results are presented below:

Number of Flashes (In Thousands)	Number of Flashes (In Thousands)	Number of Flashes (In Thousands)
15	17	20
19	22	16
14	18	15
16	9	17
12	17	16
17	16	13
16	18	15
18	17	

Obtain both a point and a 95 percent confidence interval estimate of the true, but unknown, mean number of flashes. Based on your computations, do you think the marketing manager would be able to refute Kisone's criticism of the claim made in the company's slogan?

4. As a newly hired accountant you have been assigned to a major project establishing standard costs for many of the production activities of the firm. This is of critical importance, since most of the firm's business involves unique products which are made to order. Contracts for these products are awarded on a competitive bid basis, so accurate cost estimates are essential. As a first step, you are asked to estimate the average time required to perform a welding operation. A random sample of 13 time measures to perform a welding operation is given below:

Minutes	Seconds
6	2.7
8	1.3
7	1.2
6	3.8
8	4.3

7	4.1
6	5.8
7	4.0
8	0.6
9	0.5
7	2.0
8	0.5
8	2.0

Give a 99 percent confidence interval estimate for the mean time required for the specific welding operation. Interpret your interval. How large must your sample be in order to have a probability of at least .95 that your estimate of μ will be within 15 seconds of the true value?

5. The following is an excerpt from the conversation between Linda Patell, the product manager of GroRite, a plant food manufacturing firm, and her friend A.J. Coole.

LP—You know A.J., I am pleased with the GroRite sales growth over the past eight years. Aren't you impressed?

AJ—Well, not exactly. The indoor plants are the latest fad, therefore demand for plant food had also increased. The true measure of any product's success is its market share. For example, it is possible to have increasing sales but a decreasing market share. This means that competitors are more effective in capturing the increased demand.

Ms. Patell recognized the logic of this argument and began steps to estimate GroRite's market share. She knew GroRite's sales, but there were no figures available for industry sales. For this, she selected a random sample of 300 stores that carried plant food. The average previous-week sales were 64 ounces with a standard deviation of 42 ounces. There are approximately 37,000 stores that sell plant food. Help Ms. Patell estimate the total sales of plant food for the week in question. Also obtain a 95 percent confidence interval for the true total sales.

6. Lawyers for Marshall Products are preparing a suit against Eastern Chemicals, Incorporated. Marshall Products has produced and marketed a high quality, expensive wood polish under the brand name Enfield for over 25 years. Enfield's sales have consistently been more than 75 percent of the total sales of products of this nature.

Two years ago Eastern Chemicals started distributing a wood polish under the name Infield. The product manager for Enfield started legal proceedings against Eastern Chemicals within six months after Infield appeared on the market.

The lawyers for Marshall hope to estimate the percent of individuals purchasing Infield thinking that they are buying Enfield. To do this will require a survey. However, it is difficult to get survey evidence entered into court. Therefore the lawyers are being particularly cautious. They plan to survey consumers leaving a store with a bottle of Infield. Through a series of questions that they are confident the court will accept, they will determine whether or not the consumer bought the Infield brand by mistake. Their current estimate is that 60 percent of the purchases of Infield are by mistake. They must now decide on sample size. The lawyers want to be sure they can convince the judge that their estimate is accurate. Therefore, they wish the tolerable error to be less than .1 percent with probability of .99. As interviews are expensive, they don't want to take a larger sample than necessary. Advise the lawyers.

7. Suppose that the managers of a large department store have decided that drastic action must be taken to reduce losses caused by shoplifting. Several different, expensive alternatives have been proposed. As each alternative is investigated in depth, it becomes clear that the firm needs an accurate estimate of the percentage of shoppers that engage in shoplifting. An examination of secondary sources reveals estimates of shoppers that shoplifted an item in the past week ranging from .1 percent to 9 percent. In view of the serious nature of the countermeasures proposed, the firm wants to have an estimate of the percentage of shoppers that actually shoplift an item during their visit to the store. The incidence of shoplifting will be determined by having specially trained store detectives discreetly follow a random sample of shoppers throughout their store visit, an expensive process. If the firm's management wants to be "virtually positive" that its estimate will not differ from the actual figure by more than .5 percent, how large should the sample be?

8. The administration of a state university was frequently involved in controversy with the local city government. The university was exempt from local property taxes but, in the eyes of the local government, still required substantial local services such as police, fire departments, and so forth. The controversy was not limited to this campus and threatened to produce legislation eliminating the tax-exempt status of all university property. Such a result would necessitate a rather substantial tuition increase.

   In an attempt to demonstrate the positive economic value of the university to the city, a survey of student expenditures was conducted. The survey measured *monthly* expenditures for a variety of products. Two hundred students were selected at random and each agreed to keep a diary of certain purchases for a month. The major findings are presented below.

Product	Sample Mean	Standard Deviation
Jewelry, watches	$50	12
Film	6	4
Gasoline	27	2.3
Movies	10	1.75
Gifts	15	2.2
Sports equipment	20	1.8

   Give both point and interval estimates for the mean expenditure per product for the particular month. Use the 95 percent confidence level for all product classes.

9. Suppose you are in charge of new franchises for a restaurant chain. You are continually asked by prospective franchises, "Will I be able to realize a profit by the end of the first year?" Whether a particular franchise does or does not turn a profit in its first year is a function of many things (competition, management, population, density, etc.). It would be far too complicated to try to account for all such factors. However, you think it would be informative to know what proportion of the restaurants did realize a profit in the first year of operation. Let's say you would like to estimate this proportion via sampling and you would like your estimate to be within

.03 of the true π with probability of at least .90. How large must your random sample of existing franchises be?

10. You are the controller for a medium-sized department store. Your store issues its own credit cards and thus carries a substantial level of accounts receivable. You have noticed that the total of accounts receivable has increased to an alarming level. In order to formulate an early payment discount policy, you need to know the size or magnitude of your overdue accounts. Suppose you would like to estimate the total dollar amount of receivables that are at least 90 days old. How large must your random sample be if your tolerable error is set at $10,000 with probability of .90? Assume you have 1500 separate accounts that are overdue 90 days or more, and you think the variance of these overdue accounts is less than 5000.

11. Federal Bank and Trust

    CONFIDENTIAL MEMO

    TO: *All Members, Board of Directors*

    FROM: *Ellen Malden, Attorney-at-Law*

    We operate branches within walking distance of the eight major colleges in the state. These branches serve the college students and therefore are characterized by numerous checking accounts, small balances, heavy seasonal fluctuations in activities, and a high turnover of customers. Therefore, each of these eight branches operate at a slight loss. Their existence is justified on the hereto unproven belief that the students will seek out a Federal Bank and Trust branch when they leave school and enter the work force.

    I would like to draw your attention to a published survey (copy enclosed) which shows that 40 percent of the new residents of a community selected a bank primarily because of its location, 30 percent primarily because of recommendations from friends, and only 30 percent because of past ties to another branch of the bank.

    To prove my point, I traced 60 of the campus branches' ex-accounts through their schools' alumni associations and interviewed them by telephone. Twenty-four said they were currently using a Federal branch primarily because of their experiences with the bank while in college. Therefore, in my opinion, we should seriously consider discontinuing the operation of our college branches in our next meeting.

    (SIGNED) *Ellen*

    Since you are the only data analyst on the board of the bank, your advice in these matters has always been taken seriously. Be prepared to argue for or against Ellen's proposal in the next meeting.

Chapter 8

Tests of Hypotheses

8.1 SITUATION I: WHY?

Suppose you are director of new business products for Weyerhaeuser Company. Your job consists of expanding experimental products, principally outside Weyerhaeuser's existing business, into full-scale manufacturing and marketing activities. An independent inventor has submitted an idea for a new product and has asked Weyerhaeuser to consider the production and commercialization of the product. Specifically, the product is a paper panel used for residential fencing.

The product proposal calls for the following three-piece construction of the paper panels:

1. *A 3- by 6-foot inner ply sheet of sulphite paper, approximately 0.067 inch thick.*
2. *Borders of sulphite paper approximately 0.125 inch thick bonded to both sides of the inner ply.*
3. *The entire panel, both front and back, coated with a sufficient amount of polyethylene to render the panel waterproof.*

This construction is shown in Figure 8-1. Figure 8-2 gives a graphic representation of one possible mode of fence construction. Such a product would be ideal for apartment dwellers and inhabitants of travel trailer communities in the South and Southwest.

Figure 8-1 Paper Fence Panel

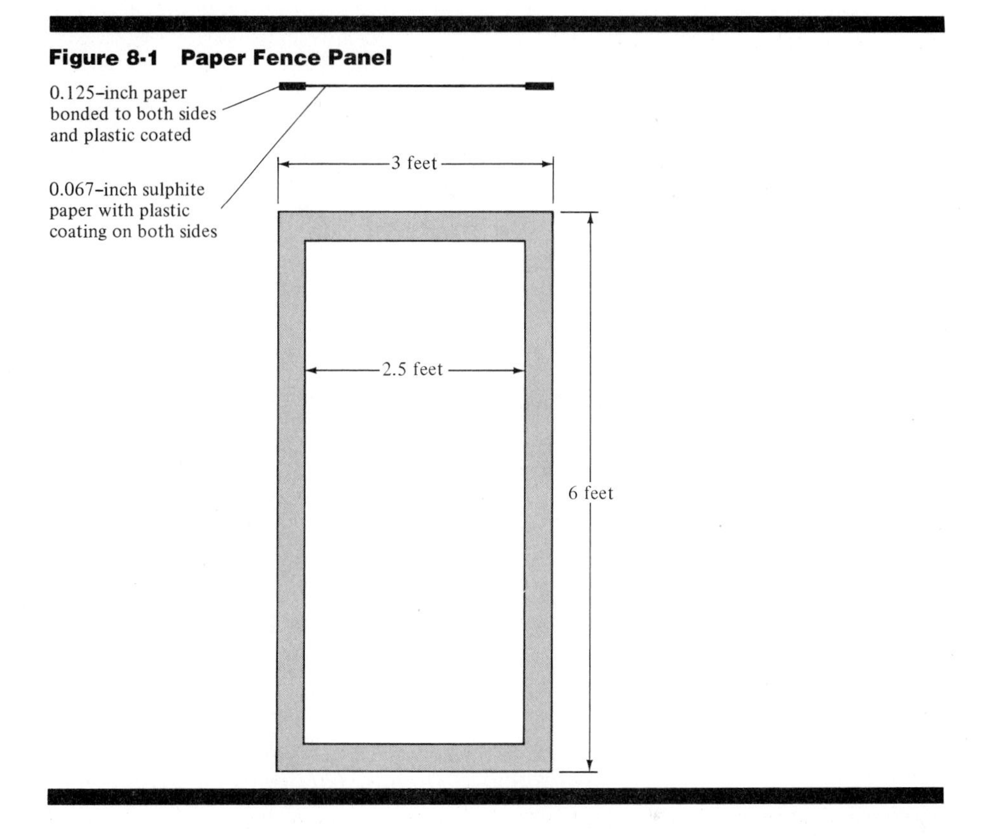

Originally fences were erected as a form of protection. Today, however, the main function of residential fencing is to provide visual privacy. This shift in product function has resulted in a change in product requirements. Strength, per se, is no longer the primary characteristic of interest. Rather, the fencing materials should be able to withstand elements of the weather. Specifically, the panels must withstand rain, moisture, and wind. Rain and moisture you can handle. After all, you do produce leakproof milk cartons that can be frozen. So the real question you are facing is this: Can you produce a paper panel that is wind resistant?

Suppose you examine the meteorological records of the areas in which you would plan to market such a product and find that wind velocity hardly ever exceeds 80 miles per hour. Thus, you conclude that if the paper panels can withstand an average wind velocity of 80 miles per hour, the product is practical.

Say you have your production manager make up 25 panels to specification and decide to test these panels to determine whether the wind-resistance strength is greater than or equal to 80 miles per hour. You take your 25 panels to the University of Washington wind tunnel in Seattle and have each panel

Figure 8-2 Paper Panel Fence Construction

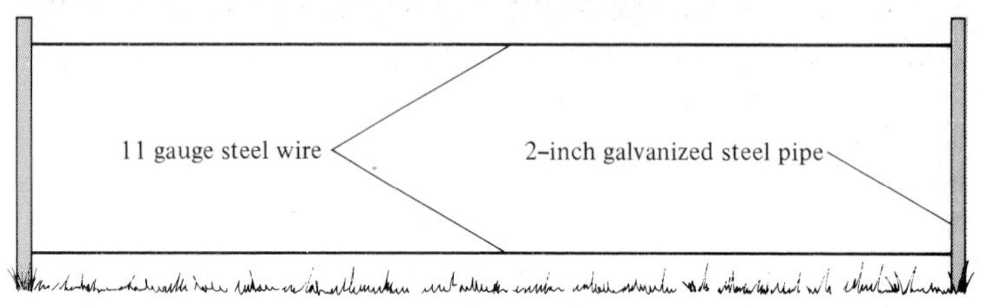

subjected to a wind test. Specifically each panel is subjected to increasing wind velocity until the panel fails (separates, rips, etc.). The wind velocity at which each panel fails is shown in Table 8-1.

Well, is the paper panel practical? We don't know the answer, yet. The first step would probably be to calculate the sample mean. So,

$$\bar{x} = 1937/25$$
$$= 77.48 \text{ mph}$$

Is 77.48 so far below 80 that you should conclude that $\mu < 80$? What do you need to be able to do? If you think $\mu \geq 80$ mph, you would judge that the paper panel is practical and therefore may want to recommend that Weyerhaeuser continue giving this product serious consideration. On the other hand, if you think $\mu < 80$, you would regard the paper panel idea as impractical for the desired market and would recommend rejection of the idea. You need to be able to decide whether you believe $\mu \geq 80$ given sample data that yields a sample mean $\bar{x} = 77.48$ mph. That is, is an $\bar{x}$ of 77.48 so far below 80 to lead you to conclude that μ is less than 80? Such a decision process is called testing hypotheses *and is, like estimation was, part of what is called* statistical inference: *drawing conclusions about populations based upon sample information.*

8.2 TESTING HYPOTHESES: HOW?

We contend that you already know how to test hypotheses. You do it every day! The only problem is that when we start dealing with abstract symbols and concepts it is easy to lose sight of what we are really doing. You *do* already know how. We'll prove it to you!

Suppose we approach you with the following proposal. We will flip a coin, say a quarter, and if the coin comes up "heads," you pay us a dollar. If, on the other hand, the quarter comes up "tails," we pay you a dollar. *Try to place yourself in this situation*. We flip the coin and it comes up heads. You pay us a dollar and we ask you if you would like to play again. You agree to play again, so we flip the coin and it comes up heads again. You pay us another dollar and we are getting excited. Will you play again? Let's suppose this process continues and the coin comes up heads *every time*. We assume that after seeing a certain number of heads and no tails, your answer to our question "Would you

Table 8-1

Panel	Wind Velocity at Failure, mph
1	82
2	91
3	72
4	71
5	68
6	71
7	82
8	94
9	106
10	74
11	72
12	76
13	82
14	88
15	62
16	51
17	59
18	71
19	81
20	88
21	59
22	89
23	79
24	68
25	101

like to play again?'' would be NO! Take a piece of paper and write down your number. That is, after seeing how many heads and no tails would you refuse to play? Don't go on until you decide what that number would have to be for you to stop playing. Ready? OK, *you have just tested a hypothesis*!

When we proposed the game and you decided to play, you actually formulated a hypothesis. *A hypothesis is a supposition about a true state of nature*. In our game your hypothesis was that the coin was fair which means that $Pr(\text{head}) = Pr(\text{tail}) = .5$. Thus your hypothesis was that the parameter π, where $\pi = Pr(\text{head})$, equals .5. You then observed sample evidence in the form of the outcomes of the various trials (flips of the coin). Why did you decide to stop playing the game? Because you decided that you do not believe that $\pi = .5$. When your answer to our question ''Would you like to play again?'' was finally NO, you reached the point at which you no longer believed that the coin was fair and rejected the hypothesis that $\pi = .5$. Why did you decide that you don't believe that $\pi = .5$? Because you observed too many heads and not enough tails in the sample. Specifically what you did was to conclude that the sample result was very unlikely to occur if the coin was indeed fair, that is, if $\pi = .5$. So you concluded that we were not playing with a fair coin and rejected the hypothesis that $\pi = .5$. Actually what you did was to decide that the probability of observing a sample result as extreme as that observed, if $\pi = .5$, was so small that you no longer believed that $\pi = .5$.

Table 8-2 shows the conditional (given $\pi = .5$) probabilities of observing n heads in n trials.

Now look back at your critical number. Table 8-2 shows how unlikely the sample result had to be for you to reject the hypothesis that $\pi = .5$.

Let's say your critical number was 5. We see in Table 8-2 that the sample outcome 5 heads in 5 flips of a *fair coin* has only 3.125 chances in 100 of occurring. *By your definition, that's too small*. After observing 4 heads in 4 flips (a sample outcome with a probability of .0625 of occurring if $\pi = .5$) you still believed that $\pi = .5$. Thus, a probability of .0625 was by your definition not too small. However, after observing 5 heads in 5 flips (probability .03125 if $\pi = .5$) you decided to reject the hypothesis that $\pi = .5$. Thus, a probability of .03125 was defined by you to be too small, so small that you no longer believed the probability of the coin coming up heads was indeed equal to .5.

The reasoning above is *precisely* the logic of *all* hypothesis-testing situations.

Logic of Testing Hypotheses

1. A hypothesis is formulated which we would like to investigate with the objective of deciding whether we believe it to be true or false.
2. Sample evidence is obtained which contains information about the truth or falsity of the hypothesis.
3. Because of the specific way in which our sample evidence is obtained and corresponding statistical theory, we are able to compute or determine the probability of

Table 8-2

n	$Pr(x = n \mid \pi = .5)$
1	.5
2	.25
3	.125
4	.0625
5	.03125
6	.015625
7	.0078125
8	.00390625
9	.00195312
10	.00097656
11	.00048828
12	.00024414
13	.00012207
14	.00006103
15	.00003051
16	.00001525
17	.00000762
18	.00000381
19	.00000190
20	.00000095

observing a sample result as extreme as that observed, or more extreme than that observed—*if the hypothesis is true.*

It is the ability to compute the conditional probability of the sample outcome when the hypothesis is true that provides the logic for the procedure.

Think of any situation where a statement or claim is made. You make a judgment about the truth or falsity of the claim by reasoning whether all your evidence—wherever it came from—supports or is consistent with the claim. ("Supports or is consistent with" means has a probability of occurring, if the claim is true, that is not too small.)

4. If the conditional probability of observing a sample result as extreme as that observed is too small, we conclude that the hypothesis is false. Either the hypothesis *is* true but we have observed what we consider an extremely unlikely sample result, or the hypothesis *is not* true and the small conditional probability is *not really the probability* of observing the particular sample result.

That's it! *This is the basic logic for all tests of hypotheses*—not just those in this chapter, but for *all tests*. Furthermore, the process is instinctive and you do it, you test hypotheses every day!

Now that we know and understand the general logic of testing hypotheses, let's move on to some specifics. The hypothesis to be tested is referred to as the *null hypothesis* and is commonly denoted by H_0. The complement of the null hypothesis is called the *alternative hypothesis* and will be denoted by H_1.

How do we formulate hypotheses in a given situation? In statistical hypothesis testing we typically dichotomize the states of nature. Our objective is to decide which of the two states we believe to be true.

The first cardinal rule of formulating hypotheses is that a *hypothesis must contain only states of nature that would uniquely lead to a single alternative*. For example, if in a given situation we would decide to do one thing if we believed μ was equal to 50 and we would choose to do a different thing if we believed μ was greater than 50, we must not formulate as one of our hypotheses $\mu \geq 50$. For, if we decide to believe this hypothesis to be true, we haven't solved our problem. If $\mu = 50$ we would want to do something different than if $\mu > 50$. Since our test may leave us believing that μ is *either* equal to *or* greater than 50, we would not know what to decide. But that's why we conducted the test in the first place—to decide.

The second cardinal rule is that *the equality must be in the null hypothesis*. This requirement is fundamental to the logic of testing hypotheses. Recall that it is the ability to compute or determine the conditional probability of observing a sample outcome "as extreme as" the sample when the hypothesis is true that provides the logic. Therefore we need to be able to specify the sampling distribution of our statistic—when the null hypothesis is true. This requires that we know where to locate the sampling distribution. If the null hypothesis does not contain the equality, we would not know where to locate the sampling distribution of our statistic; and we would not be able to compute the conditional probability of our sample results when the null hypothesis is true.

1. The null hypothesis must contain the equality.
2. No single hypothesis shall contain states of nature that lead to different decisions.

Given that we will choose one of the two hypotheses and only one hypothesis is true, there are two kinds of errors that we can make. First, we may decide to reject the null hypothesis *when it is true*. This error is referred to as a *Type I error*, and we let α be the probability of committing a Type I error. The other possible error, called a *Type II error*, is the error of accepting the null hypothesis when it is false. Typically we denote the probability of committing a Type II error by β. Table 8-3 shows the four possible outcomes of a statistical test.

There is another general issue that arises in hypothesis testing. This is whether a test should be a one- or two-tailed test. *Without exception, we will compute a statistic from our sample data and decide whether to reject or accept the null hypothesis based upon the magnitude of our computed statistic*. If the statistic could be *either* too large *or* too small for us to believe the null

Table 8-3

State of Nature	Decision	
	Accept H_0	Reject H_0
H_0 True	Correct—no error	Type I error
H_0 False	Type II error	Correct—no error

hypothesis to be true, we would want to establish two critical values. These values would be such that if our statistic was less than the smaller critical value or greater than the larger critical value we would reject the null hypothesis. If, on the other hand, our statistic fell between the two critical values, we would accept H_0.

The two critical values would be chosen such that they cut off a certain area in the upper and lower tail of the sampling distribution of our statistic. This is referred to as a *two-tailed test*.

If we would want to reject H_0 *only* when the statistic is large *or* only when it is small, we would have only one critical value for the statistic and thus the *rejection region* (interval of values of the statistic for which we would reject H_0) would be *either* the upper-tail *or* lower-tail areas—but not both—of the sampling distribution of the statistic. This test is referred to as a *one-tailed test*.

It is really quite simple to decide whether our test is one-or two-tailed. The key is the form of the alternative hypothesis. If we reject H_0 it is because our sample data seem to refute H_0 *in favor of* H_1. So what we do is ask the question, "Could the statistic get so large that we would want to reject H_0 *in favor of* H_1?" If the answer is no, we have a one-tailed test and the rejection region is for *small* values of the statistic only. If our answer is yes, we know there is a rejection region in the upper tail of our sampling distribution. We then need to ask whether we would *also* want to reject H_0 *in favor of* H_1 if we observed a small value for our statistic. If our answer is no, we have a one-tailed test. If our answer is yes, there would also be a rejection region in the lower tail of our sampling distribution—thus a two-tailed test. The three alternatives are shown graphically in Figure 8-3.

Let's address two final points before we look at a specific test. First, note the meaning of the words *accept* and *acceptance* when they relate to hypothesis testing. When we say we *accept* the null hypothesis, what we really mean is that *we are unable to reject the null hypothesis*. When we reject H_0 we are saying that there *was* sufficient information contained in the sample for us to conclude that H_0 is false. When we *accept* H_0 we are merely saying that there *was not* sufficient information in the sample to convince us that H_0 is false. This could happen because (1) H_0 is true or (2) our sample size is not large enough for us to detect that H_0 is false. Clearly, the closer H_0 is to the truth, the larger

Figure 8-3 (a) Two-tailed Test; (b) One-tailed Test, Upper Tail; (c) One-tailed Test, Lower Tail

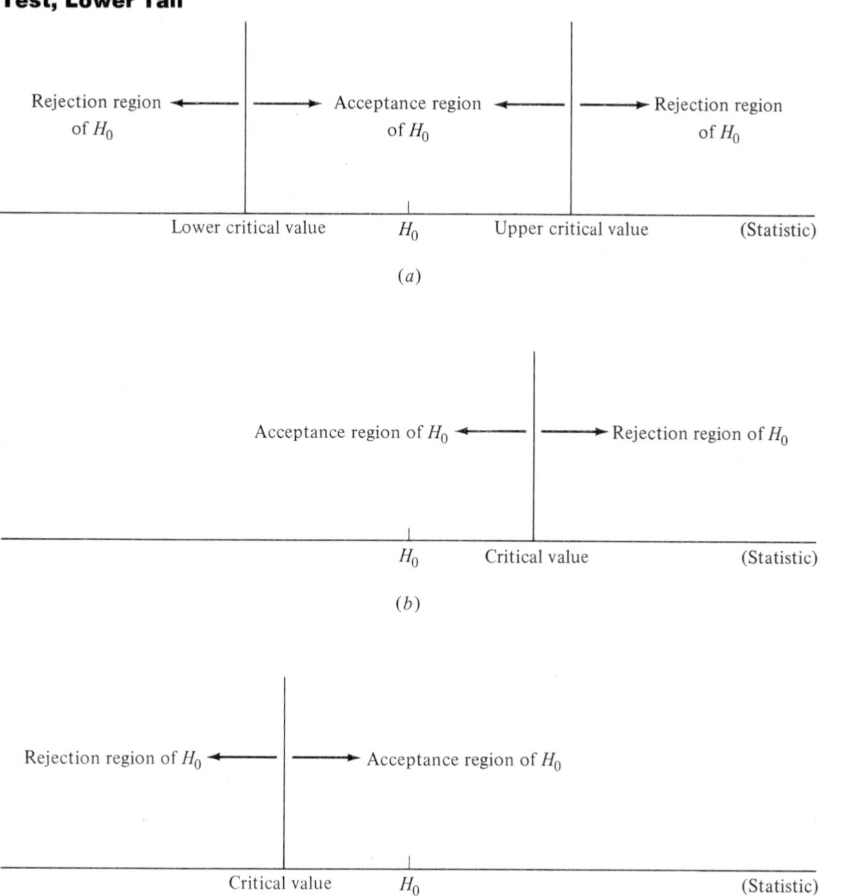

the sample size will have to be to detect that H_0 is not true. For example, if the null hypothesis is H_0: $\mu = 50$ and μ really is equal to 500, it would not require a very large sample size for us to detect that H_0 is false. However, if μ really equals 50.001, we would have to have a very large sample to have any hope of detecting that H_0 is false. So, when we *accept* H_0, it doesn't really mean that we believe it's true; it merely means that we haven't seen enough contrary evidence to reject it.

Finally, the simplest approach to testing hypotheses is to specify a value for α (the probability of rejecting H_0 when it is true); determine the rejection region or regions; compute the appropriate statistic and reject if the statistic is

in a rejection region, *accept* H_0 otherwise. The numeric value of α is known as the *level of significance*. If our sample result leads to the rejection of H_0, we say the result is *significant*.

An alternative approach is to merely compute and report the conditional probability of observing a sample result as extreme as the sample. This leaves the decision of whether the conditional probability is small enough to reject H_0 or not up to the consumer or recipient of the study. Actually, we can view α as the critical value for the conditional probability. If the conditional probability is less than or equal to α, the result is the rejection of H_0. So, when we specify α, we are defining what *too small* is for a given problem. We are defining the critical value for the conditional probability, such that if we observe a sample that has a conditional probability that is less than or equal to α, we will decide to reject H_0.

8.3 TESTING HYPOTHESES ABOUT μ: HOW?

8.3a Large Sample Size

Assuming that the population is large or perhaps infinite, and that the random sample of size n is large enough such that the sampling distribution of $\bar{x}$ is approximately normally distributed, via the central-limit theorem, we could develop a test on μ by using $\bar{x}$ as our test statistic. Since the parameter we are investigating is μ and $\bar{x}$ is an unbiased estimator of μ, it seems reasonable to focus attention on $\bar{x}$. The closer our resulting $\bar{x}$ value is to the hypothesized value for μ, the more our sample evidence would seem to support H_0. As we examine other tests we will find that it will be more convenient to actually use a transform of this statistic rather than the simple estimator itself.

Given the above assumptions, the statistic

$$\frac{\bar{x} - \mu}{s/\sqrt{n}}$$

will have as its approximate sampling distribution the standard normal. Thus,

$$z = \frac{\bar{x} - \mu}{s/\sqrt{n}} \tag{8-1}$$

and we could therefore determine various probabilities for this statistic *if we knew* μ. Since our objective is to develop a procedure that will enable us to compute probabilities when the hypothesis is true, by replacing μ in Equation (8-1) with the hypothesized value for μ denoted as μ_0, we will have the sampling distribution of a statistic when H_0 is true. There is our test. We know the sampling distribution for

$$\frac{\bar{x} - \mu_0}{s/\sqrt{n}}$$

is approximately the standard normal when H_0 is true. Therefore we are able to

determine the conditional probability of observing a sample result for $\frac{\bar{x} - \mu_0}{s/\sqrt{n}}$ as extreme as the one we actually observed. Depending upon the relative values of α and the conditional probability, we decide whether or not to reject H_0.

8.3b Small Sample Size

The only difference in the required assumptions for analyzing small sample sizes is that the population of x values is normally distributed. If this is true, then

$$\frac{\bar{x} - \mu}{s/\sqrt{n}}$$

is a student t variable with $n - 1$ degrees of freedom. So the sampling distribution of

$$\frac{\bar{x} - \mu_0}{s/\sqrt{n}}$$

is the student t with $n - 1$ degrees of freedom if $\mu = \mu_0$ (that is, when the null hypothesis is true). Let's see how to perform the test by returning to Situation I.

8.4 SITUATION I REVISITED

Recall that your problem was to decide whether you believe the wind-resistance strength of the paper panels is greater than or equal to 80 miles per hour. Since the equality must be in H_0, so must "greater than," because your decision will be the same whether you believe that $\mu = 80$ or that $\mu > 80$. Therefore our null and alternative hypotheses are as follows:

H_0: $\mu \geq 80$

H_1: $\mu < 80$

Suppose we set $\alpha = .10$. That is, we decide that we will reject H_0 if we observe a sample result that has a conditional probability of occurring (when H_0 is true) less than or equal to .10.

As n is relatively small (that is, $n \leq 30$), we must assume that the maximum wind velocity capability per panel is normally distributed. Our test statistic is $\frac{\bar{x} - 80}{s/\sqrt{n}}$ and, since only small values for $\bar{x}$ would refute H_0 in favor of H_1, only small values for $\frac{\bar{x} - 80}{s/\sqrt{n}}$ would favor H_1. Thus, we have a one-tailed test and all of α is located in the lower tail of the student t distribution with 24 degrees of freedom as shown in Figure 8-4.

Figure 8-4

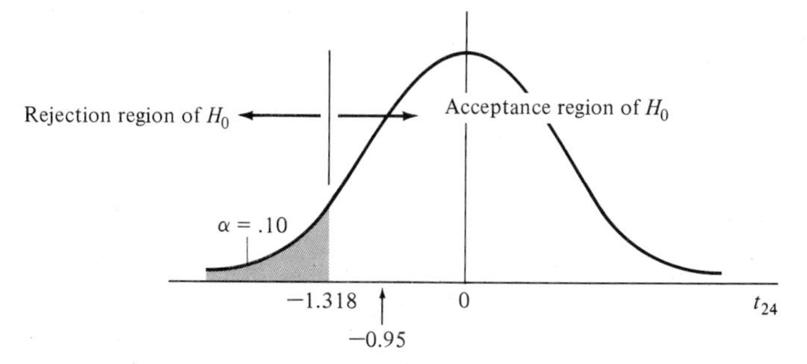

For the data in Table 8-1 we find $\bar{x} = 77.48$ and $s = 13.32$. Our computed test statistic is then

$$\frac{77.48 - 80.00}{13.32/\sqrt{25}} = -.95$$

We can see in Figure 8-4 that $-.95$ lies in the *acceptance* region of H_0, as $-.95$ is greater than the critical value for t of -1.318. Thus, the conditional probability of observing a sample outcome as extreme as ours is certainly greater than $\alpha = .10$. Our decision is to *accept* H_0 and conclude that $\mu \geq 80$ miles per hour.

Had our sample size in Situation I exceeded 30, we would have used -1.282 as the critical value for our test statistic $\dfrac{\bar{x} - 80}{s/\sqrt{n}}$ as we would use z rather than t.

8.5 *SITUATION II: WHY?*

Suppose you work in the product development division of Garrard Corporation, manufacturers of fine stereo turntables. Garrard is in the process of developing an ultralight tone arm which is designed to replace the one on their existing turntable model G-1001. The objective of the new ultralight tone arm is to reduce the weight on the diamond stylus in order to prolong both stylus and record life.

The problem is that the tone arm is so light that a special, rather expensive, diamond stylus is required to obtain maximum performance. There are two possible brands of styluses that are compatible with Garrard's new ultralight tone arm, and it is your job to select the one Garrard should purchase. Assume the two brands have basically equal availability and price. Your only concern is the life of the diamond stylus. If one brand has a longer play life, you would like to purchase it, as you would like to give your customers the most for their money.

Table 8-4 Play Life of Diamond Styluses
(In Thousands of Hours)

Brand 1	Brand 2
1.2	1.1
0.8	1.0
1.1	0.6
0.9	1.1
1.0	0.7
1.4	0.8
0.9	1.0
1.0	0.9
1.2	0.8
1.1	0.7

How would you decide if there was any real difference in play life between the two brands? Obviously the only way to tell would be to test a sample of both brands and determine the play life for each stylus.

Suppose you take a sample of 10 styluses from each brand and record the play life of each. These data are shown in Table 8-4. What do you conclude? Do you think there is a difference? Just what do you need to be able to do? You need to be able to test a hypothesis about the difference between two population means.

8.6 TESTING DIFFERENCES BETWEEN MEANS: HOW?

The parameter of interest is actually the difference between two parameters: namely, $\mu_1 - \mu_2$, where μ_1 is the mean of population 1 and μ_2 is the mean of population 2.

If the n_1 and n_2 sample observations are the result of random sampling from populations 1 and 2, respectively, it can be shown that

$$E(\bar{x}_1 - \bar{x}_2) = \mu_1 - \mu_2$$

So, a value for $\bar{x}_1 - \bar{x}_2$ close to the hypothesized value of $\mu_1 - \mu_2$ would support H_0. Our statistical question is, is $\bar{x}_1 - \bar{x}_2$ so far away from the hypothesized value for $\mu_1 - \mu_2$ that we should choose to reject H_0? The real question is, of course, how far is too far? As before, we will answer this question by determining the magnitude of the conditional probability of observing a sample result as extreme as our sample when H_0 is true.

If both populations are normally distributed and have equal variances ($\sigma_1^2 = \sigma_2^2$), the statistic

$$\frac{(\bar{x}_1 - \bar{x}_2) - (\mu_1 - \mu_2)}{\sqrt{\dfrac{(n_1 - 1)s_1^2 + (n_2 - 1)s_2^2}{n_1 + n_2 - 2} \cdot \left(\dfrac{1}{n_1} + \dfrac{1}{n_2}\right)}}$$

is student t distributed with $n_1 + n_2 - 2$ degrees of freedom. Thus, by replacing $\mu_1 - \mu_2$ with its hypothesized value, we have a statistic with known sampling distribution when H_0 is true. Our test statistic is

$$t = \frac{(\bar{x}_1 - \bar{x}_2) - D_0}{\sqrt{\dfrac{(n_1 - 1)s_1^2 + (n_2 - 1)s_2^2}{n_1 + n_2 - 2} \cdot \left(\dfrac{1}{n_1} + \dfrac{1}{n_2}\right)}}$$

where D_0 is the hypothesized value for $\mu_1 - \mu_2$.

8.7 SITUATION II REVISITED

Since we have no reason to believe that brand 1 will be superior to brand 2 or vice versa, our hypotheses are

H_0: $\mu_1 - \mu_2 = 0$

H_1: $\mu_1 - \mu_2 \neq 0$

Let $\alpha = .05$

Thus we have a two-tailed test, as $\bar{x}_1 - \bar{x}_2$ could be either too large or too small for us to believe H_0 to be true.

From the data in Table 8-4 we find

$\bar{x}_1 = 1.06 \qquad \bar{x}_2 = 0.87$

$s_1^2 = .0316 \qquad s_2^2 = .0312$

So our test statistic is

$$t = \frac{(1.06 - 0.87) - 0}{\sqrt{\dfrac{(9)(.0316) + (9)(.0312)}{18} \cdot \left(\dfrac{1}{10} + \dfrac{1}{10}\right)}}$$

$$= \frac{+.19}{.079}$$

$$= 2.405$$

For a two-tailed test we place $\alpha/2$ in the lower tail of the sampling distribution of our test statistic and $\alpha/2$ in the upper tail. Thus our critical values for t are -2.101 and $+2.101$ as shown in Figure 8-5. As our observed t value of $+2.405$ exceeds the upper critical value of 2.101, we reject H_0 and conclude that $\mu_1 - \mu_2 \neq 0$. We can see that our sample result has a conditional probability less than .05 of occurring if $\mu_1 - \mu_2 = 0$ (that is, $\mu_1 = \mu_2$).

Figure 8-5

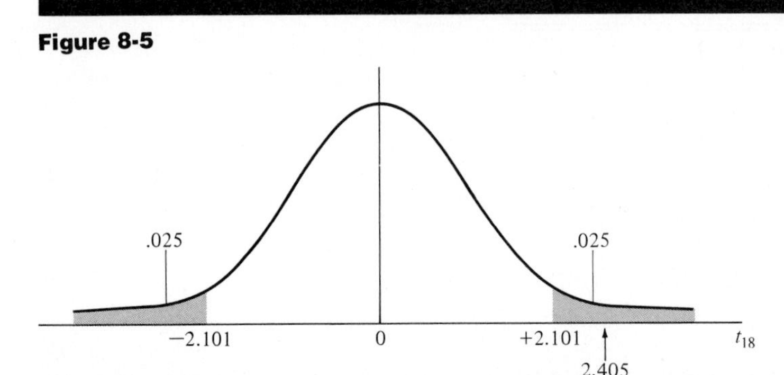

If both n_1 and n_2 had been > 30, we would have used the following approximation:

$$z = \frac{(\bar{x}_1 - \bar{x}_2) - (\mu_1 - \mu_2)}{\sqrt{\frac{s_1^2}{n_1} + \frac{s_2^2}{n_2}}}$$

This statistic does not require that $\sigma_1^2 = \sigma_2^2$ or that populations 1 and 2 be normally distributed. However, it is an approximation based upon the central-limit theorem and therefore should be used only when sample sizes are sufficiently large.

8.8 SITUATION III: WHY?

Texas Instruments Corporation has made a real impact on the pocket or small solid-state calculator market. TI not only assembles but also manufactures calculator components. The calculators contain "chips" (printed circuits designed to perform a specific operation). A separate division of TI actually makes the chips, and you are in charge of a particular assembly plant. Each calculator is tested before it is packaged for distribution.

Suppose it is typical for 10 percent of a specific model of calculator to fail the final test. These units are set aside and the necessary repair or part replacement is performed. It is somewhat expensive to repair or replace certain chips, as other circuits must be removed to get at the problem circuit. TI has found that "on the average" it takes just as long to repair such a calculator as it does to assemble one.

Suppose further that a particular assembly plant has recently experienced a rather dramatic increase in the fail rate of the final test. Upon investigation you have found that a particular chip is causing the problem. As a result of your findings, you have been asked to develop some type of screening system to "keep the defective chips out of our production line."

You realize that it would be far too expensive to test each chip prior to its entering the production process. However, it would be economically feasible to sample chips from incoming shipments to decide whether the proportion of defectives is sufficiently low. Let's say the chips arrive at your assembly plant in lots of 10,000. You decide that if $\pi \leq .04$, where π is the proportion of defective chips, you would want to accept the lot and place them in the production process. If $\pi > .04$ you would want to reject the lot and have these chips returned to the appropriate division.

You have decided to sample 125 chips from each incoming lot and decide whether to accept or reject the lot based upon the number of defective chips detected in the sample. A shipment has just arrived. You sample 125 chips and find 8 defective chips. Do you want to reject this lot?

What you need to be able to do is to test a hypothesis about π based upon sample information.

8.9 TESTING HYPOTHESES ABOUT π: HOW?

Under the assumption that the n sample observations constitute a random sample, we saw in Chapter 7 that the sample proportion is an unbiased estimator of π. We also noted in Chapter 6 that if $n\pi$ and $n(1 - \pi)$ were both greater than or equal to 5, the sampling distribution of p could be approximated by a normal distribution. This leads us to the following approximation:

$$z = \frac{p - \pi}{\sqrt{\dfrac{\pi(1-\pi)}{n}}}$$

That is, the sampling distribution of

$$\frac{p - \pi}{\sqrt{\dfrac{\pi(1 - \pi)}{n}}}$$

is approximated by the standard normal. Therefore, the statistic

$$\frac{p - \pi_0}{\sqrt{\dfrac{\pi_0(1 - \pi_0)}{n}}}$$

is approximately a standard normal random variable *if* $\pi = \pi_0$. If we let π_0 be the hypothesized value for π, we know the sampling distribution of our statistic when H_0 is true. Therefore, we have as our test statistic

$$z = \frac{p - \pi_0}{\sqrt{\dfrac{\pi_0(1 - \pi_0)}{n}}}$$

8.10 SITUATION III REVISITED

The appropriate hypotheses for your situation would be

H_0: $\pi \leq .04$

H_1: $\pi > .04$

since you would reject the lot only if you believed $\pi > .04$. Suppose you decide to set $\alpha = .05$. This means that you can expect to reject good lots 5 percent of the time. You would like to set α fairly high in this case, as it would be more expensive to accept a bad lot than to reject a good one. However, rejecting too many good lots may also cause you problems with your supplier.

Recall that you observed 8 defectives out of a sample of 125. Thus, $p = 8/125 = .064$. Is p so large that you do not believe that $\pi \leq .04$? Let's see. Since both $n\pi$ and $n(1 - \pi)$ are ≥ 5, we will use the normal approximation. Our test statistic is

$$z = \frac{p - \pi_0}{\sqrt{\dfrac{\pi_0(1 - \pi_0)}{n}}}$$

$$= \frac{.064 - .04}{\sqrt{\dfrac{(.04)(.96)}{125}}}$$

$$= 1.37$$

Figure 8-6 shows the sampling distribution of our test statistic if H_0 is true. We can see that our computed z value of 1.37 is in the *acceptance* region, and therefore the decision on this particular lot is that $\pi \leq .04$ and you would accept this shipment.

8.11 *SITUATION IV: WHY?*

Suppose you work in the product selection department for O.M. Scott and Sons Company. Your firm markets a number of lawn and lawn care products under the trade name of Scotts. A new hybrid grass seed has come to your attention, and you are charged with the job of forming a recommendation as to whether Scotts should replace their existing "all-purpose lawn" grass seed with the new

Figure 8-6

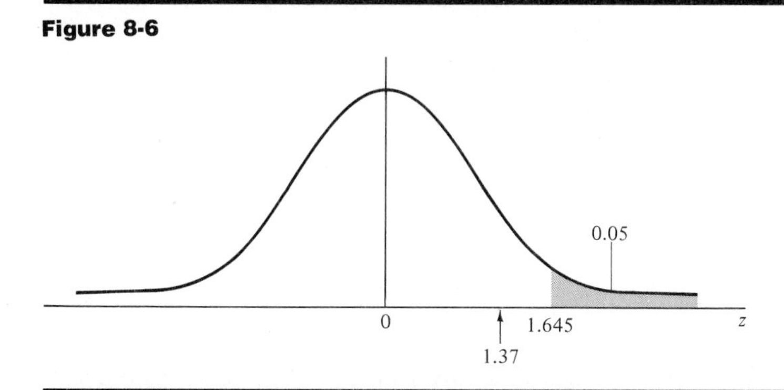

hybrid. Although the hybrid seed is slightly more expensive, the producers claim that its germination (sprouting) rate exceeds that of your present seed. If indeed the germination rate for the hybrid exceeds that of your present seed, you would like to recommend that Scotts switch seeds, as they have a company-policy "No-Quibble Guarantee" on all their products. Stated simply, if for any reason a customer is not satisfied with a Scotts product, the customer is entitled to a full refund. If a greater proportion of the new hybrid seeds will germinate, there will be less reason for customer complaints and fewer refunds required. Let's suppose that in order to determine whether the new hybrid does have a higher germination rate, you have tested a random sample of 200 seeds of both types.

You found that 150 out of the 200 seeds of your existing type germinated within 14 days. Out of the 200 new hybrid seeds, 175 germinated within the same 14-day period. Do you think the new hybrid seed has a higher germination rate? What do you need to be able to do? Specifically, you need to be able to test a hypothesis about the difference between two proportions.

8.12 TESTING HYPOTHESES ABOUT DIFFERENCES IN PROPORTIONS: HOW?

If both samples are random samples, it can be shown that

$$E(p_1 - p_2) = (\pi_1 - \pi_2)$$

Now, the form of the appropriate test statistic depends on whether we are testing that $\pi_1 = \pi_2$, which means $\pi_1 - \pi_2 = 0$, or whether we have hypothesized a specific value for $\pi_1 - \pi_2$ other than 0.

Under the assumption that n_1 and n_2 are sufficiently large, independent random samples such that the sampling distributions of p_1 and p_2 can be approximated by normal distributions, the statistic

$$\frac{(p_1 - p_2) - (\pi_1 - \pi_2)}{\sqrt{\dfrac{\pi_1(1 - \pi_1)}{n_1} + \dfrac{\pi_2(1 - \pi_2)}{n_2}}} \tag{8-2}$$

is approximately a standard normal variable.

If we are hypothesizing that $\pi_1 = \pi_2$, our null hypothesis will be

H_0: $\pi_1 - \pi_2 = 0$

If the null hypothesis is true, p_1 and p_2 are two independent unbiased estimates of the same parameter $\pi_1 = \pi_2 = \pi$. Thus, our procedure is to pool our observations to obtain the best estimate of the common value π. The pooled estimate of π is then

$$p = \frac{x_1 + x_2}{n_1 + n_2}$$

where x_i and n_i are the number of successes and sample size from population i, respectively.

Our test statistic is then

$$z = \frac{(p_1 - p_2) - 0}{\sqrt{p(1 - p)\left(\dfrac{1}{n_1} + \dfrac{1}{n_2}\right)}}$$

If H_0: $\pi_1 - \pi_2 = 0$ is true, our test statistic is approximately distributed as the standard normal. So, we again know the sampling distribution of our test statistic when H_0 is true.

If we hypothesize that $\pi_1 - \pi_2 = D_0$, where $D_0 \neq 0$, we merely replace π_1 and π_2 in the denominator of Expression (8-2) with p_1 and p_2, respectively. Our test statistic then becomes

$$z = \frac{(p_1 - p_2) - D_0}{\sqrt{\dfrac{p_1(1 - p_1)}{n_1} + \dfrac{p_2(1 - p_2)}{n_2}}}$$

8.13 SITUATION IV REVISITED

Recall that your job is to decide whether the germination rate for the new hybrid seed is greater than that for your existing seed. Since you will recommend switching types only if you believe that π for the hybrid is greater than π for the existing seed, your hypotheses will be the following

H_0: $\pi_1 - \pi_2 \geq 0$

H_1: $\pi_1 - \pi_2 < 0$

where population 1 is your existing seed type and population 2 is the new hybrid. Your sample data are shown below.

Population 1 (Existing Seed)	Population 2 (New Hybrid)
$n_1 = 200$	$n_2 = 200$
$x_1 = 150$	$x_2 = 175$
$p_1 = .750$	$p_2 = .875$

Since we are hypothesizing that $\pi_1 - \pi_2 = 0$, the pooled estimate of π is

$$p = \frac{150 + 175}{200 + 200}$$
$$= .8125$$

Our computed value for the test statistic is

$$z = \frac{(.750 - .875) - 0}{\sqrt{(.8125)(.1875)\left(\frac{1}{200} + \frac{1}{200}\right)}}$$
$$= -3.04$$

We can see by consulting Table A-4 that the conditional probability of observing a z value as small as or smaller than -3.04 is practically zero. Thus, we would reject H_0 at almost any value of α. Figure 8-7 shows that if we set $\alpha = .01$, we would reject H_0 for any computed z value less than or equal to -2.33. Your sample data strongly refute H_0 in favor of H_1, and you would want to recommend that Scotts switch to the new hybrid.

Figure 8-7

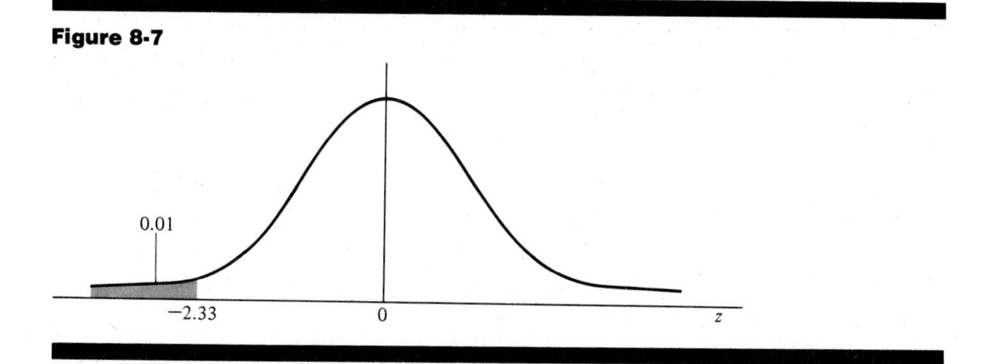

8.14 SITUATION V: WHY?

Suppose you are the general manager of A & M Vending, Incorporated, in Cleveland, Ohio. A & M is involved with many facets of the vending business, ranging from money changers to cigarette and beverage machines.

The problem on your desk at this time deals with vending machines for soft drinks. A number of your soft drink machines need to be replaced. In addition, your demand for new installations is increasing almost daily. You see the need to purchase 100 new soft drink vending machines. The question is, which brand of machine should you buy? Your present soft drink machines are all the same brand (ACE). Suppose a new brand (GEM) has come to your attention. The GEM machine is similar in both appearance and price to the ACE machine. However, there is one characteristic that interests you.

All such vending machines are equipped with an overflow container. Your service personnel verify that the overflow containers are not always empty. This indicates that there are times when your machines dispense more soft drink than the cup can hold. At the same time you have received complaints about "half-filled cups." The simple fact is that although the vending machines are adjustable, they are variable in relation to the amount of soft drink they dispense.

Let's suppose you know that the standard deviation of the fluid ounces dispensed by the ACE machines is about 0.4 ounce. You have decided that if the standard deviation for the new GEM machine is less than 0.4 ounce you would like to purchase GEM rather than ACE. After all, a smaller standard deviation should reduce both wasted product and complaints about shortages. Alternatively, you would purchase ACE if the standard deviation for GEM is equal to or greater than that for ACE. If the standard deviation for GEM is 0.4 ounce, the two brands are equivalent and stocking repair parts for two different brands would increase your costs. Thus, you conclude that you will only purchase GEM if you believe its standard deviation is less than 0.4 ounce.

Suppose you have randomly selected a GEM soft drink vending machine and have recorded the fluid ounces dispensed per cup for 15 cups. The resulting data are shown in Table 8-5. The sample variance for these data is

$$s^2 = \frac{\Sigma(x_i - \bar{x})^2}{n - 1}$$

$$= .097$$

Thus, s = .311.

Is s = *.311 so small that you would want to conclude that* $\sigma < .4$*? Let's find out.*

8.15 TESTING HYPOTHESES ABOUT σ^2: HOW?

Our statistical test is actually on the population variance σ^2 rather than on the standard deviation σ. This creates no real problem, since asking whether $\sigma < .4$ is equivalent to asking whether $\sigma^2 < .16$.

Table 8-5

Cup	Fluid Ounces
1	8.3
2	7.9
3	7.5
4	7.8
5	8.2
6	8.4
7	8.1
8	7.9
9	8.7
10	7.9
11	8.1
12	7.6
13	7.8
14	8.0
15	7.8

If the n sample observations are the result of random sampling *and* the population of values is normally distributed, the statistic

$$\frac{(n-1)s^2}{\sigma^2}$$

is chi-square distributed with $n - 1$ degrees of freedom.

The shape of the chi-square probability distribution, like that of the t distribution, depends upon the degrees of freedom associated with s^2, which is $n - 1$. Figure 8-8 shows the graphs of several chi-square distributions. We see that the chi-square is not defined below zero, and the distribution is typically asymmetric. Table A-6 gives tabulated χ^2 (chi-square) values and is constructed in the same way as Table A-5 for t. The tabulated χ^2 value is χ^2_α and represents the χ^2 value that cuts off 100α percent of the distribution's area above χ^2_α. For example, Table A-6 shows that 5 percent of the area of the chi-square distribution with 9 degrees of freedom lies above 16.9190. Similarly, we see that 2.5 percent of the area is above 19.0228 and only 1 percent above 21.666.

Since

$$\frac{(n-1)s^2}{\sigma^2}$$

is distributed as a chi-square with $n - 1$ degrees of freedom, it follows that the statistic

$$\frac{(n-1)s^2}{\sigma_0^2}$$

Figure 8-8

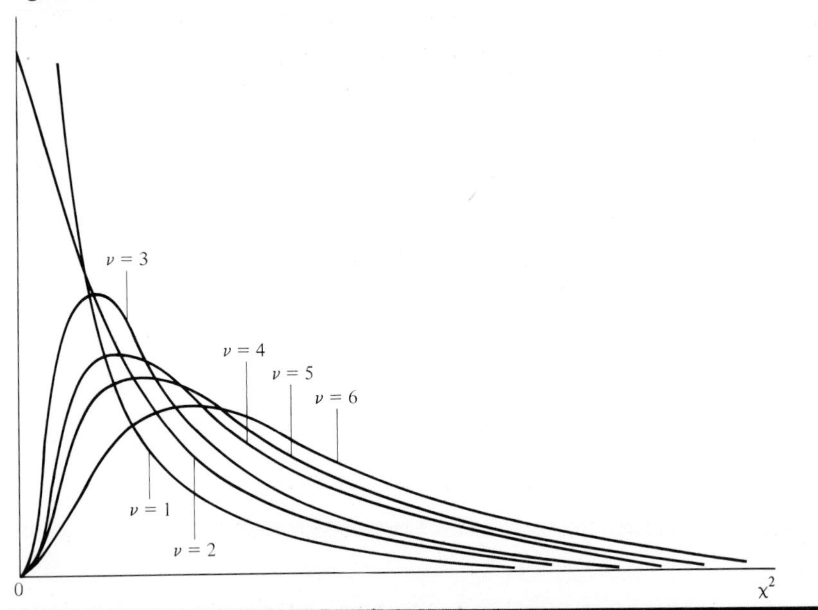

is a chi-square variable with $n - 1$ degrees of freedom *if* H_0: $\sigma^2 = \sigma_0^2$ is true. Thus, our test statistic is

$$\chi^2 = \frac{(n - 1)s^2}{\sigma_0^2}$$

8.16 SITUATION V REVISITED

Given your position and preferences as stated in Section 8.14, the appropriate hypotheses are

H_0: $\sigma^2 \geq .16$

H_1: $\sigma^2 < .16$

Suppose we choose $\alpha = .10$. We would only want to reject H_0 in favor of H_1 if we were to observe a small sample variance. Since s^2 appears in the numerator of our test statistic

$$\frac{(n - 1)s^2}{\sigma_0^2}$$

we would reject H_0 only for small values of the test statistic. Thus, we have a one-tailed test, and the rejection region is in the lower tail of the chi-square distribution with 14 degrees of freedom. Recall that $s^2 = .097$ for our sample data. Therefore

$$\chi^2 = \frac{(n - 1)s^2}{\sigma_0^2}$$
$$= \frac{(15 - 1)(.097)}{.16}$$
$$= 8.4875$$

From Table A-6 we find $\chi^2_{.90} = 7.78953$.

Since our computed $\chi^2 = 8.4875$ exceeds the critical χ^2 value of 7.78953, we accept H_0 and conclude $\sigma^2 \geq .16$. We therefore choose to purchase ACE vending machines. Figure 8-9 graphically shows our decision.

If our alternative hypothesis had been H_1: $\sigma^2 \neq \sigma_0^2$, we would choose to reject H_0 in favor of H_1 for either large *or* small values for χ^2. This would yield a two-tailed test and we would place $\alpha/2$ in the lower tail and $\alpha/2$ in the upper tail of the appropriate chi-square distribution.

8.17 SITUATION VI: WHY?

Let's return to Situation II. Recall that you were trying to decide which stylus Garrard should purchase and supply with their new ultralight tone arm. Your sample sizes were small, $n_1 = n_2 = 10$*, and you therefore applied a* t *test that was based on the assumption that* $\sigma_1^2 = \sigma_2^2$.

In any application of statistical procedures we should investigate the valid-

Figure 8-9 χ^2 **with 14 Degrees of Freedom**

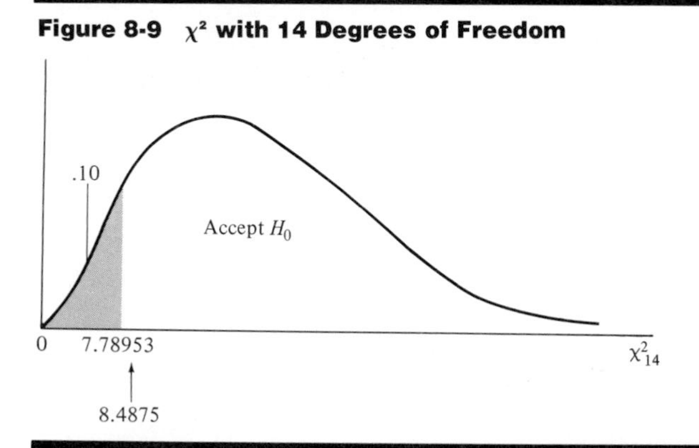

ity of the required assumptions. Was your assumption that $\sigma_1^2 = \sigma_2^2$ in Situation II valid?

Since s_1^2 is an unbiased estimate of σ_1^2 and s_2^2 is an unbiased estimate of σ_2^2, it seems reasonable that we should compare s_1^2 and s_2^2. The closer s_1^2 and s_2^2 are to being equal, the more reasonable our assumption would seem. However, if s_1^2 and s_2^2 are quite dissimilar in magnitude, we would probably choose to disbelieve that they are both estimates of the same variance. What we need is some way to decide whether $s_1^2 = .0316$ and $s_2^2 = .0312$ are close enough for us to believe that $\sigma_1^2 = \sigma_2^2$. Alternatively, are they so dissimilar as to suggest that $\sigma_1^2 \neq \sigma_2^2$?

8.18 TESTING HYPOTHESES ABOUT TWO VARIANCES: HOW?

Given the other assumptions made in Section 8.6 (that the n_1 and n_2 observations constitute independent random samples from normal populations), the statistic

$$\frac{s_1^2/\sigma_1^2}{s_2^2/\sigma_2^2} \tag{8-3}$$

is an F variable with $n_1 - 1$ degrees of freedom in the numerator and $n_2 - 1$ degrees of freedom in the denominator. F, like χ^2, is not defined below zero and is typically asymmetric. The F distribution depends on the degrees of freedom for the numerator, denoted as ν_1, and the degrees of freedom for the denominator, ν_2. Table A-7 gives tabulated F values that cut off $\alpha = .05$ of the area in the upper tail of the corresponding F distribution. Table A-8 gives the comparable F values for $\alpha = .01$.

For example, an F value of 3.33 cuts off .05 in the upper tail of the F distribution with $\nu_1 = 5$ and $\nu_2 = 10$, while an F value of 5.64 cuts off .01 in the same distribution. (See Figure 8-10.) Now, *if* the null hypothesis is true, H_0: $\sigma_1^2 = \sigma_2^2$, then $s_1^2/\sigma^2 \div s_2^2/\sigma^2$ is an F variable. All that we have done is to replace σ_1^2

Figure 8-10

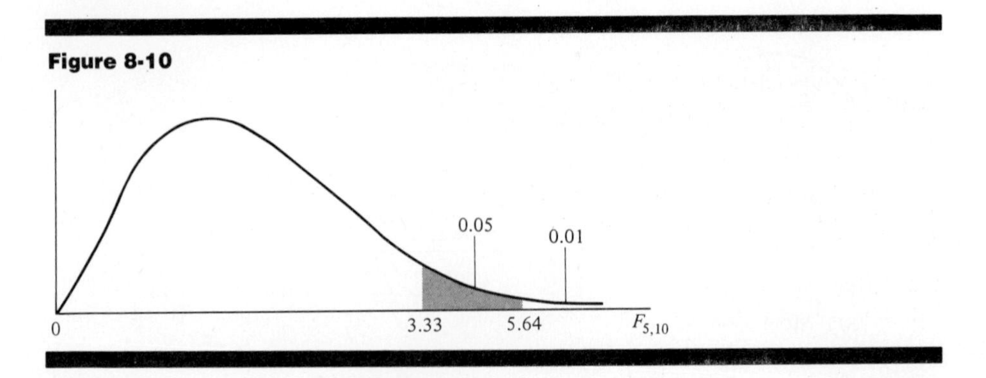

and σ_2^2 in Expression (8-3) with σ^2 since the two population variances are equal if H_0 is true. Since both the numerator and denominator are multiplied by the constant $1/\sigma^2$, we can cancel and thus $s_1^2/s_2^2 = F$ *if* H_0 *is true*.

Therefore, our test statistic for testing the null hypothesis H_0: $\sigma_1^2 = \sigma_2^2$ is

$$F = s_1^2/s_2^2$$

We have one further complication. If the alternative hypothesis is H_1: $\sigma_1^2 \neq \sigma_2^2$, we would want to reject H_0 for *either* large *or* small values for s_1^2/s_2^2. Since the F distributions are typically asymmetric, we would need the tabulated upper- *and* lower-tail F values. Unfortunately our tables are *only* upper-tail F values. It is possible to obtain lower-tail F values from corresponding upper-tail F values, but it is not as simple as just the negative of the tabulated upper-tail F value. Actually we can solve this tabulation problem another way. Since it is arbitrary which population is called 1 and which is called 2, we can place all of α in the upper tail of F *by placing the larger sample variance in the numerator*. If we adopted this convention, we would only reject H_0 for large values of our test statistic. With this convention the probability that F is greater than or equal to the tabulated F values is actually twice the α values shown in the tables. The F values given in Tables A-7 and A-8 would be the critical values for $\alpha = .10$ and $\alpha = .02$, respectively.

The discussion in the above paragraph relates to the situation where the alternative hypothesis is H_1: $\sigma_1^2 \neq \sigma_2^2$. If the alternative hypothesis is directional (H_1: $\sigma_1^2 > \sigma_2^2$ or H_1: $\sigma_1^2 < \sigma_2^2$), we place the sample variance for the population that is hypothesized to have the larger variance in H_1 in the numerator. Thus, if H_1: $\sigma_1^2 < \sigma_2^2$, our test statistic would be s_2^2/s_1^2. The rejection region for our one-tailed test would be the upper tail of the appropriate F distribution. In this case we would *not* double the tabulated probability to obtain α.

8.19 SITUATION VI REVISITED

Suppose we would like to test

H_0: $\sigma_1^2 = \sigma_2^2$

H_1: $\sigma_1^2 \neq \sigma_2^2$

at $\alpha = .10$. Adopting the above convention, we place s_1^2 in the numerator as $s_1^2 = .0316$, which is greater than $s_2^2 = .0312$.

$$\begin{aligned} F &= s_1^2/s_2^2 \\ &= .0316/.0312 \\ &= 1.013 \end{aligned}$$

The critical F for $\nu_1 = 9$ and $\nu_2 = 9$ at $\alpha = .10$ is 3.18. Since our observed F of 1.013 is less than the critical value of 3.18, we *accept* H_0: $\sigma_1^2 = \sigma_2^2$ and would then feel comfortable with our assumption of equal variances in Situation II. Be sure to note that ν_1 is the degrees of freedom for whichever s^2 we place in the numerator. In our example both n_1 and n_2 were equal. Had n_1 been 15 while n_2 was 11, since $s_1^2 > s_2^2$, ν_1 would have been 14 and ν_2 would have been 10.

8.20 COMMONALITIES OF TESTING HYPOTHESES

We argued at the beginning of this chapter that you already knew how to test hypotheses. We went on to point out that the only problem was that when we start dealing with abstract symbols and concepts, it is easy to lose sight of what we are really doing.

Well, you have seen at least six different tests, four different probability distributions, and all kinds of new terminology. It seems appropriate that we point out the commonalities among tests to help focus your picture.

Actually all these tests have followed the same sequence you went through when you rejected the hypothesis that our coin was fair in Section 8.2.

In every case we first formulated our hypotheses, adhering to our two cardinal rules.

1. The null hypothesis must contain the equality.
2. No single hypothesis shall contain states of nature that lead to different decisions.

Sample evidence was then obtained which contained information about the truth or falsity of the null hypothesis. Because of the specific way in which the sample was drawn and corresponding statistical theory, we were able to identify a test statistic with known sampling distribution if the null hypothesis was true. The test statistic was some simple transform of our unbiased estimate of the parameter under investigation. Knowing the sampling distribution of the test statistic, when H_0 is true, enabled us to determine the conditional probability of observing a sample result for the test statistic as extreme as the observed sample result. We then decided whether to *accept* or reject H_0 depending upon the magnitude of this conditional probability. "Conditional" means *when the null hypothesis is true*.

If you will recall, it was the magnitude (smallness) of the conditional probability of observing a certain number of heads in n flips of a fair coin that lead you to reject the null hypothesis that our coin was fair.

Table 8-6 provides us with a summary of the various tests presented in this chapter. (See pages 180–181.)

8.21 SELECTING α

There is one final issue that we should address. How do you select a value for α? Recall that α is the probability of committing a Type I error (rejecting H_0 when it is true). Since α is the probability of committing an error, one might think that we should set α as low as possible. We can have $\alpha = 0$ if we want by just never rejecting H_0. Clearly this is just one side of the issue. There is another kind of error which we could make. A Type II error is the error of *accepting* a false null hypothesis. Once the sample size is fixed, any reduction in the probability of committing a Type I error is at the expense of an increase in the probability of committing a Type II error.

The only way to reduce the probabilities of both errors simultaneously is by increasing the sample size. The problem of selecting α is anything but simple. One needs to take into account the economic consequences of the two types of errors for the situation at hand. If rejecting H_0 when it is true is a costly error, you would want to set α quite low. If on the other hand accepting a false null hypothesis is more serious than rejecting a true one, you would want to set α rather high in order to reduce β, the probability of accepting H_0 when it is false. Although we have not shown how, computations of β are possible. The problem is that, for all realistic situations, the computation of β is quite complex.

Suffice it to say that the economics of the given situation enter our decision process through n and α. However, once n is fixed, there is a trade-off between α and β.

8.22 IDENTIFICATION OF APPLICATIONS: WHEN?

There is, of course, a strong tie between estimating and testing hypotheses about parameters. It is the estimation and/or testing hypotheses about parameters that constitutes the area called *statistical inference*.

The only real distinction between situations where testing hypotheses would be employed and those where estimation is used is where there is a *specific value for the parameter that is pivotal in the decision*. Recall from our discussion in Section 7.14 that for many decision situations the consequences of various alternatives are a function of, or determined by, the value of a certain parameter. If there is a specific value for the parameter that is pivotal (if you believe the parameter equals the specific value you would choose to do one thing, whereas if you do not think the parameter equals the specific value you would choose to do something else), your desire would be to test the appropriate hypothesis about the parameter rather than to estimate its value.

If your decision requires that you decide whether or not a parameter equals, is greater than or equal to, or is less than or equal to a specific value, and your decision must be made using sample data, you have a testing-hypothesis problem.

PROBLEMS

1. Identify an application of testing hypotheses in your major area of study. Point out the features which make this application appropriate and of value.

Table 8-6

Parameter	Assumptions	Test Statistic		Distribution
μ	1. *a.* Random sampling. *b.* Normal population.	$\frac{\bar{x} - \mu_0}{s/\sqrt{n}}$	if $n \leq 30$ if $n > 30$	t_{n-1} z
μ	2. *a.* Random sampling. *b.* n is large enough for the sampling distribution of $\bar{x}$ to be approximated by a normal (via CLT).	$\frac{\bar{x} - \mu_0}{s/\sqrt{n}}$	$n > 30$	z
$\mu_1 - \mu_2$	1. *a.* Independent random sampling. *b.* Both populations are normal. c. $\sigma_1^2 = \sigma_2^2$	$\frac{(\bar{x}_1 - \bar{x}_2) - (\mu_1 - \mu_2)_0}{\sqrt{\frac{(n_1 - 1)s_1^2 + (n_2 - 1)s_2^2}{n_1 + n_2 - 2} \cdot \left(\frac{1}{n_1} + \frac{1}{n_2}\right)}}$	if $n_1 + n_2 \leq 30$ if $n_1 + n_2 > 30$	$t_{n_1+n_2-2}$ z
$\mu_1 - \mu_2$	2. *a.* Independent random sampling. *b.* Both n_1 and n_2 are large enough for the sampling distribution of $\bar{x}_1 - \bar{x}_2$ to be approximated by a normal (via CLT).	$\frac{(\bar{x}_1 - \bar{x}_2) - (\mu_1 - \mu_2)_0}{\sqrt{\frac{s_1^2}{n_1} + \frac{s_2^2}{n_2}}}$	n_1 and $n_2 > 30$	z
π	1. *a.* Random sampling. *b.* The population is either large compared to n or infinite.	$\frac{p - \pi_0}{\sqrt{\frac{\pi_0(1 - \pi_0)}{n}}}$	$n\,\pi_0 \geq 5$ and $n(1 - \pi_0) \geq 5$	z

		If the Null Hypothesis Is $\pi_1 - \pi_2 = 0$		
$\pi_1 - \pi_2$	1. *a.* Independent random sampling. *b.* Both n_1 and n_2 are large enough for the sampling distribution of $p_1 - p_2$ to be approximated by a normal (via CLT).	$\dfrac{(p_1 - p_2)}{\sqrt{p(1-p)\left(\dfrac{1}{n_1} + \dfrac{1}{n_2}\right)}}$ where $p = \dfrac{x_1 + x_2}{n_1 + n_2}$	$n_i p \geq 5$ and $n_i(1-p) \geq 5$ $i = 1, 2$	z
		If the Null Hypothesis Is $\pi_1 - \pi_2 = D_0$, Where $D_0 \neq 0$		
$\pi_1 - \pi_2$	2. *a.* Independent random sampling. *b.* Both n_1 and n_2 are large enough for the sampling distribution of $p_1 - p_2$ to be approximated by a normal (via CLT).	$\dfrac{(p_1 - p_2) - D_0}{\sqrt{\dfrac{p_1(1-p_1)}{n_1} + \dfrac{p_2(1-p_2)}{n_2}}}$	$n_i p_i \geq 5$ and $n_i(1-p_i) \geq 5$ $i = 1, 2$	z
σ^2	1. *a.* Random sampling. *b.* Normal population.	$\dfrac{(n-1)s^2}{\sigma_0^2}$		χ^2_{n-1}
		If the Null Hypothesis Is $\sigma_1^2 = \sigma_2^2$		
σ_1^2 and σ_2^2	1. *a.* Independent random sampling. *b.* Both populations are normal.	$\dfrac{s_i^2}{s_j^2}$ where $s_i^2 > s_j^2$		F_{ν_1, ν_2} $\nu_1 = n_i - 1$ $\nu_2 = n_j - 1$

2. Bill Schwick, vice president of Travis Savings and Loan Association, recently read in an article in *Business Week* that the national figure for the rejection rate of Savings and Loans was 28 percent in the past year. He was immediately curious as to how close his own association was to this figure. If it was rejecting substantially more than the national average, it would strengthen his suspicion that his association's advertising strategy was attracting too many high-risk borrowers.

   Schwick calls you into his office and requests that you compare the reject rates of Travis with those of the nation as a whole. He concludes your meeting with this statement: "I've suggested several times that our rejection rates are higher than those experienced by other S & L's. However, I have not been able to convince the other managers. I want to be certain that our rate is higher than the national average before I discuss this at our next meeting. Look at a sample of our loan applications over the past year and tell me how we stand compared to the nation as a whole." From a random sample of 50 applications, you find 18 rejections. Do your findings indicate that Travis Savings rejects a greater proportion of loan applications than the national average? (Use $\alpha = .01$.)

3. Morning House markets a large variety of household items by catalog and door-to-door sales. The door-to-door approach has been utilized for only four years and is slowly expanding on a region-by-region basis. Initial research has shown that customers who purchase an item from a direct contact are very likely to make follow-up purchases from the catalog which is left by the salesperson. Therefore, the firm's management is interested in maximizing the percent of initial sales calls that result in sales.

   The firm recently decided to test the effectiveness of a preliminary television campaign prior to sending its sales force into a new territory. Two similar territories were selected. One received the television campaign and one did not. The percent of successful calls in the territory with the television campaign was 14 percent compared to 10 percent in the territory without the campaigns. Approximately 10,000 calls were made in each territory. In order to be economical, the cost of local advertising campaigns of this nature requires a 3 percent increase in the sales-call success ratio. If π_T and π_N denote the percent of successful calls in television and nontelevision campaign territories, do these data provide sufficient evidence to indicate that the cost incurred in local advertising campaigns is justified? (Use $\alpha = .05$.) *Hint*: Test the hypothesis that $\pi_T - \pi_N \geq .03$.

4. The snack-foods market in the United States is highly competitive. Thus, a good marketing manager always likes to know the factors that are associated with "successful" products. As a newly hired vice president of marketing you wanted to have a feel for these factors. Accordingly, you selected two products—one successful and one unsuccessful—and had a separate random sample of 100 customers evaluate each, using an interval scale with a range of 1 to 100. The study measured a number of beliefs about the product. The respondents' evaluation of the "nutritious" feature of the successful brand was 60 (standard deviation 1) while the unsuccessful brand scored 59 (standard deviation 3, much, much larger than the other). You directed your statistician to test whether the nutritious feature of the successful brand 1 was greater than that of unsuccessful brand 2 by using a z test and a level of significance of 1 percent. What was the finding of your statistician?

5. Sarah Warner, comptroller of Beck's Inc., recently read an industry survey based on an analysis of 10,000 accounts receivable records collected from 75 firms. Among

other things, the survey indicated that the average time between billing and payment was 15.7 days with a standard deviation of 4. Ms. Warner feels that this is substantially shorter than the time experienced by her company. This disturbs her very much. She always prides herself as "one of the best comptrollers in business." In order to investigate whether her company is really performing below the national average, she takes a random sample of 40 accounts. These accounts reveal the information shown below:

Account Number	Days between Billing and Payment
1	17
2	21
3	15
4	9
5	16
6	17
7	32
8	18
9	16
10	14
11	17
12	13
13	16
14	15
15	19
16	11
17	11
18	18
19	23
20	14
21	13
22	24
23	17
24	12
25	10
26	20
27	14
28	17
29	12
30	18
31	13
32	17
33	19
34	31
35	18
36	15
37	10
38	14
39	22
40	16

What would you conclude? (Let $\alpha = .05$.)

6. Suppose DuPont is testing the chip resistance of two new paint components designed for automobiles. DuPont's engineers believe that one type, called Polydura, is superior to Resista, the second type. However, Polydura is substantially more expensive to produce. Therefore, they are interested in knowing whether Polydura is really superior. To test the two paints, several cars are painted with each and the chipping point is measured in 15 trials at different angles on each type of paint. For Polydura the average chipping point is at 31.62 psi with a standard deviation of 2.63. For Resista, the figures are 27.84 with a standard deviation of 3.14. Do these data provide sufficient evidence to conclude that Polydura is superior to Resista? (Use $\alpha = .01$.)

7. Daniel Thurn recently agreed to serve as chairman of the alumni fund-raising campaign of his college. Thurn examined the appeals used in the past and decided that they did not ask enough of the alumni. In particular, he felt it was necessary to increase the "suggested minimum contribution" from $10 to $25. However, the alumni fund-raising committee expressed concern that the net effect would be a reduction in total revenue generated. The committee was very reluctant to alter a campaign which had worked reasonably well in the past. Therefore, Thurn decided to test his idea.

   A random sample of 15 alumni received the request with the $25 minimum and a second random sample of 15 received the request with the $10 minimum. The results of the test are given below.

$10 Request		**$25 Request**	
Individual	Amount	Individual	Amount
1	0	1	25
2	10	2	0
3	10	3	0
4	10	4	0
5	25	5	25
6	0	6	25
7	50	7	0
8	10	8	0
9	100	9	100
10	5	10	25
11	0	11	15
12	0	12	0
13	10	13	0
14	10	14	50
15	0	15	10

   Do these data present sufficient evidence to change the suggested minimum contribution from $10 to $25? (Use $\alpha = .05$.)

8. Steelrite Manufacturing Company makes carburetors for small gasoline engines. It produces most of the parts for these engines. These engines require small needlelike valves for producing jets which are critical to the functioning of carburetors. These valves are of varying diameters, and the machine which produces them is supposed to be acceptable if it produces valves which result in jets with a variance of less than

5 percent of the specified diameter. A manufacturer who is trying to sell his machine to Steelrite claims that his machine meets the acceptance criterion. A trial run of 101 jets with a diameter set at 0.3 centimeters produces an average diameter of 0.297 centimeters and a variance of 0.02 centimeters. Applying the test for a given population variance, do you think that this new machine could be acceptable to Steelrite? (Use $\alpha = .05$.)

9. Cleveland Iron and Supply specializes in producing custom-made axles for a wide variety of vehicles, particularly off-road military vehicles. It recently had an opportunity to bid on a subcontract for axles that were to be used in the landing gear of aircraft. The general manager of Cleveland Iron was hesitant to bid: "We haven't seen the detailed specifications for the contract yet, but I'm afraid our manufacturing process is just too variable. We haven't developed the precision required to make axles of this nature."

   The general manager's view was not held by several members of the management staff. Prior to investigating the upcoming contract, they decided to estimate the precision of the firm's production process. They randomly selected 51 axles from the last order the firm produced. These axles were to have a standard deviation of 0.05 centimeters. The data for axle diameter in centimeters are:

   5.11, 5.01, 4.96, 5.03, 5.01, 5.02, 5.03, 5.04, 5.03, 5.02, 5.0, 5.12, 5.10, 5.16, 5.12, 5.12, 5.12, 5.11, 5.0, 5.11, 5.09, 5.08, 5.06, 5.09, 5.02, 5.03, 5.02, 5.11, 5.02, 5.13, 5.21, 5.11, 5.12, 5.14, 4.26, 4.28, 4.99, 5.01, 4.96, 4.93, 4.85, 4.79, 4.89, 4.88, 5.02, 5.02, 4.89, 4.99, 5.08, 5.16, 5.20

   Based on these data, do you conclude that the variance is the same as that desired by the management staff? (Use $\alpha = .01$.)

10. Doris Spiegel, a staff member of the finance department of Amac Inc., is currently evaluating a request from the production department to replace the firm's existing lathes with a newer model. One month ago, the production department acquired six of the new lathes on an experimental basis. The new lathes are easier to operate than the old ones but cost substantially more. If they reduce direct labor hours more than 4 minutes, they should be purchased. Otherwise, the firm should continue with the less expensive lathe.

    Six existing lathes are selected at random and the performance of the two groups is monitored. Each group of lathes is involved in the production of 900 items. The average direct labor time per item with the existing lathes is 38.6 minutes (standard deviation of 3.8). The average direct labor time with the new lathes is 33.5 minutes (standard deviation of 4.1). Should Spiegel recommend approval of the request? Why or why not?

11. As a new graduate with a degree in real estate, you have been hired by a firm which specializes in site-location analyses. Your first project is to assist Brian Kerrigan, a local retailer, in selecting one of two potential sites for a new store. Kerrigan specializes in expensive gift items. Approximately 60 percent of his business comes from customers who actively seek out his firm. Since the two potential outlets are equally accessible, Kerrigan is indifferent to this aspect of the choice.

    The remaining 40 percent of his sales comes from "unplanned" purchases made by upper-income individuals. The two sites are several blocks apart in the downtown area. Pedestrian traffic counts for the two outlets reveal virtually identical numbers of pedestrians per each outlet each day. Since the outlet on 16th Avenue rents for 5 percent less a month than the one on 22d Avenue, Kerrigan is tempted to sign a lease. However, after a lengthy discussion you convince him to examine the composition of the traffic, that is, what percent of the traffic by each store is in an upper-income bracket. Kerrigan concludes that the 22d Avenue outlet would have to have 10 percent more high-income traffic than 16th Avenue outlet before it would be worth the additional rent. A random sample of 100 pedestrians is taken from in front of each store. Twenty-four percent of those in front of the 16th Avenue outlet are in the upper-income categories, while 36 percent of those from the 22d Avenue store are from these categories.

What would you recommend to Kerrigan? Why?

12. Abbot Gifts is a small firm that sells expensive gift items by placing small ads in such magazines as *Sunset* and *Better Homes and Gardens*.

    Robin Abbot, president of Abbot Gifts, is considering a major change in the firm's advertising approach. The basic change would be to increase the size of the advertisements from 2 column inches to a quarter page. If the new approach increases the percent of orders to circulation by at least .0005 points, it should be adopted.

    To test the effectiveness of the new approach, a split-run arrangement has been made with one of the magazines frequently used by Abbot Gifts. Every other copy of an issue has carried a quarter page ad, while the remaining copies have carried the smaller ad. The two ads have utilized different addresses so that the percent responding to each could be measured. The larger ad produces a .0032 order rate from the 600,000 copies in which it has appeared. The smaller ad produces an order rate of .0024 from the 600,000 copies in which it has appeared.

    Should the larger size ad be used regularly? Why or why not?

13. Disco Records

    MEMO

    TO: *Bob Singer, Consultants Inc.*

    FROM: *Susan Byerlyn, Manager, Disco Records*

    As you might have read in a recent *Rolling Stone* article, our firm's market share is declining due to some sour deals we made recently. I am convinced that our best chance for survival is to specialize to a greater extent than is common in the industry. My investigations have convinced me to focus on the 10- to 12-year-old "preteen" group and design records particularly to their taste. Therefore, my main concern is with the differences if any between male and female record-purchasing patterns at this age.

    A magazine recently conducted a study of preteen behavior that included some coverage of record-purchasing behavior. The study was based on a random sample of 100 preteenage boys and 100 preteenage girls. With respect to records it found that the boys spent an average of $0.60 per year on records with a standard deviation of .4. The girls spent $0.80 per year with a standard deviation of .5. In the same study the magazine also found out that 62 percent of the records purchased by boys had male vocalists, while 74 percent of the records purchased by girls had male vocalists.

    What recommendations, if any, would you make based on these findings?

    (SIGNED) *Susan Byerlyn*

14. Suppose you are the production manager for the Coca-Cola Bottling Company of Houston. On your last trip to the supermarket you took time to look over your product and the competition. Upon examination you were concerned about one fact. The volume content of your 32-ounce bottle seemed to be more variable than that of all the competitors.

    The neck of your 32-ounce bottle is quite narrow and minor variations in volume are easily detected. A variation of 0.1 ounce is visually obvious! You reason that such obvious variations may not be beneficial for customer opinion of your product quality.

    You are aware of a rather expensive adapter that is compatible with your filling machines. The adapter is adjustable and its manufacturer claims that the adapter substantially increases the accuracy of the fluid ounces dispensed. If indeed the adapter would reduce the variance of dispensed fluid ounces, you would like to recommend the installation of an adapter on each of your fill lines.

    Suppose you have arranged to have one adapter installed for test purposes. You have obtained the following data. What do you conclude?

Fluid Ounces Dispensed

With Adapter	Without Adapter
32.02	32.01
31.97	31.96
31.98	32.07
32.01	31.91
31.97	31.86
32.03	32.09
32.05	32.11
31.96	31.92
32.03	32.12
	32.07

Chapter 9

Decision Theory

9.1 SITUATION I: WHY?

Suppose you have recently been named production manager of the Vick Chemical Company. Your predecessor was removed from the position because of his tendency to order excessive amounts of certain chemicals. This has resulted in considerable losses due to waste when the chemicals lose their effectiveness and have to be discarded.

As you assume your new position, you face a problem caused by your predecessor. Vick Chemical has in its inventory a substantial amount of doxylamine succinate, which is used in its Formula 44 cough syrup and its NyQuil cold remedy. In fact, the amount is sufficient for the next quarter's production of the two products. However, the chemical is subject to spoilage while being stored and its quality is suspect at this point. You are trying to decide whether to use the doxylamine succinate you have or to discard it and purchase more for this quarter's production.

If you process the chemical you have, one of three possibilities will be observed. First, you might observe that the products are totally ineffective. Second, the resulting medicines might be somewhat effective but less effective than usual. The third possibility is that the medicines produced would be of normal quality. If the products were totally ineffective, the entire production would have to be destroyed. Further, it would necessitate Vick's purchase of finished products through another drug firm, since time would not permit another production run. Such purchases would be at the same price at which Vick sells its products, so no profit would be realized on those transactions.

If the products were somewhat effective, they would not be marketed under the Vick's brand name. So, again, Vick would have to purchase finished products from the other firm. However, the less effective medicines could be sold at half price to offset some of the costs incurred.

The important costs and prices in this problem are summarized below:

Cost of purchasing doxylamine succinate	\$ 185,000
Cost of processing doxylamine succinate	700,000
Value of finished products	\$1,320,000

You do some quick calculations to determine if the problem warrants careful analysis. If you discard the chemical on hand, Vick would incur costs of

\$185,000 + \$700,000 = \$885,000

and receive \$1,320,000 for a net gain of

\$1,320,000 − \$885,000 = \$435,000

on the production of Formula 44 and NyQuil. If you process the chemical you have in inventory and the products exhibit their normal effectiveness, Vick would incur only the processing cost of \$700,000 while receiving \$1,320,000. This amounts to a profit of \$620,000, or a 42.5 percent increase over the \$435,000 realized if the chemical were discarded. On the other hand, if the products are totally ineffective, the processing costs of \$700,000 would be lost. Thus, processing the chemical on hand could increase profit by \$620,000 or decrease it by \$700,000. With the possibility of such substantial gains, the alternative to process the chemical in inventory cannot be discarded without further analysis. The very real possibility of substantial losses dictates that this action cannot be taken without careful consideration.

You have been informed by one of the chemists on the staff that you can perform the following test to get some idea of the essential properties of the doxylamine succinate. Under normal conditions, the chemical exhibits certain density characteristics when cooled. The chemical can be cooled and its density can be observed through sound and vibration readings. The cost of this test will be about \$1000, and time will allow only one such test.

This test will indicate one of the three conclusions. First, it may indicate that the doxylamine succinate has deteriorated. Second, the test may indicate that the chemical is of sufficient quality for successful production. Finally, the test may be inconclusive.

You ask the chemists on the staff to develop probability distributions for the outcomes of the test so that you can use statistical analysis in your decision. They submit a report to you which contains the following information:

1. If the doxylamine succinate has deteriorated:

   Pr(test indicates deteriorated chemical) = .80

Pr(test indicates chemical of sufficient quality) = .05
Pr(test is inconclusive) = .15

2. If the doxylamine succinate has lost some of its quality:

Pr(test indicates deteriorated chemical) = .30
Pr(test indicates chemical of sufficient quality) = .20
Pr(test is inconclusive) = .50

3. If the doxylamine succinate has maintained all its essential properties:

Pr(test indicates deteriorated chemical) = .05
Pr(test indicates chemical of sufficient quality) = .80
Pr(test is inconclusive) = .15

Finally, the chemists' report indicates that the quality of the doxylamine succinate translates directly to the effectiveness of the cough syrup and the cold remedy. That is, deteriorated doxylamine succinate will produce totally ineffective products; some loss of essential properties will result in the products' being only somewhat effective. Retention of the essential properties of doxylamine succinate will mean the Formula 44 and NyQuil produced is marketable under Vick's brand name.

Your original thought was to employ hypothesis testing in your analysis. However, several complications have been introduced which seem to preclude hypothesis testing. First, there are three "hypotheses"—the doxylamine succinate has deteriorated, has somewhat deteriorated, or has not deteriorated. Second, the probability distributions are quite different; your judgment seemingly cannot be linked to parameters. Third, the diversity of the values that could be realized by Vick Chemical in connection with the various possibilities seems to demand explicit consideration. This decision warrants analysis, but what can you use?

Situation I is a quite complicated decision problem. In addition to the uncertainty, there are several important relationships which have a bearing upon the final outcome. A great deal of information is presented, but it's not entirely clear where it all fits in. Let's begin by sorting out the key elements.

9.2 COMPONENTS OF DECISION PROBLEMS

In Chapter 8, we introduced the concepts and techniques of hypothesis testing. That form of statistical inference is sometimes called *statistical decision making*. On the basis of the information as interpreted with the sampling distribution of the test statistic, the statistician attempts to choose between accepting or rejecting the null hypothesis.

Hypothesis testing is a simple example of a decision-making problem. Decision theory represents an attempt to extend the concepts and techniques of

statistical inference to most or all decision problems. To see this extension, we examine the hypothesis-testing problem in an effort to identify those elements which are common to more general decision-making problems.

In hypothesis testing, the statistician has a choice between two alternatives—one to accept the null hypothesis, the other to reject the null hypothesis. Every decision involves alternatives. We can extend the concept of hypothesis-testing alternatives. First, we need not confine our analysis to alternatives which are expressed in terms of parameter values or forms of probability distributions. Second, we need not be restricted to decision problems with only two alternatives. In more general settings, then, the decision maker must choose one of two or more possible alternatives which represent courses of action.

DECISION THEORY IS THE APPLICATION OF PROBABILISTIC AND STATISTICAL CONCEPTS TO GENERAL DECISION-MAKING PROBLEMS.

Business decision making is done to improve the performance of the firm, in whatever terms that performance might be measured. Because the alternatives have effects on performance, any analytical approach to decision making must consider them explicitly. In the vocabulary of decision theory, the set of all alternatives is called the *action space*, and each alternative is called an *action*. Actions are within the control of the decision maker in that he or she can select which action is taken.

There are other elements in every decision problem which cannot be controlled and yet which influence the eventual value of the performance measure. These uncontrollable elements are called *states of nature*. In the hypothesis-testing problem, the states of nature are two: one standing for the truth of the null hypothesis and the other representing the falsity of that hypothesis.

States of nature can be expressions for the possible environments in which the decision will be made. These environments might be particular formations of consumer preferences or market conditions. States of nature can be used to identify the actions of competitors or the rulings of federal regulatory agencies. In short, the states of nature identify any elements beyond the control of the decision maker which affect the eventual result of the action taken.

Let's clarify one point about the states of nature immediately. Very often, we equate the inability to control with the lack of accountability. That is, we often hear statements like, "That was beyond his control; there was nothing he could do about it." But there are two different assertions between the quotation marks—one concerning control and one about the impossibility of action. An administrator with a fire department cannot control where fires break out; but, if the department cannot respond to a fire alarm within its jurisdiction, the administrator is certainly held responsible. On the other hand, should a fire start on the upper floors of a skyscraper and the necessary equipment has not been developed, the administrator would not be held accountable.

The existence of more than one state of nature implies uncertainty, for the decision maker does not know which state of nature is present. As we mentioned in Chapter 4, there are two ways in which decision makers are uncertain. First, decision makers are uncertain about events that have yet to occur. The states of nature here would describe the various possible outcomes which might happen. The second uncertainty comes from a lack of knowledge. In this case, the states of nature include all the possibilities which could be learned.

Very often in business, uncertainty is either not perceived or simply ignored. In these cases, only one state of nature would be identified by the decision maker. Of course, there is no need to specify the state of nature in this case. In this way, the decision maker has constructed what we called a *decision under certainty* in Chapter 4. We feel that this happens so often because uncertainty is an uncomfortable state to be in, and most decision makers are not familiar enough with the concepts and uses of probability and statistics to be convinced of their value. We can only say of this situation that the decision maker is not solving the real problem he or she faces. The consequences of such a strategy can be extremely costly.

A third component of a decision problem is the *payoff*. Of the components we discuss, only the payoff is not explicitly expressed in hypothesis-testing problems. Generally, we behave as if correctly identifying the true hypothesis costs us nothing while failing to identify the true hypothesis incurs a cost. We can think of the relative costs of Type I and Type II errors as dictating which levels of significance (α values) are "small" and which are not. However, such direct considerations of costs, or payoffs in general, are missing from the hypothesis-testing procedures.

In more general decision making, of course, consideration of payoffs is critical. Payoffs can be measured in terms of cost, or profit, or sales, or any of the other usual measures of a firm's performance or an individual decision maker's performance. It is not always easy to measure the payoffs. For example, how should the fire chief measure the performance of a particular squad's response to a house fire? What payoff should be used to evaluate the acquisitions made by a librarian?

Even if we measure payoffs in more conventional terms, such as profit or cost, there are some problems which arise. For instance, suppose you are the manager of a department store, and you are considering the possibility of having a clearance sale on children's shoes. In measuring the payoff from the sale, what cost should you assign to the children's shoes you have in inventory? With the amount of information given, the question cannot be answered. The cost assigned should be what is called the *decision cost*; that is, the cost incurred as a direct result of the action taken. If the children's shoes will be replaced in inventory after the sale, then the cost of replacement should be assigned, because the decision to have the sale is also the choice to incur this cost. On the other hand, if the clearance sale marks the end of your store's carrying children's shoes and the shoes will not be replaced, no cost should be assigned. The payoff should reflect only the costs incurred and the benefits realized as a result of the action taken.

The final component of decision problems is *information*. In this context, information is defined as the outcomes of random variables whose probability distributions depend upon the prevailing state of nature. The word "information" is used interchangeably with the term "data." In the hypothesis-testing problem, the value of the test statistic constitutes information or data. The test-statistic value is the outcome of a random variable, and the sampling distribution of that random variable depends upon which hypothesis is true.

Data are gathered in the problem to reduce the uncertainty present. Consequently, without uncertainty, there is no need to collect data. But surely there are measurements to be made and values to be calculated. Our point here is that, in the vocabulary of decision theory, these measurements and values do not constitute data. Let's reiterate: *Data are defined as the outcomes of random variables whose probability distributions depend upon the prevailing state of nature.*

The components of decision problems are:

1. The existence of actions or alternatives
2. The states of nature
3. Payoffs
4. Information

To solidify these concepts, let's look at two brief illustrations of decision problems and identify in each the components we have just enumerated. In Situation I, you can choose to process the doxylamine succinate in inventory or you can discard it. The action space can be expressed as the collection of actions a_1 and a_2, where

a_1 represents processing the chemical
a_2 represents discarding the chemical

The states of nature enumerate the various levels of quality of the chemical in inventory. You might use s_1 to represent deteriorated chemical, s_2 for partially deteriorated chemical, and s_3 to indicate chemical of normal quality. You would choose to express the payoffs in terms of profit since the necessary values are available. Finally, for information you have the opportunity to perform the density test described by the chemist. Because of the dependence of the outcome of this experiment upon the existing state of nature, this qualifies as information in the decision-theory sense. The time limitation mentioned in Situation I precludes the possibility of more tests.

As another example, suppose you work for a publishing company and are deciding upon the number of copies to be printed this year of your company's latest novel. A likely choice as the performance measure would be profit; thus payoffs would be expressed as profit. Production requirements are such that books are printed in lots of 10,000 copies. If your estimate of the maximum

demand for the novel is 55,000 copies, your action space would be {10,000, 20,000, 30,000, 40,000, 50,000, 60,000}. The states of nature are the possible values for total demand in the year. For purposes of simplification, suppose you group the states into the classes 5000–15,000, 15,000–25,000, 25,000–35,000, 35,000–45,000, and 45,000–55,000. Finally, each state could be named by its class mark. The states of nature would then be listed as {10,000, 20,000, 30,000, 40,000, 50,000}. Data could be gathered by a market survey; interviewers might describe the novel to individuals who were selected at random. This sample information could take the form of the percentage of those interviewed who expressed an interest in reading the novel.

9.3 SUMMARIES OF DECISION PROBLEMS

A graphic illustration of a decision problem is provided by a *decision tree*, which consists of nodes and branches. Nodes are points at which choices are made or possibilities are selected or realized. Each branch connects one node to another and corresponds to an alternative which might be chosen or an outcome which might be observed.

There are three kinds of nodes—each with its own symbol in the legend of the diagram. First, there is the *decision node*, which is identified by a small square □. The decision node indicates that, at that juncture in the decision problem, the decision maker chooses an action. Each course of action available to the decision maker has associated with it a branch of the tree emanating from the decision node.

The second kind of node is the *random node*, which is denoted by a small circle ○. The random node symbolizes uncertainty, which we will think of as a process or experiment. Each branch emanating from a random node corresponds to one of the possible outcomes of the process.

The final kind of node corresponds to a payoff realized. The node is represented by a dot • and is called a *terminal node*. There is also a number identifying the value of the payoff listed at a terminal node.

The decision tree stretches from left to right and, in that way, indicates the time flow from action taken to payoff realized. This time flow is dictated by what the decision maker perceives or will perceive as the situation unfolds. To illustrate, the decision maker first chooses a course of action and second observes the state of nature or realizes a payoff. This is the time sequence observed by the decision maker, and it may be in conflict with the actual chain of events.

A decision tree for a simple decision problem might look like the diagram in Figure 9-1. This decision problem has three actions (a_1, a_2, and a_3), two states of nature (s_1 and s_2), and payoffs (in profit, say) as listed on the right. These components have been labeled in the diagram. For each action the tree shows a branch leaving the decision node; for each state, a branch emanates from the random nodes.

Figure 9-1

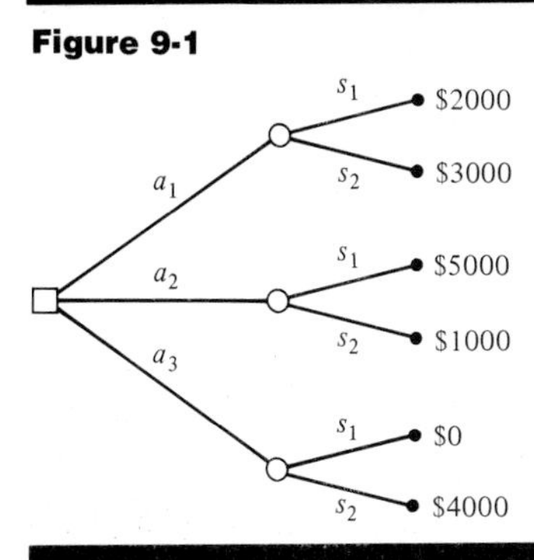

Information is incorporated in the tree by producing a new action—that of gathering data. The branch for that action leads to a random node representing the process which produces the data. From that point, there are several repeats of the decision node of Figure 9-1. If the information can take two possible values, the new decision tree contains the branch diagrammed in Figure 9-2. There are no numbers associated with the terminal nodes at this point. Whatever the cost of information, that cost is now incorporated in the payoffs.

Figure 9-2

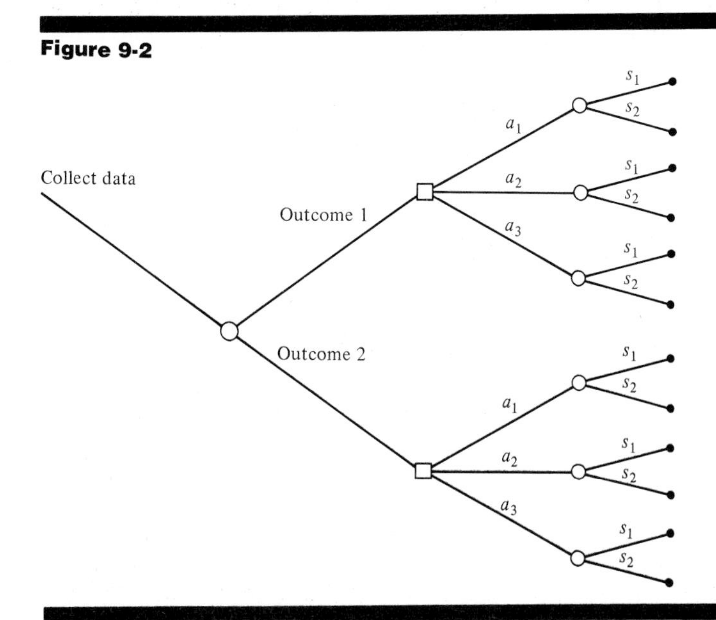

Table 9-1 Payoff Table for the Decision Problem of Figure 9-1

		States	
		s_1	s_2
Actions	a_1	2000	3000
	a_2	5000	1000
	a_3	0	4000
		Payoff (in Dollars)	

A second and alternative summary of a decision problem is provided by the *payoff table*. For each state of nature and each action, there is a payoff to the decision maker. A two-dimensional array of numbers then is quite appropriate for the summarization and display of the payoffs. One dimension corresponds to the states of nature; the other corresponds to the actions.

We prefer to label the columns by the states of nature and the rows by the actions. This is a matter of personal preference and is by no means universally accepted.

The payoff table in Table 9-1 summarizes the decision problem diagrammed in Figure 9-1. The payoff for each combination of action and state is listed in the appropriate position in the table. For instance, if the decision maker chooses action a_1 and state of nature s_1 is present, the payoff will be \$2000. If action a_3 is taken and state s_2 is present, the decision maker realizes a payoff of \$4000.

The payoff table includes all the components of the decision problem with the exception of information. Of course, we could incorporate information by producing a new action as we did with the decision tree. However, we choose to treat information differently, as we will see later in Section 9.5.

In practice, the payoff table is used less often than the decision tree. However, it is easily the more beneficial format for discussing and illustrating decision-theory techniques and principles. Consequently, for the remainder of our presentation, we use the payoff table as our summarization of decision problems.

9.4 ANALYSIS: HOW?

An important part of decision making is the determination of the action space. The superior decision makers are better able to recognize present actions and generate new actions. Similar statements can be made about recognizing or anticipating states of nature. But decision theory focuses on the analysis of decision problems, not on the formulating or structuring of those problems. Consequently, we begin our treatment with the actions and states given. Also taken as given are the payoffs. In other words, our treatment of the techniques of decision theory begins with the payoff table.

The analysis of the payoff table proceeds by assigning a value to each action. After the value is assigned, it's a simple matter to determine which value is best and therefore which course of action to choose. Let's use the payoff table of Table 9-1 as a reference. Each action represents a gamble for the decision maker. If action a_1 is chosen, the payoff is either \$2000 or \$3000. Action a_2 leads to \$5000 or \$1000, and action a_3 leads to either \$0 or \$4000. What we need to do is come up with one number which describes these possibilities.

Our earlier remarks about control and accountability suggest that the decision maker must explicitly consider the role of the states of nature in producing a payoff. One must plan for the various possibilities. Individuals and business enterprises often plan for "the unexpected" or the uncontrollable by purchasing insurance. This planning generally incorporates two considerations—the consequences of the uncontrollable event and the likelihood of the uncontrollable event.

In a similar fashion, decision theory suggests incorporation of these same two considerations when dealing with the states of nature. The consequences of the various states of nature are captured in the payoff table; so, when we analyze the payoff table, we will be incorporating the consequences of the various states. The likelihoods of the states of nature are captured by means of a probability distribution. For the decision problem of Table 9-1, this requires the assignment of two probabilities—one to state s_1 and the complementary probability to state s_2.

A brief word of caution is called for at this point. The probabilities assigned to the states of nature should accurately reflect the decision maker's uncertainty. The process of assigning accurate reflections is quite difficult for most individual decision makers. Research psychologists have documented the results of numerous experiments which identify many of the problems encountered. This phase of the analysis should be carried out with the help of an expert who specializes in probability assignment anytime the magnitude of the problem necessitates the careful development of numerical inputs.

Once probabilities are assigned, each action leads to an experiment which produces the payoff. The experiment is described by various payoffs and probabilities associated with them through the states of nature. The value assigned to a particular action is the expected value of the experiment associated with it. Returning to the payoff table of Table 9-1, suppose we assign probabilities of .6 to state s_1 and .4 to state s_2. The evaluation of the three actions then proceeds by computing the expected payoff for each action. Recall that in Chapter 5, Section 5.7, we introduced the concept of *mathematical expectation* and gave the expected value for a discrete random variable as

$$E(x) = \Sigma x f(x)$$

In our present situation the discrete random variable is payoff. Therefore, we can compute the *expected payoff* for each action by summing the products of the various payoffs and their probabilities of occurrence. These calculations are shown below.

$$a_1\text{: } .6(\$2000) + .4(\$3000) = \$2400 \tag{9-1}$$

$$a_2\text{: } .6(\$5000) + .4(\$1000) = \$3400$$

$$a_3\text{: } .6(\$0) + .4(\$4000) = \$1600$$

Since payoffs are expressed in terms of profit, \$3400 is the most preferred value assigned to an action. Since action a_2 is associated with this highest expected payoff, a_2 is the action prescribed by the analysis.

At the beginning of this discussion, we mentioned that the analysis must incorporate both the consequences and the likelihoods of the uncontrollable elements. In this analysis, the uncontrollable elements are the states of nature, and their consequences and likelihoods have been incorporated in the evaluation of the actions. *So, given a payoff table and probabilities on the states of nature, we compute the expected payoff for each action and choose the action with the most preferred expected payoff.* Note that this is without purchasing additional information about the states of nature.

9.5 SAMPLE INFORMATION

Information is gathered to reduce the decision maker's uncertainty. This is accomplished by using the sample information to revise the probabilities assigned to the states of nature. These "new" probabilities are used as the "old" ones were to determine the most desirable action; that is, we calculate the expected payoff for each action. The payoff table remains unchanged as data are gathered.

These ideas require formalization. Let y be a random variable. If observations on y are to constitute data, there must be a fundamental relationship between the probability distribution of y and the prevailing state of nature. This can be captured in the notation for the distribution of y. We could write $Pr(y|s)$ to represent the probability function of y when s is the state of nature. Suppose Table 9-2 gives the probability distributions for the sample decision problem of Table 9-1.

The concept of conditional probability plays a major role in this analysis.

Table 9-2 Conditional Probabilities for y

		States	
		s_1	s_2
Outcomes	y_1	.7	.2
	y_2	.3	.8

Table 9-3 Conditional Probability Tables

	States		
	s_1	s_2	s_3
y_1	.4	.7	.9
y_2	.6	.3	.1

(A)

	States		
	s_1	s_2	s_3
y_1	.4	.4	.4
y_2	.6	.6	.6

(B)

The use of the conditional probability notation suggests that we do not have the same probability distribution for y for every state of nature. For instance, Table 9-3 gives two tables of conditional probability distributions for y for various states of nature. Table 9-3(A) shows a case in which information is available in y, while Table 9-3(B) gives an example of a situation in which y does *not* constitute information. This is so because in Table 9-3(B) all the probability distributions for y are the same; there will be no way of distinguishing between the various states of nature based upon observations on y. This is the essence of the meaning of information in the vocabulary of decision theory. The conditional probability notation also suggests that, if we know the prevailing state of nature, we have learned something about the distribution of y. That is also embodied in the meaning of information.

We denote by $Pr(s_i)$ the probability assigned to the state of nature s_i. The availability of information in the form of an observation on y allows us to revise $Pr(s)$ to $Pr(s|y)$. So, we get a new probability distribution on the states of nature for each outcome of the random variable y. This is accomplished via Bayes' rule. The probability assignment $Pr(s)$ is called the *prior distribution* on the states of nature because it is formulated *before* sampling. The conditional probability distribution $Pr(s|y)$ is called the *posterior distribution* because it is calculated *after* data are observed.

In Chapter 4, we stated Bayes' rule in terms of an event A and a collection of events $B_1, B_2, \ldots, B_n$, as

$$Pr(B_i|A) = \frac{Pr(A|B_i)Pr(B_i)}{Pr(A|B_1)Pr(B_1) + Pr(A|B_2)Pr(B_2) + \cdots + Pr(A|B_n)Pr(B_n)}$$

Here the states of nature play the roles of the B_i's and the sample data constitute A. If we let B_i represent the state s_i and A the information y, we are able to replace

$Pr(B_i|A)$ with $Pr(s_i|y)$

$Pr(B_i)$ with $Pr(s_i)$

$Pr(A|B_i)$ with $Pr(y|s_i)$

and so on. So, if we have n states of nature $s_1, s_2, \ldots, s_n$, the probability of state s_i, after the particular value y is observed, is

$$Pr(s_i|y) = \frac{Pr(y|s_i)Pr(s_i)}{Pr(y|s_1)Pr(s_1) + Pr(y|s_2)Pr(s_2) + \cdots + Pr(y|s_n)Pr(s_n)} \tag{9-2}$$

The posterior distribution on the states of nature is given by $Pr(s|y)$.

We had better get some numbers in here. In Table 9-2, $Pr(y_1|s_1) = .7$, with $Pr(y_2|s_1) = .3$; and $Pr(y_1|s_2) = .2$, with $Pr(y_2|s_2) = .8$. Let's assume $Pr(s_1) = .6$ and $Pr(s_2) = .4$. Now, *if y_1 is observed*, then the posterior distribution on the states of nature is calculated as

$$Pr(s_1|y_1) = \frac{Pr(y_1|s_1)Pr(s_1)}{Pr(y_1|s_1)Pr(s_1) + Pr(y_1|s_2)Pr(s_2)} \tag{9-3}$$

$$= \frac{.7(.6)}{.7(.6) + .2(.4)} = \frac{.42}{.50} = .84$$

$$Pr(s_2|y_1) = \frac{Pr(y_1|s_2)Pr(s_2)}{Pr(y_1|s_1)Pr(s_1) + Pr(y_1|s_2)Pr(s_2)}$$

$$= \frac{.2(.4)}{.7(.6) + .2(.4)} = \frac{.08}{.50} = .16$$

Notice that

$$Pr(s_1|y_1) + Pr(s_2|y_1) = .84 + .16 = 1.0$$

so that the posterior distribution *is* a probability distribution on the states of nature. This will always be the case, because the numerator is one of the terms of the denominator, and there is a state of nature and a probability for each term in the denominator. That is, the denominator is of the form

$$A + B + \cdots + K$$

and the posterior probabilities calculated are

$$\frac{A}{A+B+\cdots+K}, \quad \frac{B}{A+B+\cdots+K}, \quad \ldots, \quad \frac{K}{A+B+\cdots+K}$$

Once the posterior distribution is developed, the analysis proceeds as before. That is, the probability distribution on the states of nature is applied to determine an expected payoff for each action. The appropriate probability distribution is now the posterior distribution for the observed sample data. Under the supposition that y_1 is our observed sample outcome, we evaluate the action as follows:

$$a_1\text{: } .84(\$2000) + .16(\$3000) = \$2160 \tag{9-4}$$

$$a_2\text{: } .84(\$5000) + .16(\$1000) = \$4360$$

$$a_3\text{: } .84(\$0) \quad + .16(\$4000) = \$\ 640$$

Since \$4360 is the largest expected profit, action a_2 is the prescribed choice.

This completes the analysis in the event that y_1 is the observed sample outcome.

If, on the other hand, our experiment results in the observation of y_2, then we would calculate the posterior distribution as follows:

$$Pr(s_1|y_2) = \frac{Pr(y_2|s_1)Pr(s_1)}{Pr(y_2|s_1)Pr(s_1) + Pr(y_2|s_2)Pr(s_2)} \tag{9-5}$$

$$= \frac{.3(.6)}{.3(.6) + .8(.4)} = \frac{.18}{.50} = .36$$

and

$$Pr(s_2|y_2) = 1 - Pr(s_1|y_2) = .64$$

This probability distribution is utilized as before to determine the prescribed action in the event that the sample outcome is y_2. Thus, *if* y_2 *is the observed outcome*, the expected payoffs for the actions are calculated as:

$$a_1\text{: } .36(\$2000) + .64(\$3000) = \$2640 \tag{9-6}$$

$$a_2\text{: } .36(\$5000) + .64(\$1000) = \$2440$$

$$a_3\text{: } .36(\$0) + .64(\$4000) = \$2560$$

This time the greatest expected profit is associated with action a_1, so the decision maker should take course of action a_1.

The information received in this simple example leads to the selection of a different preferred action depending upon whether y_1 or y_2 is observed. Recall from Equations (9-1) that action a_2 was the prescribed choice prior to the acquisition of the sample information. Action a_2 is also the choice if our sample information is y_1. However, if we observe y_2 as the outcome from our experiment, we should choose action a_1. Since this sample information would alter our choice, the sample information should have some value. That value can be seen by examining the expected payoff before the data and the expected payoff after. The difference in these two values can be viewed as the *value of information*.

The expected payoff *before* observing data is easily seen from Equations (9-1). It is \$3400. *After* observing the data, we have from Equations (9-4) and (9-6) the expected payoff

\$4360 if y_1 is observed

\$2640 if y_2 is observed

To determine the expected payoff if the data are gathered, we must determine two probabilities: namely, $Pr(y_1)$ and $Pr(y_2)$. The expected payoff with sampling will then be

$$\$4360\ Pr(y_1) + \$2640\ Pr(y_2)$$

These probabilities have already been computed above. The denominator of

Bayes' rule for $Pr(s_1|y_1)$ is exactly $Pr(y_1)$. Recall that, if y_1 occurs, it occurs in conjunction with state s_1 or with state s_2. The calculation is

$$\begin{aligned} Pr(y_1) &= Pr(y_1s_1) + Pr(y_1s_2) \\ &= Pr(y_1|s_1)Pr(s_1) + Pr(y_1|s_2)Pr(s_2) \\ &= .7(.6) + .2(.4) = .50 \end{aligned} \quad (9\text{-}7)$$

Also,

$$Pr(y_2) = 1 - Pr(y_1) = 1 - .50 = .50$$

Again, the formula used in Equation (9-7) is just the denominator from the calculations of Equation (9-3). A similar computation for $Pr(y_2)$ would give the denominator of Equation (9-5). [We should point out that it is only coincidental that $Pr(y_1) = Pr(y_2)$ in this calculation, as you will see in subsequent examples.]

The expected payoff after data is

$$.50(\$4360) + .50(\$2640) = \$3500$$

Thus the value of information, the difference between the expected payoffs before and after data, is

$$\$3500 - \$3400 = \$100$$

This value, \$100, can now be compared with the cost of gathering the information to determine if data collection is warranted in the problem at hand. Clearly we would only gather information if its value exceeded its cost. Thus, if the cost of observing the random variable y is less than \$100, we would choose to sample. Otherwise, we would not.

9.6 SITUATION I REVISITED

Now you have the tools to handle the problem of the doxylamine succinate. First, you develop the payoff table. You have already calculated some of the numbers which go into the table. The table will have two rows—one for each of the possible courses of action. The table will have three columns—one for each of the possible states of nature. That means that six payoffs will be displayed in the payoff table. These are

1. The payoff when deteriorated chemical is processed
2. The payoff when somewhat deteriorated chemical is processed
3. The payoff when the chemical on hand is processed and is not deteriorated
4. The payoff when the deteriorated chemical is not processed
5. The payoff when the somewhat deteriorated chemical is not processed
6. The payoff when the chemical on hand is not processed but is not deteriorated

Payoffs 4, 5, and 6 are all the same because they do not depend upon the state of nature. These payoffs were calculated to be $435,000. Payoff 1 was calculated to be −$700,000. Payoff 3 was calculated to be $620,000. Only Payoff 2 remains to be calculated. Payoff 2 is computed as follows:

Processing the chemical on hand	−$ 700,000
Selling the somewhat effective medicines (at half price)	660,000
Purchasing finished products	− 1,320,000
Selling finished products	1,320,000
Payoff	−$ 40,000

The payoff table is then

		States		
		s_1 (Not Effective)	s_2 (Somewhat Effective)	s_3 (Normal Quality)
Actions	a_1 (process)	−700	−40	620
	a_2 (discard)	435	435	435

Payoff (Profit in Thousands of Dollars)

Next, you ask the chemists on the staff to provide you with probabilities that you might assign to the states of nature. Their response is expressed in the prior distribution

$Pr(s_1) = .2 \qquad Pr(s_2) = .4 \qquad Pr(s_3) = .4$

Since you are going to consider the possibility of conducting the test, you must first determine the expected payoff without the data.

a_1: $.2(-\$700{,}000) + .4(-\$40{,}000) + .4(\$620{,}000) = \$\ 92{,}000$

a_2: $.2(\$435{,}000) + .4(\$435{,}000) + .4(\$435{,}000) = \$435{,}000$

Action a_2 has the highest expected payoff at this stage, and its expected profit is $435,000.

The chemists' earlier report provides you with the conditional probability table for the interpretation of the test results.

		States		
		s_1	s_2	s_3
Test Results	Indicates deteriorated chemical	.80	.30	.05
	Indicates no deterioration	.05	.20	.80
	Inconclusive	.15	.50	.15

Conditional Probabilities

These probabilities will be used via Bayes' rule to determine the posterior probabilities of the states of nature.

For the case in which the test indicates that the doxylamine succinate has deteriorated (call this outcome D), the probabilities on the states of nature are revised in the following way:

$$Pr(s_1|D) = \frac{Pr(D|s_1)Pr(s_1)}{Pr(D|s_1)Pr(s_1) + Pr(D|s_2)Pr(s_2) + Pr(D|s_3)Pr(s_3)}$$

$$= \frac{.80(.2)}{.80(.2) + .30(.4) + .05(.4)} = \frac{.16}{.30} = .533$$

$$Pr(s_2|D) = \frac{.30(.4)}{.30} = \frac{.12}{.30} = .40$$

$$Pr(s_3|D) = 1 - .533 - .40 = .067$$

With these probabilities, you evaluate the two actions as

a_1: .533(−\$700,000) + .40(−\$40,000) + .067(\$620,000) = −\$347,560

a_2: .533(\$435,000) + .40(\$435,000) + .067(\$435,000) = \$435,000

(It should be becoming clear at this point that action a_2 will always have this expected payoff.) In this case, action a_2 is prescribed, and its expected payoff is \$435,000.

For the case in which the test indicates a chemical of sufficient quality, SQ, the probabilities are

$$Pr(s_1|SQ) = \frac{Pr(SQ|s_1)Pr(s_1)}{Pr(SQ|s_1)Pr(s_1) + Pr(SQ|s_2)Pr(s_2) + Pr(SQ|s_3)Pr(s_3)}$$

$$= \frac{.05(.2)}{.05(.2) + .20(.4) + .80(.4)} = \frac{.01}{.41} = .024$$

$$Pr(s_2|SQ) = \frac{.20(.4)}{.41} = \frac{.08}{.41} = .195$$

$$Pr(s_3|SQ) = 1 - .024 - .195 = .781$$

With these probabilities, the expected payoffs for the two actions are

a_1: .024(−\$700,000) + .195(−\$40,000) + .781(\$620,000) = \$459,620

a_2: \$435,000

The prescribed action in this case is a_1; that is, the probability that the doxylamine succinate in inventory is of sufficient quality has attained a level high enough to warrant processing of the chemical on hand. The expected payoff of processing under these conditions is \$459,620.

The final possible outcome of the test is that the results are inconclusive, NC. Under this condition, the probabilities for the states of nature are revised more simply by making use of the fact that the denominators of the two previous calculations have given you the following probabilities:

$Pr(D) = .30$ and $Pr(SQ) = .41$

Consequently, the denominator for the next set of probabilities you compute will be

$1 - .30 - .41 = .29$

since that denominator will be the probability of inconclusive results from the test. The calculations are

$$Pr(s_1|NC) = \frac{.15(.2)}{.29} = .103$$

$$Pr(s_2|NC) = \frac{.50(.4)}{.29} = .690$$

$$Pr(s_3|NC) = 1 - .103 - .690 = .207$$

Using these posterior probabilities, you calculate

a_1: .103(−\$700,000) + .690(−\$40,000) + .207(\$620,000) = \$28,640

a_2: \$435,000

Under the condition that the test is inconclusive, the prescribed action is a_2 with an expected profit of \$435,000.

Well, let's see where you are now. You know that without data the prescribed choice is to discard the chemical on hand and the expected payoff for this action is \$435,000. If you decide to perform the test, which costs \$1000, you will select an action based upon the outcome of the test. The three outcomes lead to three expected payoffs. The expected profits are \$435,000, \$459,620, and \$435,000. The probabilities associated with these expected payoffs are .30, .41, and .29, respectively. So, the expected profit for the problem with the sample information is

.30(\$435,000) + .41(\$459,620) + .29(\$435,000) = \$445,094.20

The value of the information to be gained from the test is

\$445,094.20 − \$435,000 = \$10,094.20

Since the test costs only about \$1000, it is certainly a valuable device for aiding in this decision.

Your final decision can be summarized as follows:

Proceed with the test on the doxylamine succinate on hand. If the test indicates that the chemical has retained its essential properties, then process the chemical on hand. Otherwise, discard the chemical on hand and purchase a new supply for your processing purposes.

Now is a good time to recap what we've accomplished. We set out in Section 9.2 to simplify a complex decision problem. To do that, we explored the elements or components common to all decisions, and we identified these components for the decision problem of Situation I. Once that was done, we were in a position to complete the analysis and eventually make a recommendation. There is often a tendency to underestimate the value of this identification process. Suffice it to say that until the components are adequately specified, the valuable analysis cannot proceed. One advantage of decision theory is that it forces the decision maker to identify these components at the outset.

The steps involved in decision-theory applications include:

1. Identification of the actions, the states of nature, the payoffs, and the information available
2. Assignment of probabilities to the states of nature
3. Evaluation of the actions by computing the expected payoff for each
4. Determination of the expected value of the information
5. Selection of the prescribed action by comparing the expected payoffs

Step 4 makes frequent use of Bayes' rule for the calculation of posterior probabilities. The selection mentioned in step 5 includes the possibility of choosing to collect data and acting in accordance with the findings from the sample data.

9.7 EXAMPLES

(1) Let's return to the book publisher, who is now considering the printing of the company's latest college-level history text. The first-year demand for the text will be below 15,000 copies. The printing must be done in lots of 5,000 copies. The publisher has arranged the demand possibilities into the classes:

s_1: demand below 5,000 copies
s_2: demand between 5,000 and 10,000 copies
s_3: demand between 10,000 and 15,000 copies

The courses of action available to the publisher are

a_1: print 5,000 copies
a_2: print 10,000 copies
a_3: print 15,000 copies

Suppose the payoff table is

		States		
		s_1	s_2	s_3
Actions	a_1	−4.5	12.5	12.5
	a_2	−34.5	8.0	25.0
	a_3	−64.5	−22.0	20.5

Payoff
(Profit in Thousands of Dollars)

The publisher assigns probabilities of .1, .5, and .4, respectively, to the states of nature.

As with other texts, the publisher has sent brochures on the new text to a large number of college faculty. The response rate on these mailings has been observed to have the following relationship with the first-year sales:

		States		
		s_1	s_2	s_3
Response Rate	< 5%	.7	.3	.1
	5%–20%	.2	.5	.1
	> 20%	.1	.2	.8
		Conditional Probabilities, *Pr*(response|s)		

The rate of response has already been observed for this mailing as 11 percent. So, the posterior probabilities on the states of nature are given by

$$Pr(s_1|5\%\text{–}20\%) = \frac{Pr(5\text{–}20\%|s_1)Pr(s_1)}{Pr(5\%\text{–}20\%|s_1)Pr(s_1) + Pr(5\%\text{–}20\%|s_2)Pr(s_2) + Pr(5\%\text{–}20\%|s_3)Pr(s_3)}$$

$$= \frac{.2(.1)}{.2(.1) + .5(.5) + .1(.4)} = \frac{.02}{.31} = .065$$

$$Pr(s_2|5\%\text{–}20\%) = \frac{.5(.5)}{.31} = \frac{.25}{.31} = .806$$

$$Pr(s_3|5\%\text{–}20\%) = 1 - .065 - .806 = .129$$

Using this posterior distribution, the publisher evaluates the actions as

a_1: .065(−\$4500) + .806(\$12,500) + .129(\$12,500) = \$11,395

a_2: .065(−\$34,500) + .806(\$8000) + .129(\$25,000) = \$7,430.50

a_3: .065(−\$64,500) + .806(−\$22,000) + .129(\$20,500) = −\$19,280

Since \$11,395 is the highest expected profit, the prescribed course of action is to print 5,000 copies.

(2) A life insurance company has purchased land to begin construction on a new building. The board of directors is studying two different plans for the structure. The smaller building would cost \$120 million and have 175 offices and suites for lease. The larger building would cost \$200 million and have 250 offices and suites to be used for rental income. Each building would provide similar facilities for the insurance company operations.

The demand for office space in the area will ultimately determine the degree of success achieved. The board assigns three levels for the demand for office space: low, moderate, or high. The payoff table is developed as

		States		
		s_1 Low	s_2 Moderate	s_3 High
Actions	a_1 Smaller Building	0.2	2.5	3.0
	a_2 Larger Building	−0.1	1.9	4.1

Payoff (Annual Profit in Millions of Dollars)

The board has solicited probabilities from real estate experts in the area and has arrived at

$Pr(s_1) = .1 \qquad Pr(s_2) = .6 \qquad Pr(s_3) = .3$

Without data, the expected payoffs are

a_1: $.1(0.2) + .6(2.5) + .3(3.0) = 2.42$

a_2: $.1(-0.1) + .6(1.9) + .3(4.1) = 2.36$

The smaller building would lead to the higher expected payoff and so is the prescribed choice. The expected payoff is $2,420,000.

For $20,000, the board can commission a rental market survey which will identify demand as "marginal" or "substantial." The relationship between the states of nature and the survey findings is captured in the conditional probability table:

	States		
	s_1	s_2	s_3
M = marginal	.7	.5	.4
S = substantial	.3	.5	.6

Conditional Probability Table

To reach a decision, the board must determine the value of information. If the survey report suggests a "marginal" demand,

$$Pr(s_1|M) = \frac{Pr(M|s_1)Pr(s_1)}{Pr(M|s_1)Pr(s_1) + Pr(M|s_2)Pr(s_2) + Pr(M|s_3)Pr(s_3)}$$

$$= \frac{.7(.1)}{.7(.1) + .5(.6) + .4(.3)} = \frac{.07}{.49} = .143$$

$$Pr(s_2|M) = \frac{.5(.6)}{.49} = \frac{.30}{.49} = .612$$

$$Pr(s_3|M) = 1 - .143 - .612 = .245$$

In this case, the expected payoffs are

a_1: .143(0.2) + .612(2.5) + .245(3.0) = 2.2936

a_2: .143(−0.1) + .612(1.9) + .245(4.1) = 2.153

The prescribed choice in this event is action a_1, with expected annual profit of $2,293,600.

If a "substantial" market is indicated,

$$Pr(s_1|S) = \frac{.3(.1)}{.3(.1) + .5(.6) + .6(.3)} = \frac{.03}{.51} = .059$$

$$Pr(s_2|S) = \frac{.5(.6)}{.51} = \frac{.30}{.51} = .588$$

$$Pr(s_3|S) = 1 - .059 - .588 = .353$$

The actions are now evaluated as

a_1: .059(0.2) + .588(2.5) + .353(3.0) = 2.5408

a_2: .059(−0.1) + .588(1.9) + .353(4.1) = 2.5586

Action a_2 is now prescribed, and the expected payoff is $2,558,600.

The expected annual profit with the information available is

.49($2,293,600) + .51($2,558,600) = $2,428,750

This tells us that the information is valued at only

$2,428,750 − $2,420,000 = $8,750

But the information costs the board of directors $20,000. The board, then, should not purchase the information. Further, the board should pursue the construction of the smaller building.

(3) The chief legal counsel for a food processing firm has hired Dr. Jean Robinson, a statistical analyst, as a consultant in a case. A few months ago, the firm recalled several cases of cream-style corn because of the presence of bacteria. Unfortunately, one family had already eaten some of the product before the notice was broadcast on radio and television. Two children became ill. Both recovered after being hospitalized. But now the family is suing the food processor.

The suit asks for real damages (doctors, hospital costs, etc.) of approximately $75,000, punitive damages of $100,000, and approximately $825,000 for "mental anguish." The total sought in the suit is just over $1 million.

During pretrial negotiations, the attorney for the family offered to settle out of court for $250,000. Dr. Robinson is to help the chief counsel make a decision about the settlement offer. The chief counsel can elect to accept the settlement offer, a_1, or pursue the court proceedings, a_2. There is also the possibility of making a counteroffer, but Dr. Robinson suggests omitting that action for the present.

Dr. Robinson works with the chief counsel to identify the following states of nature:

s_1: court ruling in the firm's favor

s_2: court ruling in favor of the family

s_3: an intermediate ruling

After deliberations, they assign the probabilities .1, .1, and .8, respectively, to the states of nature.

The payoffs are expressed as costs to the food processing firm and are presented in the table:

	States		
	s_1	s_2	s_3
a_1	250,000	250,000	250,000
a_2	10,000	1,000,000	150,000

Payoff (Cost in Dollars)

The actions are then evaluated. Since all the payoffs associated with action a_1 are \$250,000, its expected cost is \$250,000. (This is similar to what we saw in analyzing the decision problem of Situation I. There, the expected payoff for one of the actions was always \$435,000, regardless of the probability distribution on the states of nature.) The evaluation of action a_2 is

$$a_2: .1(\$10,000) + .1(\$1,000,000) + .8(\$150,000) = \$221,000$$

Since action a_2 has the *lower* expected *cost*, it is the prescribed choice at this stage.

Dr. Robinson now returns to the possibility of submitting a counteroffer. The expected cost for action a_2 can now serve as the largest settlement offer the firm will accept. A settlement offer any larger than \$221,000 is unacceptable because the expected cost of the court proceedings is \$221,000. Dr. Robinson suggests that the chief counsel present a counteroffer somewhat below \$221,000 in order to allow for more negotiation should that counteroffer be rejected.

(4) A manufacturer of airbags for automobiles is considering the expansion of its production facilities. The expansion would be in anticipation of a sharp increase in demand which would result if Congress passed the pending legislation requiring airbags in all new automobiles.

The manufacturer is considering not only expansion now but also the possibility of beginning modifications so that expansion can be done more easily next year, if needed. The action space, then, consists of

a_1: expand now

a_2: modify now

a_3: take no action at this time

The states of nature express the possibilities of congressional action:

s_1: Congress passes the pending legislation
s_2: Congress delays action on the bill
s_3: Congress defeats the legislation

The probabilities assigned to these states are .3, .3, and .4, respectively. The payoff table is expressed in profit over the next two years.

		States		
		s_1	s_2	s_3
Actions	a_1	105	20	−60
	a_2	75	80	25
	a_3	40	40	35

Payoff
(Profit in Thousands of Dollars)

The evaluations of the actions are

a_1: $.3(105) + .3(20) + .4(-60) = 13.5$
a_2: $.3(75) + .3(80) + .4(25) = 56.5$
a_3: $.3(40) + .3(40) + .4(35) = 38$

Since action a_2 has the highest expected profit, it is the prescribed choice. There was no source of information, and therefore no revision of prior probabilities on the states of nature.

9.8 UTILITY

In the evaluation of each action, we have used the expected value of the payoffs. In the case of Situation I, you were urged to process the chemical on hand if the test was positive. However, processing the chemical on hand would result in one of three payoffs: a gain of $620,000, a loss of $40,000, or a loss of $700,000. Few decision makers would feel comfortable turning their backs on a certain $435,000 (which was the payoff for discarding the chemical) for a chance at $620,000 with the accompanying risk of losing $700,000.

Let's consider some other possibilities—all with the same expected payoff.

		States	
		s_1	s_2
Actions	a_1	50	50
	a_2	0	100
	a_3	−100	200

Payoff
(Profit in Thousands of Dollars)

If each of the states is assigned a probability of .5, then each action has an expected payoff of $50,000. According to our decision criterion, each of these actions is equally desirable, or, in other words, equivalent. *But these three alternatives represent very different opportunities.*

In Chapter 2, we used the mean or expected value to describe a collection of numbers. In subsequent chapters, we have seen the importance of the expected value in statistical inference. That importance stems from the meaning of the expected value. As the number of observations increases, the relative frequencies of the outcomes become closer and closer to their probabilities. So, the average of the outcomes will be close to the expected value.

Consequently, if the decision maker faces the same decision problem a large number of times, then taking a particular action will return an average payoff close to the expected payoff. But specific decision problems are usually faced only once. Even if the problem setting is the same, the states of nature change, the possible courses of action change, the payoffs change, or the probabilities change. *Decision problems are usually unique.* The expected payoff, then, can be misleading in evaluating the actions.

Some executives would argue that the three actions in the above table are indeed equivalent. These decision makers are described as having *linear utility*; that is, each additional dollar of payoff produces the same change in their total satisfaction. Each dollar of payoff is equally desirable to them whether it is the first dollar, the one-hundredth dollar, or the one-millionth dollar. For these decision makers, the expected payoff is appropriate in evaluating the actions.

Utility is a concept used to deal with the uniqueness problem. In each decision problem, we can establish a substitute for each payoff. The substitution procedure incorporates the riskiness of an action into the evaluation of that action. The substitute for the payoff is called the *utility* of the payoff.

In any decision problem, the best payoff is assigned a utility of 1, and the worst or least desirable payoff is given a utility of 0. Every other payoff gets a utility value between 0 and 1, which reflects the decision maker's preference for that payoff relative to the best and the worst. It is very similar to a rating system on "a scale of 1 to 10." (It should be clear that the more desirable payoff has the greater utility.)

Just as we cautioned you about the difficulties of assigning probabilities to the states of nature, we must caution you here about the difficulties of developing utilities. Again, this phase of the analysis of the decision problem should be performed with the help of a trained specialist. This is particularly true when the importance of the problem dictates careful measurements of the quantitative inputs.

The utility of a payoff can be interpreted as a probability of winning the best or most desirable payoff. This probability is substituted for the payoff in the analysis. For instance, if the utilities of $40,000 and $20,000 are .9 and .5, respectively, the payoff table

		States	
		s_1	s_2
Actions	a_1	10	50
	a_2	40	20

Payoff
(Profit in Thousands of Dollars)

is replaced by the utility table

		States	
		s_1	s_2
Actions	a_1	0	1
	a_2	.9	.5

Utility

The evaluation of the actions is accomplished in the same way as with payoffs; that is, each action is evaluated by its expected utility. It can be shown mathematically that this expected utility can also be interpreted as the probability of winning the best payoff. So each action is evaluated by a probability of winning. Selecting the action with the largest expected utility is selecting the action with the largest probability of winning.

If the prior distribution on the states is given by

$Pr(s_1) = .4$ and $Pr(s_2) = .6$

then, using payoffs, we get

a_1: $.4(\$10,000) + .6(\$50,000) = \$34,000$

a_2: $.4(\$40,000) + .6(\$20,000) = \$28,000$

However, using utility, we have

a_1: $.4(0) + .6(1) = .6$

a_2: $.4(.9) + .6(.5) = .66$

Consequently, while action a_1 has the higher expected payoff, action a_2 has the higher expected utility.

It is unfortunate that expected payoff and expected utility may lead to different prescribed actions. It is unfortunate because finding the prescribed course of action with payoffs is easier than with utilities. But the additional information contained in the utility table makes the analysis using utility preferred. So, in the conflict above, action a_2 is prescribed.

As an example, let's reconsider Example 3 in which Dr. Robinson was advising the chief counsel for a food processing firm. The payoffs in that problem were expressed in terms of cost. Since $10,000 was the least-cost payoff, it is assigned a utility of 1. The greatest-cost payoff, $1,000,000, is assigned a utility of 0. Suppose a utility of .4 is assigned to the $150,000 cost, and a utility of .3 is assigned to the $250,000 cost. Then, the payoff table

		States		
		s_1	s_2	s_3
Actions	a_1	250,000	250,000	250,000
	a_2	10,000	1,000,000	150,000

Payoff (Cost in Dollars)

is replaced by the utility table

		States		
		s_1	s_2	s_3
Actions	a_1	.3	.3	.3
	a_2	1	0	.4

Utility

The evaluations of the actions in terms of expected utility are

a_1: .3

a_2: $.1(1) + .1(0) + .8(.4) = .42$

Action a_2 is prescribed. Consequently, this serves as an example in which the prescribed choice using payoffs and the prescribed choice using utilities are the same.

9.9 IDENTIFICATION OF APPLICATIONS: WHEN?

When we began our discussion of decision theory, we talked in terms of extending the concepts of hypothesis testing to all decision problems. We looked at the various components of a decision problem—not all of which were explicitly considered in hypothesis testing. Every decision problem presents the decision maker with two or more possible courses of action. Every realistic decision situation holds some uncertainties for the decision maker. Decisions are made to accomplish some purpose; the accomplishments are measured in the payoffs. Finally, most decision problems afford the decision maker the opportunity to collect data.

In our discussion we've kept the number of actions and the number of states small. This allowed our presentation to be more understandable because the problems and illustrations remained tractable. Much larger and more complex problems have been analyzed via decision theory with very satisfactory results. Such applications rely heavily on computers.

It is also possible to develop parameter estimates with decision-theory methods. In these cases, the states of nature are all the possible true values for the parameter. The action space consists of all the possible "guesses" available to the statistician. (Procedures are available for dealing with an infinite number of actions and an infinite number of states.) These problems are more often encountered by statisticians than business decision makers and so have been omitted from our presentation.

Because of the many difficulties involved with probability and utility assignments as well as the careful development of payoff numbers, there is frequently a substantial cost figure attached to decision-theory applications. Consequently, direct application of decision-theory techniques is carried out on major problems involving large sums of money. Some examples include space exploration, airport location problems, and corporate policy formulation.

However, the approach of decision theory to decision problems carries over to all decisions, large and small. In any decision problem, we can think in terms of identifying the states of nature and the actions, of roughly estimating payoffs and consequences, and of guessing at likelihoods. So, even on the smallest of problems, the concepts of decision theory guide the formulation and analysis for decision making. In that sense, decision theory can always be used in decision making.

PROBLEMS

1. Identify an application of decision theory in your major field of study. Point out the unique characteristics of the situation which make this application appropriate and of value.

2. Suppose the National Science Foundation (NSF) invited bids to conduct a study of information utilization by consumers. As a recently hired analyst of Market Consultants Incorporated, you want to answer the question "to bid or not to bid." It will cost your firm $8,500 to prepare the bid, and you estimate that the firm will realize $50,000 profit if it receives the contract. Given the level of competition, you estimate the probability of receiving the contract to be .15. Would you advise your firm to bid on the contract?

3. Atlantic Coast Corporation (A.C.C.), which manufactures throwaway bottles, is deciding whether to spend $1 million in a campaign to help defeat a state referendum requiring returnable bottles. If the referendum is successful and returnable bottles are required, it will cost A.C.C. approximately $12.5 million in lost profits. If the referendum fails, the firm's profit figure will be unaffected (except for any funds it spends to advertise against required returnable bottles). It is currently estimated that the bill has a 50-50 chance of passing. Several political analysts contacted by A.C.C. agree that the firm's advertising campaign will reduce the odds of the referendum's being successful from 50-50 to 45-55. Should the firm engage in the advertising (from a financial standpoint)?
Hint: Use a decision tree.

4. Suppose General Electric is considering locating a plant in a developing country that

has a history of political instability. The plant appears to have substantial potential, as it will serve to give G.E. an advantage in a rapidly developing area. At the same time G.E. wants to know the expected payoff before it decides to locate the plant. G.E. analysts break the numerous possible outcomes down to the following five major outcomes with the attached probabilities of occurrence and payoffs. One, the plant can be operated as specified in the contract, .3, $37,500,000. Two, the plant can be partially nationalized with adequate compensation, .2, $28,300,000. Three, the plant can be completely nationalized with adequate compensation, .2, $63,500,000. Four, the plant can be nationalized with inadequate compensation, .2, $6,000,000. Five, the plant can be nationalized with no compensation, .1, −$63,500,000. Based on your expected payoff computation, would you advise G.E. to build the plant?

5. Recently, you were hired by Blue Sheet, a real estate investment company, as an investment analyst. On the first day of your job Aaron Stowe, the owner, needed your advice on the following two real estate investments. One is a large office complex being built near the edge of the city's expanding south side. The discounted cash flow that this investment will generate depends on the economic conditions in the local area. In the very likely event (a probability of .8) that the currently healthy local economy continues in its present state, the investment will return $1,355,000 current value. If the local economy falters (a .2 probability), the investment will return $815,000.

   The second investment involves purchasing a large piece of real estate 25 miles southeast of town. The purchase is purely speculative. There is a chance (estimated at 1 in 10 by knowledgeable sources) that the government will construct a military base near this site. If the government does build, the current value of the gain will be $25,000,000. However, since land prices have already risen because of the speculation, a decision by the government to build elsewhere would result in a loss of $1,200,000.

   a. Which investment would you recommend based on the expected monetary value of each investment?

   b. Would you use expected monetary value as the criterion in this case? Why?

6. The Federal Drug Administration (FDA) is taking strong steps to enforce "truth in advertising" due to many false claims in television advertisements of over-the-counter drugs. Naturally, Emmett Watson, Standard Pharmaceutical's controller, and Laura Ellsworth, marketing manager, were concerned with the issue of whether or not to engage in television advertising of its over-the-counter birth control devices. Both wanted to avoid the adverse public reaction to such advertising. The marketing and controller groups eventually agreed that the infinite number of possible public reactions could be broken down into five representative positions: (1) strongly unfavorable, (2) unfavorable, (3) neutral, (4) favorable, or (5) strongly favorable. After two more meetings which lasted a total of seven hours, both groups agreed on the following net cash payoffs for each of the five outcomes: (1) −$10,000,000 (2) −$4,000,000 (3) $1,000,000 (4) $5,000,000, and (5) $6,000,000.

   Unfortunately the two groups were unable to agree on the likelihood of each outcome. The faction headed by Emmett Watson assigned the following probabilities: (1) .4, (2) .2, (3) .05, (4) .3, and (5) .05. The second faction, led by Laura Ellsworth, argued for the probabilities: (1) .1, (2) .15, (3) .2, (4) .35, and (5) .2.

What is the expected value of the advertising campaign to the Watson group? To the Ellsworth group?

7. In an attempt to resolve the conflict presented in Problem 6, a laboratory test of the advertising campaign was proposed and an independent consulting firm was contacted. The firm stated that their testing service, which consisted of monitoring key changes in the physiology of a large random sample of consumers, would cost $125,000. They stated that, based on numerous previous studies, they would identify either extreme reaction correctly 90 percent of the time. If an extreme reaction occurred and was misidentified, it would be identified as the adjacent category. An actual extreme reaction would never be miscategorized as neutral or on the opposite side of the continuum. An actual reaction of unfavorable or favorable would be correctly identified 70 percent of the time. Mistakes would be equally divided between the neutral category and the extreme category. An actual neutral reaction would be correctly identified 60 percent of the time. Mistakes would be divided with 15 percent going into each adjacent category and 5 percent going into each extreme.

   a. Is the proposed research a "good buy" from Watson's point of view? From Ellsworth's?

   b. If the research indicated a strongly favorable reaction, would the two groups agree to introduce the campaign?

   c. If the research indicated a neutral reaction, what would each group recommend concerning the campaign? How can you explain this?

   d. If the results indicated an unfavorable reaction, would both groups agree not to run the campaign?

8. To curb its recent slump, Atlantic Airways petitioned the Civil Aeronautics Board (CAB) with regard to several changes in fares and schedules. The CAB gave permission to Atlantic to add a daily Student Special flight on its New York–Paris–Rome route. The Student Special will provide minimum comforts, i.e., no frills, and more crowded seating than the regular flights. Fares on this flight will be 25 percent lower than the standard fares. If at least 75 percent of the passengers on these flights are passengers who would *not* have flown on a regular Atlantic flight, the firm will realize a $2 million profit. Atlantic's executive committee felt that there was a .7 probability that this would be the outcome. If less than 75 percent of the passengers are added passengers, Atlantic expected to lose $1.4 million.

   a. Should the flight be added?

   In response to the situation described above, the research department proposed an extensive survey of current Atlantic passengers, college students who had flown to Europe in the past two years, and college students in general. The research director estimated that the survey would predict whether or not 75 percent of the passengers were added passengers with 80 percent accuracy. The cost of the proposed research is $100,000.

   b. Should the survey be conducted?

9. Athens Rope Corporation is the world's third largest producer of rope, twine, netting, and kindred products. Originally all of its production was based on sisal, a plant fiber grown in arid regions. All of Athens' supply has been purchased in Africa for the past 32 years. Although 60 percent of the firm's products are now based on synthetic fibers, sisal products still account for the remaining 40 percent of sales.

The profits derived from the sisal-based products have varied widely from year to year due to fluctuation in the price of African sisal. The price of the finished products could not be altered to meet these shifts since the finished products were in direct competition with products made from synthetic fibers as well as sisal fibers from other parts of the world.

The trade minister of the recently independent African nation from which Athens secures 60 percent of its sisal supply recently approached Athens with a proposal. His government was taking over the marketing of his country's sisal output. The wide price fluctuations had proved damaging to the country's economy in the past and therefore he was willing to enter into a fixed-price agreement with Athens. Basically, the proposal called for Athens Rope to purchase all of its sisal requirements (or 250,000 tons, whichever was larger) from this country for the next 10 years. The price of the sisal would be the average of the prices for the past 5 years.

The company was impressed with the proposal. However, the length of the contract was of some concern. It was possible that new synthetic fibers might render sisal obsolete. This was certainly not likely to happen in the first 5 years, but it might occur in years 6 to 10. The research and development department and accounting department developed the following probability estimates and payoffs.

	Year					
	6	7	8	9	10	After 10
Pr	.05	.07	.12	.14	.20	.42
Payoff (in millions of dollars)	−14	−8	−2	+4	+10	+16

Should Athens Rope sign the contract?

10. As a plant manager for Herlon, Inc., you are worried about upcoming labor negotiations. The plant's labor contract expires in three months. Current inventories are extremely low and the plant's two-shift, six-day-a-week operation is not sufficient to build up the inventories to any significant degree. However, due to seasonal fluctuations, demand will diminish in about four months. In the meantime, the two-shift operation is adequate.

    Adding a third shift for a short period of time would be a very expensive operation. However, if labor negotiations fail and the firm is caught in a strike with limited inventories, it will suffer heavy losses. The chief labor negotiator for the firm feels that there is a .6 probability that there will be no strike, a .25 probability that there will be a minor, one-week strike, and a .15 probability that there will be a longer strike with more serious consequences.

    The estimated profits with the third shift given the three possibilities stated above are −$650,000, −$100,000, and +$1,700,000, respectively. Without the third shift, the estimated profits are $1,200,000, $300,000, −$2,700,000. Should the third shift be added?

11. Rutherford Manufacturing has decided to buy an additional manufacturing plant in a Southern state. The only decision is whether to buy this spring or wait a year until next spring. The firm will receive a 10 percent tax credit (valued by the firm at $250,000) if it buys this year. There is an 80 percent chance that this tax credit will

also be available next year. A complicating factor is defense spending for conventional weapons. If it stays the same or increases, the new plant will operate profitably at once. However, if defense spending in this area drops by 5 percent or more, the plant will have to remain idle for the first year. Rutherford's management estimates that this would cost the firm $750,000. However, if spending remains at or above current levels, the firm will realize an additional profit of $600,000 during this year. Management estimates the odds as 50-50 on a decrease in spending of 5 percent or more.

When should Rutherford buy the plant?

12. After his return from a two-week vacation in Hawaii, Bob Pickett found the following memo on his desk.

MEMO: From the desk of Roberta Blueforte, Production Manager

TO: Bob Pickett, Chief Analyst

Hope you had a nice vacation. In your absence, we completed the evaluation of your brainchild "Angels-on-the-Run" doll. The evaluation committee is unanimous in assigning the doll a .8 probability of success, which means a profit of $800,000. If the doll is not accepted, the firm can expect to lose $120,000. Not introducing the product would generate neither profits nor losses. What is your opinion now?

(SIGNED) *Blueforte*

Pickett replied with the following:

MEMO: From the desk of Bob Pickett, Chief Analyst

TO: Roberta Blueforte, Production Manager

I agree with the evaluation committee's report, but I feel that the doll should be tested at the upcoming trade show to ascertain buyers' interests.

(SIGNED) *Pickett*

cc: Rick Matlowe, Production Engineer
Kay Warner, Marketing Manager

Pickett was interested in this demonstration of the doll at the trade show because he knew that these shows correctly predict product successes and failures 80 percent of the time. The cost of this demonstration would be $2500.

The next morning Pickett received the following comments from Rick Matlowe and Kay Warner. Rick Matlowe commented, "We're 80 percent sure we've got a winner. Your test is only 80 percent accurate, so we gain nothing by conducting it, since we are already at the 80 percent level." Kay Warner added, "Nothing your test could show would alter our decision. That is, even if the test is negative, the best decision would be to introduce the product. Therefore, why conduct the test?" How should Pickett respond to their comments?

Chapter 10

Simple Linear Regression

10.1 SITUATION I: WHY?

Suppose that you have just been hired by one of the large commercial banks in your state. Your boss calls you into her office and tells you that there were many applicants for your position; however, you were selected because certain key individuals within the bank noticed that you have had a course in statistics. The bank has had a problem for some time and it is believed that the application of statistical techniques may actually provide the solution to the problem—thus your selection. It turns out that the problem is branch bank location. In the past the decision as to whether or not to locate a branch bank in a particular neighborhood was merely committee consensus. It seems as though the committee has made a few bad branch bank location decisions lately and there is considerable interest in what you can come up with. Your boss continues to fill you in. There is a metropolitan area in your state of approximately 250,000 people. Your bank presently has 15 branches in this area. However, there has been some new building (residential housing) in the northeast section, and you do not have a branch located in that area. The immediate question is whether or not to build a branch in this northeast locality. You are informed that your boss and the committee expect a recommendation report from you, relative to this particular decision, within the month!

After the initial shock of your assignment wears off, you realize that you are in need of more information. At your request, you are told (1) that your bank has a policy that all branches must be of equal staff size and that the *buildings must be of approximately equal cost, and (2) that for the most part, branch bank profitability depends upon total deposits (both time deposits and demand deposits). It has been found that if a branch's total deposits are greater than or equal to \$2.5 million, the branch is able to show a profit that is sufficient for the bank committee to be happy about that particular branch.*

At this point it occurs to you that what you really want to do is to decide whether or not the total deposits for the proposed branch in the northeast section of the metropolitan area will equal or exceed \$2.5 million. If you think so, you will want to recommend that the branch be built. Otherwise, your recommendation should be negative. What do you think? Would the total deposits for the proposed branch equal or exceed \$2.5 million? *If you can answer the question at this point, then you know something that we don't. We are trying to predict or estimate something that has not happened. Therefore, the value that we seek is not observable* at this time. *However, there* may *be another variable that is observable at the present and that may be related in some manner to what the future total deposits of the proposed branch would actually be. Suppose that after some thought, we arrive at the conclusion that total affluence of the relevant neighborhood would more than likely be related to total deposits. Let us assume that we can define the relevant neighborhood for any branch bank by examining a map of the area and the location of the nearest alternative branch. Looking for a measure of total affluence of a particular neighborhood, we decide that a viable measure might be the total assessed value of residential units. This measure is chosen because we expect income (which is not readily available) to be strongly related to the assessed value of the home or apartment in which a family resides. The assessed value of residential units is available at the county assessor's office. Therefore, we define the* quantitative variable x *to be the total assessed value of residential units in that area.*

Now that we have an observable quantitative variable that is believed to be related to total deposits, we need some means of linking the two, or estimating just what the relationship is. In order to begin, we need observations on both total deposits of a branch bank, which we will call y, *and our defined variable* x. *Believing that the 15 existing branches in the particular metropolitan area are the most relevant, we would obtain both values,* x *and* y, *for each of the existing branches. The data are shown in Table 10-1.*

As we look at the data in Table 10-1, it becomes apparent that values of x *and* y *indicate the relationship suspected. That is, it appears that branches with the larger affluence values* x *also have the larger total deposits* y. *But, how could we estimate the apparent relationship? No doubt the best first step would be to plot the data on a two-dimensional grid as shown in Figure 10-1.*

The scatter diagram graphically depicts the degree (strength) and nature (form) of the relationship—should one exist. We can see from Figure 10-1 that (1) a relationship appears to be present and (2) the relationship looks to be linear. That is, the pattern shown in Figure 10-1 seems to suggest that total deposits y *and total assessed value of residential units* x *are linearly related. In*

Table 10-1 Total Deposits and Affluence of Relevant Areas for Each of the Existing Branch Banks (In Millions of Dollars)

Branch	Total Deposits y	Total Assessed Value of Residential Units x
1	$3.1	$41.1
2	4.0	66.0
3	2.6	35.1
4	2.3	14.0
5	2.9	47.9
6	3.9	77.9
7	3.3	57.8
8	2.7	30.6
9	3.1	36.0
10	4.3	72.4
11	3.5	64.2
12	2.2	22.5
13	3.8	70.0
14	3.3	42.2
15	3.7	53.0

order to use the apparent relationship to predict, we must first estimate the relationship. The estimation of the relationship will be accomplished by somehow fitting a line to the points that appear in the scatter diagram. We could, of course, visually fit the line by moving a ruler about on the scatter diagram until we thought we had found the "best" line. And, the visually fit line may be quite satisfactory. However, each person would fit or draw his or her line, and no two lines would be exactly the same. From an objective point of view, this situation would seem to be lacking. What we need is some objective method of estimating a relationship between quantitative variables.

REGRESSION ANALYSIS IS A TECHNIQUE DESIGNED TO ASSIST IN THE EMPIRICAL ESTIMATION OF FUNCTIONAL RELATIONSHIPS BETWEEN QUANTITATIVE VARIABLES.

10.2 FITTING A LINE: HOW?

Consider for a minute just what property you would examine if you were to visually fit or draw a line through the data points. As we are really trying to predict the value of y given a specific value of x, the important measure would be the vertical distance between an observation and the fitted line. Note that this vertical distance is really error—the difference between the actual value of y and the predicted value using the estimated relationship. These distances or errors are shown in Figure 10-2 for our 15 branch bank observations.

You may recall that the equation of a straight line is

$$y = a + bx$$

where a = the y intercept

b = the slope

For our purpose we will denote the equation of the line *fitted* to sample data as

$$\hat{y} = b_0 + b_1x$$

where b_0 = the y intercept

b_1 = the slope

$\hat{y}$ = the point on the fitted line which corresponds to a specific x value

The vertical distance (deviation) between the ith observation of y and the fitted line would then be $y_i - \hat{y}_i$. One might expect that a "good" line is the one that minimizes the sum of the vertical distances for the sample data; namely,

$$\sum_{i=1}^{n} (y_i - \hat{y}_i)$$

However, *any* line that passes through the point with coordinates $(\bar{x}, \bar{y})$ will

Figure 10-1 Scatter Diagram of Branch Total Deposits and Area Affluence

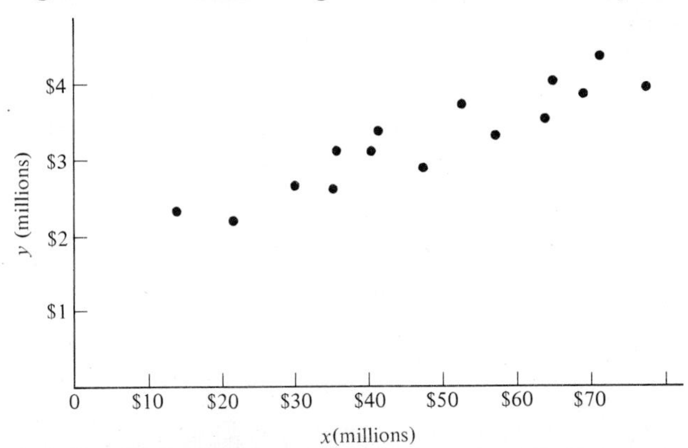

Figure 10-2

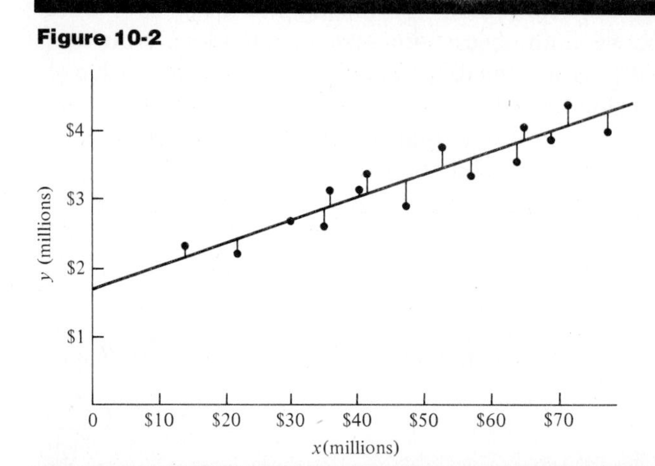

have a sum of the deviations equal to zero. It turns out that for any such line, the sum of the positive deviations exactly cancels out the sum of the negatives. We can get around this problem if we square the deviations before we sum. There is, of course, a direct relationship between the magnitude of a deviation and the magnitude of a deviation squared. Therefore, we might accept as our criterion for fitting a line to data points the minimization of the sum of the squared deviations.

Symbolically, we would like to find the equation of the line that would minimize

$$\sum_{i=1}^{n} (y_i - \hat{y}_i)^2$$

This is precisely what the *method of least squares* does.

> THE METHOD OF LEAST SQUARES DETERMINES THE EQUATION OF THE LINE THAT MINIMIZES THE SUM OF THE SQUARED DEVIATIONS OF THE SAMPLE OBSERVATIONS ABOUT THE FITTED LINE.

Method of Least Squares

Let us denote

$$\sum_{i=1}^{n} (y_i - \hat{y}_i)^2$$

as SSE (sum of squares error). Recall that $y_i - \hat{y}_i$ is the error or deviation of the observed value y_i about its predicted value $\hat{y}_i$. We then seek the values for b_0 and b_1 that will minimize SSE for a given set of observations. The procedure for deriving the equation for b_0 and b_1 is outlined below.*

That is, through differential calculus we are able to determine the equations for both b_0 and b_1 such that SSE will be minimized for a given set of data. The solution for b_1 is as follows:

$$b_1 = \frac{n\Sigma xy - \Sigma x\,\Sigma y}{n\Sigma x^2 - (\Sigma x)^2} \tag{10-1}$$

As the least squares line *must* pass through the point $(\bar{x}, \bar{y})$, it follows that,

$$\bar{y} = b_0 + b_1\bar{x}$$

Solving for b_0 we obtain

$$b_0 = \bar{y} - b_1\bar{x} \tag{10-2}$$

Thus we compute b_1 using Equation (10-1) and force b_0 by means of Equation (10-2). The results (b_0 and b_1) are the coefficients of the least squares line. It is the line that best fits the sample data. "Best" means that of all possible lines, the fitted line has the *minimum* SSE (sum of the squared errors or deviations). The fitted least squares line is known as the *estimated regression equation*.

10.3 SITUATION I REVISITED

We are now in a position to determine the line best fitted to the data for branch banks. Then we can use the regression equation (fitted line) to predict total deposits for the proposed branch. In order to use the regression equation to predict total deposits for the proposed branch, we would need the value of x, total assessed value of residential units in the proposed area. Let us suppose we have found this to be $28 million.

The necessary calculations are given in Table 10-2. By substituting the

$$\begin{aligned}
{}^*\text{SSE} &= \Sigma\,(y - \hat{y})^2 \\
&= \Sigma\,(y - (b_0 + b_1x))^2 \\
&= \Sigma\,y^2 - 2b_0\Sigma y - 2b_1\Sigma xy + nb_0^2 + 2b_0b_1\Sigma x + b_1^2\Sigma x^2
\end{aligned}$$

$$\frac{\partial \text{SSE}}{\partial b_0} = -2\Sigma y + 2nb_0 + 2b_1\Sigma x$$

$$\frac{\partial \text{SSE}}{\partial b_1} = -\,2\Sigma xy + 2b_0\Sigma x + 2b_1\Sigma x^2$$

If both partial derivatives are equated to zero and the resulting equations solved simultaneously for b_0 and b_1, we obtain the pair of values for b_0 and b_1 that will minimize SSE for the given set of observations.

Table 10-2

Branch	y	x	y^2	x^2	xy
1	3.1	41.1	9.61	1,689.21	127.41
2	4.0	66.0	16.00	4,356.00	264.00
3	2.6	35.1	6.76	1,232.01	91.26
4	2.3	14.0	5.29	196.00	32.20
5	2.9	47.9	8.41	2,294.41	138.91
6	3.9	77.9	15.21	6,068.41	303.81
7	3.3	57.8	10.89	3,340.84	190.74
8	2.7	30.6	7.29	936.36	82.62
9	3.1	36.0	9.61	1,296.00	111.60
10	4.3	72.4	18.49	5,241.64	311.32
11	3.5	64.2	12.25	4,121.64	224.70
12	2.2	22.5	4.84	506.25	49.50
13	3.8	70.0	14.44	4,900.00	266.00
14	3.3	42.2	10.89	1,780.84	139.26
15	3.7	53.0	13.69	2,809.00	196.10
	48.7	730.7	163.67	40,768.73	2,529.43

appropriate sums from Table 10-2 into Equation (10-1) we obtain

$$b_1 = \frac{n\Sigma xy - \Sigma x\Sigma y}{n\Sigma x^2 - (\Sigma x)^2} \tag{10-1}$$

$$= \frac{(15)(2{,}529.43) - (730.7)(48.7)}{(15)(40{,}768.73) - (730.7)^2}$$

$$= \frac{2{,}356.36}{77{,}608.46}$$

$$= +.03$$

Next, we divide Σy and Σx by n to obtain $\bar{y} = 3.25$ and $\bar{x} = 48.71$. With $\bar{y}$, $\bar{x}$, and b_1, we are able to substitute in Equation (10-2) to get

$$b_0 = \bar{y} - b_1\bar{x} \tag{10-2}$$

$$= 3.25 - (.03)(48.71)$$

$$= 3.25 - 1.46$$

$$= +1.79$$

Thus, the least squares (regression) equation is

$$\hat{y} = 1.79 + .03x \tag{10-3}$$

Figure 10-3 shows both the scatter diagram of our 15 observations and the fitted regression equation. Looking at Figure 10-3 we see that our fitted equation is reasonable. Note that the data points or sample observations tend to cluster about the least squares line. Figure 10-3 also enables us to obtain a feel for the graphic meaning of intercept and slope. Our estimate is that as x increases 1 unit, y increases .03 units. Or, for a \$1 million increase in total assessed value of residential units in an area, total deposits increase an average of (.03)(\$1,000,000) = \$30,000.

Now, knowing that the total assessed value of residential units for the proposed area x is \$28 million, our prediction of total deposits y is

$$\begin{aligned}\hat{y} &= 1.79 + (0.03)(28.0)\\ &= 1.79 + .84\\ &= \$2.63 \text{ million}\end{aligned}$$

Are you ready to make your recommendation? If total deposits for the proposed branch would equal or exceed \$2.5 million, your bank would like to go ahead on the branch. Our estimate or prediction is that the branch would attract total deposits of \$2.63 million. Thus, our analysis would indicate a positive recommendation. But, how sure are we that total deposits would equal or exceed \$2.5 million? What we have is a point estimate based upon an estimated linear relationship, which itself is based upon sample observations. *How "good" our point estimate is depends on (1) whether or not* x *and* y *are linearly related, (2) if they are related, the strength of the linear relationship, and (3) the sample size.*

There are really two questions we would like to ask at this point. First, is there really a linear relationship between x and y? Realize that the method of least squares will give a fitted line for any set of data, whether the variables are related or not. Second, if we decide there is a linear relationship, can we give an

Figure 10-3 Fitted Regression Equation

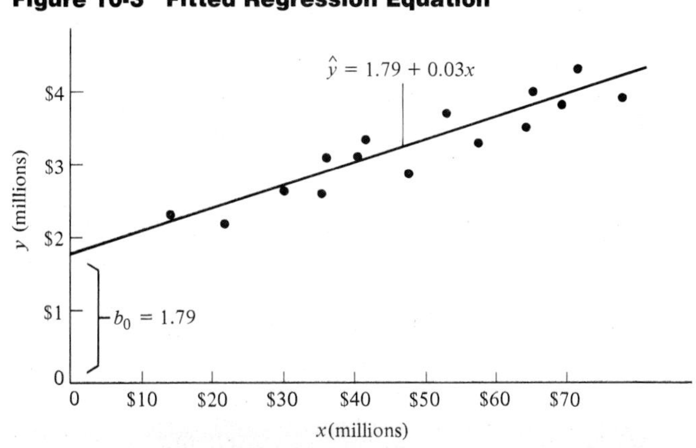

interval estimate along with the point estimate for total deposits of the proposed branch?

In order to answer these two questions, we must make certain assumptions about the "true" relationship between x and y.

10.4 REGRESSION-MODEL ASSUMPTIONS

Mathematically, the regression model is

$$y_i = \beta_o + \beta_1 x_i + \epsilon_i \tag{10-4}$$

where the random variable ϵ_i is called the *error term*. The error term is normally distributed with mean zero and variance σ^2 and errors in successive observations are statistically independent. What this really means is that for any given value of x, say x_p, the distribution of the y_i values is normal with mean $E(y) = \beta_0 + \beta_1 x_p$ and variance equal to σ^2. Figure 10-4 graphically depicts the simple linear regression model. Note that the distribution of y for any given value of x is normal with its mean *on* the regression line and that all such distributions have the *same* variance. In effect, the regression model can be viewed as a band of normal distributions with equal variances aligned in a straight line.

Assuming this model, we can now return to our two questions.

10.5 ARE x AND y LINEARLY RELATED?

Note that β_1 is the average change or effect on y for a 1-unit increase in x. If x and y are *not linearly related*, the average effect on y for a 1-unit increase in x

Figure 10-4

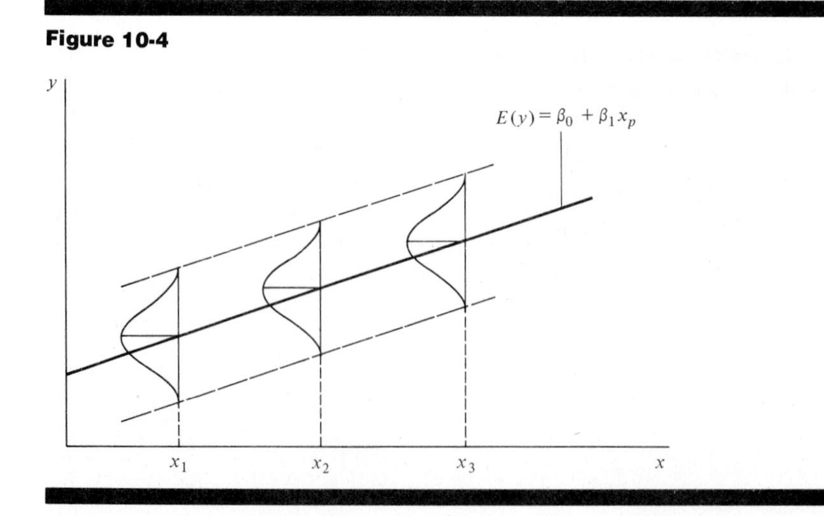

would be zero, $\beta_1 = 0$. Therefore, in order to investigate our question, "Are x and y linearly related?" we actually want to test the null hypothesis H_0: $\beta_1 = 0$.

If our population meets the assumptions of the regression model as given in Section 10-4, we have the following two results:

1. $E(b_1) = \beta_1$, which means that the slope of the fitted line is an unbiased estimate of the "true" regression slope (β_1).
2. $$\frac{b_1 - \beta_1}{s_{y \cdot x} \Big/ \sqrt{\sum_{i=1}^{n} (x_i - \bar{x})^2}}$$

   which is a random variable, follows the student t distribution with $n - 2$ degrees of freedom.

The term $s_{y \cdot x}$ in the above expression is known as the *standard error of the estimate* and is defined as

$$s_{y \cdot x} = \sqrt{\sum_{i=1}^{n} (y_i - \hat{y}_i)^2 \Big/ (n - 2)} \qquad (10\text{-}5)$$

The statistic $s_{y \cdot x}$ can be viewed as a measure of the standard deviation of the sample points about the fitted line. Computationally, we obtain the value of $s_{y \cdot x}$ by the following formula:

$$s_{y \cdot x} = \sqrt{(\Sigma y^2 - b_0 \Sigma y - b_1 \Sigma xy)/(n - 2)} \qquad (10\text{-}6)$$

Knowing the sampling distribution of the statistic

$$\frac{b_1 - \beta_1}{s_{y \cdot x} / \sqrt{\Sigma (x - \bar{x})^2}}$$

enables us to determine whether or not the conditional probability (given H_0 is true) of obtaining a sample result b_1 as extreme as the one we observed is less than or equal to a predetermined level α. Thus, we have a logical procedure for testing the null hypothesis.

Let us apply this procedure to our branch bank data in order to determine whether or not we believe total assessed value of residential units x and total deposits of a branch y are linearly related.

If we chose $\alpha = .05$, we will reject H_0: $\beta_1 = 0$ if our statistic

$$\frac{b_1 - 0}{s_{y \cdot x} / \sqrt{\Sigma (x - \bar{x})^2}}$$

is either less than -2.160 or greater than $+2.160$, since a t value of 2.160 cuts off .025 in the right tail of the student t distribution with 13 degrees of freedom.

If our hypothesis is true,

$$t = \frac{b_1}{s_{y \cdot x} / \sqrt{\Sigma(x - \bar{x})^2}}$$

follows the student t distribution with $n - 2$ or 13 degrees of freedom. This distribution is shown below.

Student t Distribution with 13 Degrees of Freedom

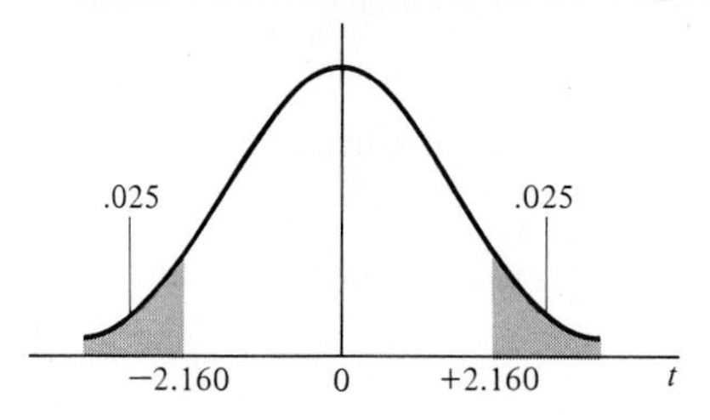

Here comes our test:

H_0: $\beta_1 = 0$

H_1: $\beta_1 \neq 0$

Let $\alpha = .05$

$$t = \frac{b_1}{s_{y \cdot x} / \sqrt{\Sigma(x - \bar{x})^2}} = \frac{+.03}{.217/\sqrt{5173.9}} = \frac{+.03}{.217/71.9} = +9.94$$

We have the value for b_1. However, we need to compute both $s_{y \cdot x}$ and $\Sigma(x - \bar{x})^2$. By Equation (10-6) we obtain

$$s^2_{y \cdot x} = (\Sigma y^2 - b_0 \Sigma y - b_1 \Sigma xy)/(n - 2)$$

$$= [(163.67) - (1.79)(48.7) - (.03)(2529.43)]/13$$

$$= .047$$

$$s_{y \cdot x} = \sqrt{.047}$$

$$= .217$$

Algebraically

$$\Sigma(x - \bar{x})^2 = \Sigma x^2 - (\Sigma x)^2/n$$

Thus for our data,

$$\Sigma(x - \bar{x})^2 = 40768.73 - 730.7^2/15$$

$$= 5173.9$$

As $+9.94 > +2.160$, we reject the null hypothesis H_0: $\beta_1 = 0$ and conclude that x and y *are* linearly related.

If our computed value of t had been in the acceptance region $-2.160 < t < +2.160$, we would have concluded that x and y *were not* linearly related, and it would make no sense to use the estimated relationship when we believe there is none.

10.6 PREDICTION INTERVAL

Now that we have decided that total assessed value of residential units x and total deposits of a branch y are linearly related, we should feel comfortable about using our estimated linear relation to predict the "would-be" total deposits for the proposed branch. Recall that, using the fitted regression equation, our prediction was \$2.63 million. However, this is merely a point estimate and you will recall from our discussion on estimation in Chapter 7 that a point estimate generally does not provide information on its precision. The standard operating procedure would be to accompany the point estimate with an *interval* estimate. The interval estimate is designed to indicate the *precision* or *goodness* of the estimate.

Expression (10-7) provides an interval estimate for a particular value of y given that x is equal to the specific value x_p.

$$\hat{y} \pm t(\alpha/2;\ n - 2)s_{y \cdot x}\sqrt{1 + \frac{1}{n} + \frac{(x_p - \bar{x})^2}{\Sigma(x - \bar{x})^2}} \tag{10-7}$$

Some statisticians refer to an interval resulting from Expression (10-7) as a *confidence* interval, while others consider the interval as a *prediction* interval. The interval is constructed exactly the same way that confidence intervals are formed. However, the interpretation of the interval is somewhat different. Intervals resulting from Expression (10-7) are not designed to capture a parameter as such. Rather, the interval is designed to capture a single observation of a random variable. No doubt the interval, which we will call a *prediction interval*, does in some way indicate the precision of our point estimate. Clearly the wider the prediction interval, for a given value of α, the less precise our estimate. And, the greater the difference we might expect between the actual value of y and our prediction.

Now, by applying Expression (10-7), let us generate a 90 percent prediction interval estimate for the future total deposits of our proposed branch, given that the total assessed value of residential units in the area is \$28 million.

$$\hat{y} \pm t(\alpha/2;\ n - 2)s_{y \cdot x}\sqrt{1 + \frac{1}{n} + \frac{(x_p - \bar{x})^2}{\Sigma(x - \bar{x})^2}}$$

$$2.63 \pm (1.771)(.217)\sqrt{1 + \frac{1}{15} + \frac{(28.00 - 48.71)^2}{5173.9}}$$

$$2.63 \pm .41$$

$$2.22 \leq y_i \leq 3.04$$

OK, what is your recommendation? Unfortunately our 90 percent prediction interval contains values on *both sides* of the crucial \$2.5 million figure. Had the lower limit of the prediction interval exceeded \$2.5 million, you would have made a positive recommendation. On the other hand, had both the point estimate and the upper limit of the interval been less than \$2.5 million, a negative recommendation would have been in order. However, as the problem turned out, our "best" estimate $\hat{y} = \$2.63$ million does exceed the crucial \$2.5 million but it is quite possible that the actual value could fall short. Perhaps the best recommendation would be to defer the decision until additional residential units are realized.

10.7 STRENGTH OF A LINEAR RELATIONSHIP

In most situations where we apply simple linear regression, we are also interested in a measure of the *strength* of the linear relationship. The magnitude of b_1 does not contain information on, or provide a measure of the strength of, the linear relationship. Note Figure 10-5, which shows two cases with identical regression equations but rather different strengths of linear relationship.

There is no doubt that the *standard error of the estimate* ($s_{y \cdot x}$) does indicate the strength of the linear relationship in the sample data. The closer $s_{y \cdot x}$ is to zero, the stronger is the linear relationship. However, the magnitude of $s_{y \cdot x}$ is dependent upon the scales of measurement for x and y. If you change the respective scales, the magnitude of $s_{y \cdot x}$ will change. This is a rather undesirable characteristic.

To overcome this problem a separate statistic has been defined.

$$r = \frac{n\Sigma xy - (\Sigma x)(\Sigma y)}{\sqrt{[n\Sigma x^2 - (\Sigma x)^2][n\Sigma y^2 - (\Sigma y)^2]}} \tag{10-8}$$

The statistic r is called the sample *coefficient of correlation*. And, r has the following properties:

1. The magnitude of r is *independent* of the scales of measurement for x and y.
2. $-1.0 \leq r \leq +1.0$

   a -1.0 value indicates a *perfect* (all points lie on a straight line) *inverse* linear relationship between x and y.

Figure 10-5

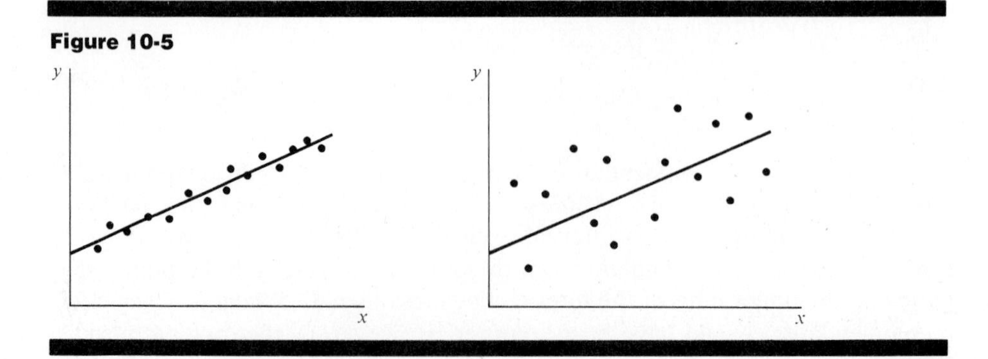

Figure 10-6 Scatter Diagrams and the Associated Correlation Coefficients

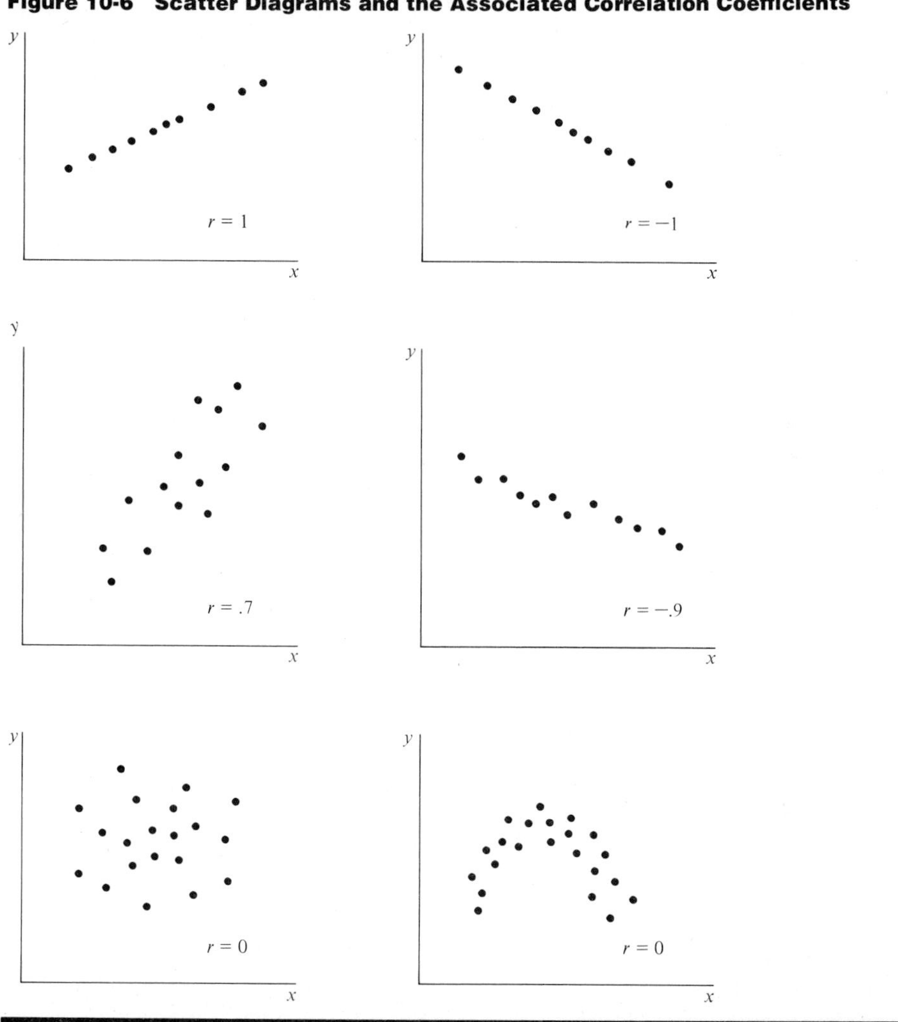

a +1.0 value indicates a *perfect direct* linear relationship between x and y.

a value of zero indicates *no* linear relationship between x and y.

3. r and b_1 must have the same sign. This follows directly from the fact that neither the denominator of r or b_1 can be negative, and r and b_1 have the same numerators. Thus, they must have the same sign. A result of this third property is that $r = 0$ when $b_1 = 0$.

Figure 10-6 shows a selection of scatter diagrams and the associated correlation coefficients.

Let us use Equation (10-8) to compute the sample coefficient of correlation for our branch bank data.

$$r = \frac{n\Sigma xy - (\Sigma x)(\Sigma y)}{\sqrt{[n\Sigma x^2 - (\Sigma x)^2][n\Sigma y^2 - (\Sigma y)^2]}} \qquad (10\text{-}8)$$

$$= \frac{(15)(2{,}529.43) - (730.7)(48.7)}{\sqrt{[(15)(40{,}768.73) - 730.7^2][(15)(163.67) - 48.7^2]}}$$

$$= \frac{2{,}356.4}{2{,}543.5}$$

$$= +.926$$

An r value of this magnitude would indicate a *strong positive linear relationship*.

Actually, the square of the coefficient of correlation r^2 has an interesting interpretation. The coefficient of correlation squared is called the *coefficient of determination* and represents the proportion of the total variation in y that is *explained*, or accounted for, by the fitted regression equation. The total variation in y is expressed as

$$\sum_{i=1}^{n} (y_i - \bar{y})^2$$

and can be partitioned into two components:

1. $\sum_{i=1}^{n} (\hat{y}_i - \bar{y})^2$

   which is the variation *explained* by the fitted equation

2. $\sum_{i=1}^{n} (y_i - \hat{y}_i)^2$

   which is the variation *unexplained* by the fitted equation

Figure 10-7 shows this partitioning of the total variation of y into its two parts or components.

Although Equation (10-8) does not look like it, it is algebraically equivalent to

$$\sqrt{\frac{\Sigma(\hat{y}_i - \bar{y})^2}{\Sigma(y_i - \bar{y})^2}}$$

Thus the coefficient of determination r^2 equals the ratio of the sum of the explained deviations squared to the sum of squared total deviations. In our example, $r^2 = +.926^2$, which is .857. Therefore, we can say that 85.7 percent of the variation in y is explained by the variable x.

Needless to say, as $-1.0 \leq r \leq +1.0$, then $0 \leq r^2 \leq +1.0$.

Figure 10-7 Partitioning of Variation of a Point into Explained and Unexplained

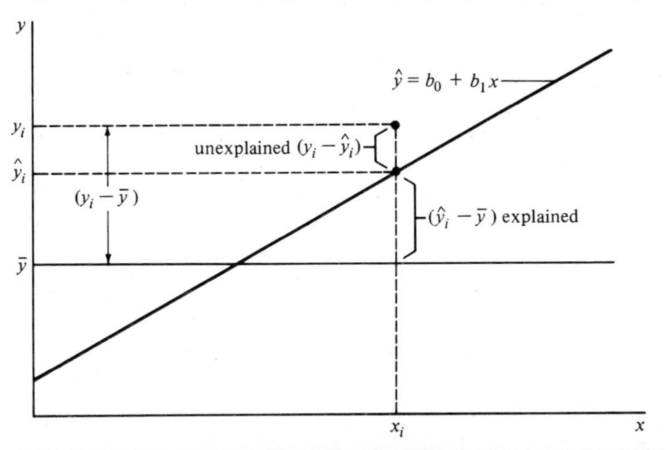

10.8 A COUPLE OF QUALIFICATIONS

This chapter has been restricted to simple linear regression and correlation. One of the first steps was to plot the data points on a scatter diagram to see if a line would be an appropriate fit to the sample data. In most situations it is unrealistic to expect the relationship between two variables to be *exactly linear*. However, this does not negate the value of linear regression. Consider the "true" relationship shown in Figure 10-8. Even though the true relationship is curvilinear, a linear fit would do quite well as an estimate of the relationship for x values in the interval $10 \leq x \leq 40$. Actually any relationship can be approximated by a straight line if the interval of x is sufficiently small.

This condition brings us to one of the dangers of applying linear regression. *Our statistical conclusions (inferences) are relevant only to the range of data in our sample*. Extrapolating beyond the range of data in our sample should be done *only with extreme caution*.

Consider Figure 10-8. Suppose we were to draw samples where $10 \leq x \leq 40$ and fit a linear regression equation to the sample observations. The fitted line might quite accurately represent the relationship between x and y, so long as we restricted its use to situations where $10 \leq x \leq 40$. However, if we were to extrapolate the estimated relationship and use the fitted line to predict a value of y given $x = 100$, our error of prediction would be enormous. When we extrapolate we make the assumption that the underlying relationship over a certain range, which we have estimated, continues in a region where we have no observations—mighty risky.

Figure 10-8

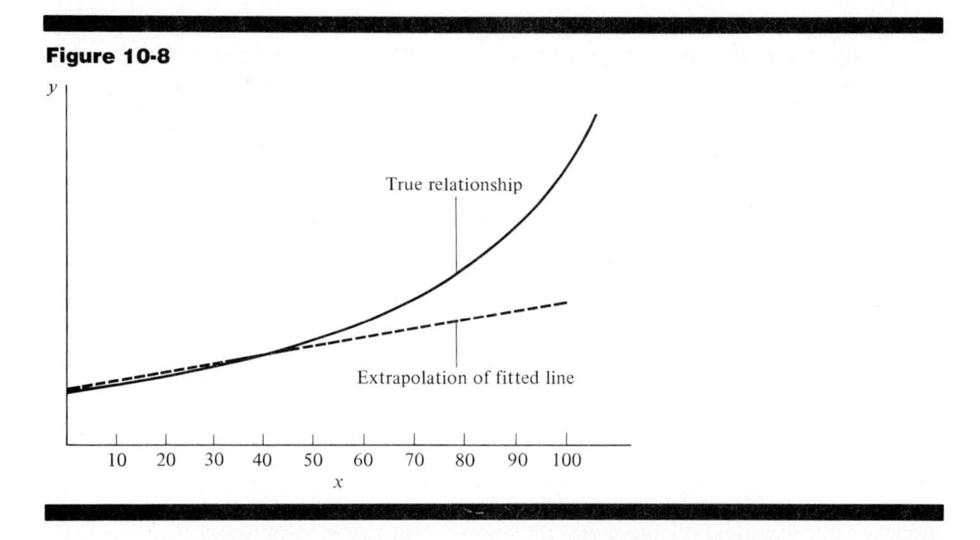

The second qualifier is that we must realize that rejecting the null hypothesis $\beta_1 = 0$ does *not prove* that x causes y. We do not statistically determine causality (cause and effect). Our statistical conclusion, if we reject H_0: $\beta_1 = 0$, would be that x and y appear to *move together*, but the apparent movement or relationship may be due to some other factor.

10.9 REGRESSION ESTIMATORS FOR FINITE POPULATIONS

We saw, in Situation I, how simple linear regression could be used to predict the value of one variable by capitalizing upon a second, related (we hope), auxiliary variable. Another interesting application of linear regression is the estimation of the mean and/or total of a finite population.

Suppose you are the controller for a small beer and wine distributing company. Among your duties is the preparation of the quarterly balance sheet and income statement. Because of state regulations relating to inventory tax, a physical count of inventory must be made on December 31. As a result, the inventory figure which appears on the year-end balance sheet (and which is used to determine the cost-of-goods-sold figure on the income statement) is based on the *actual* number of units on hand as of that date.

It is now March 31 and time for you to prepare the first quarter reports (balance sheet and income statement). Your firm carries 1000 items in inventory. An example of a single item would be Gallo Pink Chablis—half gallons. As it is time-consuming to physically count each unit for all 1000 items, and such a physical count is not required for reports other than year-end report, you decide to "estimate" cost of inventory on hand by drawing only a sample of the 1000 items. Let us suppose you draw a random sample of 25 items. For each of the 25 items, you do physically count the units on hand. Returning to the office, you determine the cost per item in your sample by multiplying the number of units on hand times your cost-per-unit figure for that item. The results of your sample are shown in Table 10-3.

Your objective is to estimate total cost for the 1000 items (total inventory) based upon your sample of 25 items. At this point there are two ways to proceed. First, recall from Chapter 7 that you could estimate the total cost by merely applying Equation (7-9).

Estimator of the Population Total τ:

$$\hat{\tau} = N\bar{y} = \frac{N\sum_{i=1}^{n} y_i}{n} \tag{7-9}$$

Table 10-3 Cost of Inventory per Item Sampled, March 31, 197X

Item	Cost y	y^2
1	$ 620	384,400
2	1,431	2,047,761
3	794	630,436
4	141	19,881
5	293	85,849
6	4,021	16,168,441
7	314	98,596
8	129	16,641
9	184	33,856
10	72	5,184
11	15	225
12	1,841	3,389,281
13	724	524,176
14	308	94,864
15	842	708,964
16	2,016	4,064,256
17	600	360,000
18	53	2,809
19	1,980	3,920,400
20	0	0
21	241	58,081
22	390	152,100
23	100	10,000
24	68	4,624
25	424	179,776
	$17,601	32,960,601

Estimated Variance of $N\bar{y}$:

$$\text{var}(N\bar{y}) = N^2 \frac{s^2}{n}\left(\frac{N-n}{N}\right) \tag{7-10}$$

where $$s^2 = \frac{\sum_{i=1}^{n}(y_i - \bar{y})^2}{n-1} = \frac{\Sigma y_i^2 - (\Sigma y_i)^2/n}{n-1}$$

N = the size of the finite population

If the n sample observations are the result of random sampling, Equation (7-9) gives an unbiased estimate of the true value of τ. The result of this application is as follows:

$$\hat{\tau} = (N\bar{y}) = \frac{N\Sigma y_i}{n} \tag{7-9}$$

$$= \frac{(1{,}000)(17{,}601)}{25}$$

$$= \$704{,}040$$

$$\text{var}(N\bar{y}) = N^2 \frac{s^2}{n}\left(\frac{N-n}{N}\right) \tag{7-10}$$

$$= 1{,}000^2 \frac{(857{,}033)}{25}\left(\frac{1{,}000 - 25}{1{,}000}\right)$$

$$= 33{,}423{,}980{,}000$$

where $$s^2 = \frac{32{,}960{,}601 - 17{,}601^2/25}{24}$$

$$= 857{,}033$$

An alternative approach would be to try to identify an auxiliary (related) variable, and to capitalize on whatever information the auxiliary variable contains to aid in the estimation. In our inventory example you might expect the cost of inventory on hand December 31 for a particular item to be related (correlated) with the cost of inventory on hand March 31 for the same item. Not only would this information be available for the 25 selected items on March 31, but the total for this auxiliary variable would be the cost-of-inventory figure which appeared on the December 31 balance sheet. Thus, let us suppose you return to your December 31 inventory records and retrieve the inventory-cost figures for the 25 selected items. These data and the appropriate sums are shown in Table 10-4. We can capitalize upon the information in x (December 31 cost of inventory) to aid our estimation of y (March 31 cost of inventory) by applying Equation (10-9).

Regression Estimator of the Population Total τ:

$$\hat{\tau}_{reg} = N[\bar{y} + b_1 (\mu_x - \bar{x})] \tag{10-9}$$

Estimated Variance of $\hat{\tau}_{reg}$.

$$\text{var}(\hat{\tau}_{reg}) = N^2\left(\frac{N - n}{N}\right) s^2 (1 - r^2)\left[\frac{1}{n} + \frac{(\mu_x - \bar{x})^2}{\Sigma(x - \bar{x})^2}\right] \tag{10-10}$$

where b_1 = Equation (10-1)

r = Equation (10-8)

$$\mu_x = \sum_{i=1}^{N} x_i/N$$

Table 10-4 Cost of Inventory per Item Sampled
(x = December 31 Cost, y = March 31 Cost)

Item	y	x	
1	$ 620	$ 510	
2	1,431	1,340	
3	794	601	
4	141	150	
5	293	308	
6	4,021	4,225	
7	314	107	Σx = 17,342
8	129	0	Σy = 17,601
9	184	300	Σx^2 = 34,718,316
10	72	0	Σy^2 = 32,960,601
11	15	0	Σxy = 33,119,147
12	1,841	2,500	
13	724	1,020	
14	308	280	
15	842	493	
16	2,016	1,542	
17	600	341	
18	53	200	
19	1,980	1,851	
20	0	500	
21	241	184	
22	390	521	
23	100	38	
24	68	10	
25	424	321	
	$17,601	$17,342	

In order to proceed, certain calculations are necessary. Suppose your total-cost-of-inventory figure for December 31 was $753,240. Then,

$$\mu_x = \frac{\$753{,}240}{1{,}000}$$
$$= \$753.24$$

$$b_1 = \frac{n\Sigma xy - \Sigma x \Sigma y}{n\Sigma x^2 - (\Sigma x)^2} \qquad (10\text{-}1)$$
$$= \frac{(25)(33{,}119{,}147) - (17{,}342)(17{,}601)}{(25)(34{,}718{,}316 - 17{,}342^2)}$$
$$= \frac{522{,}742{,}133}{567{,}212{,}936}$$
$$= .92$$

$$r = \frac{n\Sigma xy - (\Sigma x)(\Sigma y)}{\sqrt{[n\Sigma x^2 - (\Sigma x)^2][n\Sigma y^2 - (\Sigma y)^2]}} \qquad (10\text{-}8)$$
$$= \frac{(25)(33{,}119{,}147) - (17{,}342)(17{,}601)}{\sqrt{[(25)(34{,}718{,}316) - 17{,}342^2][(25)(32{,}960{,}601 - 17{,}601^2]}}$$
$$= .968$$

$$\sum_{i=1}^{n} (x_i - \bar{x})^2 = \Sigma x_i^2 - (\Sigma x_i)^2/n$$
$$= 34{,}718{,}316 - 17{,}342^2/25$$
$$= 22{,}688{,}518$$

Now, by applying Equations (10-9) and (10-10), we obtain the following:

$$\hat{\tau}_{\text{reg}} = 1{,}000\,[704.04 + .92(753.24 - 693.68)]$$
$$= \$758{,}840$$

$$\text{var}(\hat{\tau}_{\text{reg}}) = 1{,}000^2 \left(\frac{1{,}000 - 25}{1{,}000}\right) (857{,}033)(1 - .968^2) \left[\frac{1}{25} + \frac{(753.24 - 693.68)^2}{22{,}688{,}518}\right]$$
$$= 2{,}114{,}086{,}000$$

This regression estimator, Equation (10-9) is generally biased: That is, $E(\hat{\tau}_{\text{reg}}) \neq \tau$. However, the bias

1. Decreases as n is increased
2. Decreases as r approaches ± 1
3. Is zero if the relationship between y and x is linear

In our example one might expect the bias to be negligible as $r = +.968$. An r value of this magnitude would indicate a strong *linear* relationship.

We have seen two different ways of estimating the same parameter (total cost of inventory on hand March 31). The first approach was without the aid of an auxiliary variable. The second method utilized the information in an auxiliary variable through simple linear regression. Yes, the second approach (regression estimator) does require considerably more computations. And, when we use the regression estimator, we face the possibility of bias.

At this point you must ask "why?" Why go through the extra computations and worry about bias? The answer is in the precision (variances) of the two estimators. Look at the ratio of the two estimated variances:

$$\frac{\text{var}(N\bar{y})}{\text{var}(\hat{\tau}_{\text{reg}})} = \frac{33{,}423{,}980{,}000}{2{,}114{,}086{,}000}$$

$$= 15.8$$

The estimated variance of the simple estimate (without the aid of the auxiliary variable) is approximately 15.8 times as large as the estimated variance for the regression estimate. In other words, it would require a sample of approximately $(15.8)(25) = 395$ using the simple estimator Equation (7-9) to attain the same precision (variance) as 25 observations using the auxiliary variable and Equation (10-9).

Now that is impressive! When observations are costly, an n of 25 versus an n of 395 would be important!

The reason for such a large gain in precision is the strong ($r = +.968$) linear relationship between x and y. Had the coefficient of correlation been zero, there would have been *no* gains. If $r = 0$, then $b_1 = 0$, and under this condition Equation (10-9) becomes $\hat{\tau}_{\text{reg}} = N\bar{y}$, which is Equation (7-9). It's when r approaches ± 1 that *real* gains result.

10.10 IDENTIFICATION OF APPLICATIONS: WHEN?

Ordinarily the variable of interest, which we have called y, is known as the *dependent variable*, whereas our auxiliary variable x is referred to as the *independent variable*. The important characteristics of a situation where simple linear regression might be applied are:

1. Both dependent and independent variables are *quantitative*.
2. The plot of the sample data on a scatter diagram gives the appearance of a *linear* relationship.

In Situation I simple linear regression was used to "predict" an unobservable value of one variable (total deposits of a proposed branch bank) by identifying an observable variable (total assessed value of residential units in the relevant area) which appeared to be linearly related. In the final example in this chapter (controller for a beer and wine distributing company) linear regression was used as an aid in estimating the total of a quantitative variable for a finite population.

Another rather classic type of situation where regression would be of value is where the dependent variable (the variable of interest) is expensive to measure because the measurement is *time-consuming* or perhaps *destructive*. If an auxiliary variable, which is relatively inexpensive to measure, can be found that is highly correlated with the variable of interest, it may be possible to obtain a more precise estimate (smaller variance) for a fixed cost. This may be accomplished by reducing the size of the sample of the expensive measurement and capitalizing upon the inexpensive information in the auxiliary variable through regression.

Through regression we extract, capitalize on, or use the information contained in the independent variable to aid or assist in predicting or estimating the value of the dependent variable. This is accomplished by estimating the underlying relationship.

PROBLEMS

1. Identify an application of simple linear regression analysis. What conditions are present in your identified situation that would make this application appropriate and of value?

2. Morning House is a mail-order firm which carries a wide range of rather expensive art objects for homes and offices. It operates by advertising a particular item either in selected magazines or in a direct-mail program. Suppose the sales response varies widely by item and the firm's management has been unable to predict in advance which items will sell well and which will not. Consequently the firm frequently experiences either stock-outs or excessive inventories.

   For many of the products Morning House sells, it is possible to order a limited amount for inventory and to place a subsequent order for delivery within two weeks. Thus, if the firm could make an early prediction of the ultimate sales of a product, its inventory problems would be greatly reduced. Since it takes approximately six weeks to receive 90 percent of the response to a given campaign, an accurate prediction of total sales made as late as the end of the first week of receiving orders would be useful.

   The first week's sales and total sales of the last 12 campaigns of the firm are shown below. Can the first week's sales be used to predict total sales?

	First Week's Sales	Total Sales
1	32	167
2	20	91
3	114	560
4	66	335
5	18	70
6	125	650
7	83	401
8	65	320
9	94	470
10	5	15
11	39	210
12	50	265

a. Plot a scatter diagram.

b. Obtain the regression equation of total sales on the first week's sales.

c. Test the null hypothesis H_0: $\beta_1 = 0$ at the .05 level. What is the use of this test?

d. Obtain the sample coefficient of correlation.

e. Give a 95 percent prediction interval for the total sales for a specific campaign or item that had sales of 40 for the first week.

f. Based on your findings in parts *a* through *e*, what is your opinion about using the first week's sales to predict total sales?

3. Assume that shortly after acquiring your undergraduate degree in finance, you were hired by Macy's Department Stores as a trainee in the finance department. After a six-month training period you are assigned a special project involving credit card customers. In order to more closely monitor cash flow as well as to effectively use direct-mail programs aimed at credit card purchasers, it is necessary to have an accurate estimate of the first year's purchases of new credit card holders.

   It would be fairly simple to compute the average first-year purchases of new credit card holders based on historical data. However, past attempts using this approach have often been less accurate than desired. Furthermore, it is desirable to be able to predict individual purchase levels in advance. Large purchasers could then be sent relatively frequent direct-mail promotions.

   You decide to investigate the relationship between two possible known variables and an individual's future credit card purchases. These variables are total household income taken from the credit card application form and total credit purchases during the first month of card ownership. A simple random sample of 25 credit card holders that have owned the card for at least 12 months is drawn and the following data recorded.

Total 12-month Purchases	Total Household Income (After Tax) (In Thousands of Dollars)	Total First-Month Purchases
$ 1,735	$14.2	$141
6,381	18.5	534

Total 12-month Purchases	Total Household Income (After Tax) (In Thousands of Dollars)	Total First-Month Purchases
2,972	17.6	248
6,439	25.0	535
7,356	15.3	613
5,014	16.8	406
2,349	17.0	187
7,018	18.9	568
9,284	20.2	844
7,498	18.3	623
696	16.9	058
9,057	13.8	727
2,036	15.9	176
9,549	16.0	933
8,921	21.3	811
10,560	21.3	921
5,639	14.6	449
3,576	15.9	298
11,844	26.3	987
1,768	17.8	136
591	19.3	047
3,304	20.0	263
3,575	21.8	287
1,918	22.0	154
7,562	23.1	612

a. Identify the dependent (response) and independent (explanatory) variables.

b. Plot scatter diagrams using each of the independent variables.

c. Obtain the functional relationships between response and each of the explanatory variables separately.

d. Test for the linear relationships in each case (that is, in each case test the null hypothesis that slope is zero). Use $\alpha = .05$.

e. Obtain coefficients of determination in each case.

f. If total after tax household income is 20,500, what would be the total 12-month purchases?

g. If first-month purchases of an individual total $200, what would this person's total 12-month purchases be?

h. Based on your findings in parts *a* through *g*, which, if any, of these variables appear to be useful predictors?

4. Suppose the University of Oregon, which is funded based on student FTE (full time equivalent, i.e., 15 credit hours equals 1 FTE), is attempting to predict total FTE

based on early applications. If this prediction can be made accurately, staffing and supply decisions can be made much more efficiently, and the university can effectively argue about its budget requirements before the legislature. Based on the number of applications received by March 1 and the number of FTE for the next fall for the previous eight years (as shown below), do you think that early applications serve as a useful predictor of subsequent FTE?

Year	No. of Applications Received by March 1	No. of Student FTE for the Next Fall
1	18,000	14,400
2	19,550	16,100
3	20,150	17,040
4	23,200	19,290
5	24,600	20,728
6	25,500	21,600
7	26,750	22,630
8	28,000	24,920

The following points will be helpful in organizing your thinking:

a. Find the appropriate functional relationship between these variables.

b. Plot the points and graph the estimated functional relationship.

c. Check whether b_1 is statistically significant. (Use $\alpha = .05$.)

d. Obtain the coefficient of correlation.

Now you can answer the question posed to you earlier. Give a 95 percent prediction interval estimate of the number of student FTE for next fall if the number of applications received by March is 27,550.

5. The paint department of Hempwell's department store has two major sales per year, one in the spring and one in the fall. The price of the paint has been reduced from 10 to 30 percent during these sales. The firm's management has decided to reduce the price of paint by 50 percent for this spring's sale. Although Hempwell's will not show a profit on the paint at this discount level, the firm's marketing manager hopes to generate enough traffic to ensure a strong profit on sales of auxiliary items such as brushes, masking tape, and paint thinner.

As a newly hired staff member in the acounting department, you are eager to impress the firm's price-policy-making committee. One of the members has raised the following point: "Nothing irritates a customer more than going to a sale and finding that the advertised item has been sold out. We must be sure that we have an adequate supply of paint. However, we don't want to have an excessive supply or our inventory costs over the summer will offset any profits from the sale."

Each member of the committee is asked to examine the past sales information and to report back to the full committee the following week.

Examine the following data (arranged according to discount level) and prepare a report for the committee.

Percentage Discount	Percentage Change in Sales
10	3.43
12	3.75
14	4.52
16	5.13
18	5.94
20	6.35
22	7.99
24	8.23
26	9.46
28	10.35
30	12.00

The chief accountant believes in a clean and organized report. Therefore, you should analyze the data in the following order.

a. Get an idea of the relationship between the two variables using a scatter plot.

b. Using the technique you have learned, estimate the functional relationship.

c. Find out whether the relationship is appropriate at a prechosen significance level.

d. Also find out how much variation is explained by your regression equation.

Now you are ready to prepare the report for the committee. Do not forget to make your own recommendations which might be useful to the price-policy committee.

6. J. C. Penney's has, say, 1600 retail outlets in the United States. Suppose that this firm conducted a major sales-training effort designed to improve the effectiveness of in-store sales personnel. Prior to the training program, traffic counts of the number of individuals passing through selected points in a store were conducted throughout the chain. Based on the traffic counts, an estimate of "sales per shopper" was calculated. Immediately after the sales-training program, the "sales per shopper" was recalculated. An analysis of the results indicated that the sales training had been effective.

   Six months have passed since the completion of the second measurement. As a staff member in the personnel department, you have been asked to determine the long-term effect, if any, of the training program. Unfortunately, you were told that "the first two studies cost a fortune, you must minimize expenses on this follow up, without giving up the reliability (precision) of your estimate."

   Total and per-outlet sales figures are readily available. The problem is to determine traffic counts, an expensive process. Finally, you take a sample of 20 stores and commission a traffic count in each. The results of these counts as well as the results of the previous count for each of these stores and the total count from the previous study are presented below.

Store	Current Traffic per Hour	Previous Traffic per Hour
1	63	56
2	56	51
3	36	26
4	49	32

Store	Current Traffic per Hour	Previous Traffic per Hour
5	52	38
6	51	43
7	45	31
8	39	28
9	46	31
10	47	35
11	36	30
12	28	21
13	42	38
14	46	35
15	53	43
16	60	54
17	62	55
18	31	21
19	42	31
20	21	16

Mean traffic per hour for *all* 1600 stores at the previous time was 36.8.

a. Estimate the total traffic per hour without the use of the previous data.

b. Estimate the variance of your point estimator in part *a*.

c. Estimate the total traffic per hour with the use of the previous data.

d. Obtain the estimated variance for your estimator in part *c*.

e. Which estimator would you use and why?

7. The director for a chain of self-service hardware stores was very concerned about the wide variation in profitability among the firm's retail outlets. She asked you to look into the matter. As a first step you talk with a number of people in the firm and find that there are two very different models or theories of why some stores are more successful than others.

One group believes that location, particularly as it relates to the flow of traffic by the store, is the key variable. Its basic model is that more traffic causes more total sales which causes more profits. A second group in the firm thinks that sales effort on the part of the sales clerks is the key variable. Its basic model is that sales effort by the sales clerks causes a higher average sale per customer, which leads to increased profits.

The two theories call for different types of managerial efforts. The location hypothesis suggests concentrating on opening new outlets in ''good'' locations and closing or selling outlets that are currently unprofitable. The sales-effort hypothesis suggests emphasizing the recruiting and training of store managers and sales clerks.

To determine which if either of the theories is valid, the average monthly sales for the past two years of each of the 17 outlets located in the home-office city are collected. From the municipal traffic department, you are able to secure average daily traffic counts from the past year for the streets running by 14 of the 17 outlets.

You also collect cash-register tapes from each of these 14 outlets and calculate the average purchase size. These data are presented below. What should you recommend to the director?

Store	Average Daily Auto Traffic (In Hundreds)	Average Purchase Size (In Dollars)	Average Monthly Sales (In Thousands of Dollars)
1	14.2	15.20	691.79
2	10.4	14.28	279.82
3	22.3	13.86	851.79
4	24.1	14.98	994.40
5	42.1	13.68	904.68
6	37.5	14.36	786.02
7	77.9	15.23	1712.40
8	99.5	16.21	1959.50
9	96.3	16.20	1974.03
10	89.5	15.20	1388.67
11	85.4	15.12	1568.50
12	28.9	15.16	866.30
13	63.5	13.22	1032.63
14	09.4	15.12	283.21

8. Household Products, Inc., sells a complete line of household products (cleaners, light bulbs, brushes, etc.) on a door-to-door basis. Salespeople are assigned exclusive sales territories. For several years the firm has been concerned about its inability to accurately evaluate the sales performance of individual sales representatives. Total sales per salesperson vary widely. However, the sales manager feels that at least some of this variation is due to differences in sales potential among the various territories.

   As a member of the accounting department, you were recently involved in establishing standard costs for many of the items that Household Products manufactures itself. Now your boss asks that you develop some standards for evaluating sales performance.

   By examining census-tract data you obtain estimates of total household income in a random sample of 20 sales territories. You also obtain last year's total sales in each of these territories. Can these data be used to develop standards for evaluating sales performance?

Territory No.	Total Sales (In Millions of Dollars)	Total Household Income (In Millions of Dollars)
1	2.12	110.2
2	13.26	1002.3
3	.48	16.3
4	2.89	116.8
5	13.82	1139.2
6	6.83	500.8
7	14.56	1468.3
8	15.00	1500.8
9	3.83	124.3
10	5.26	463.2
11	4.82	150.8

Territory No.	Total Sales (In Millions of Dollars)	Total Household Income (In Millions of Dollars)
12	3.89	132.3
13	10.06	931.4
14	9.83	821.3
15	1.36	68.2
16	11.29	888.8
17	22.85	2132.6
18	16.89	1782.9
19	.48	12.3
20	1.68	98.3

9. As a trainee in the financial department of Greenwall Plastics, a medium-sized producer of plastic products used primarily in the construction industry, you are involved in analyzing the results of an experiment that has caused some controversy among the firm's top-level managers.

For the past six months, the firm has utilized a variety of different payment requirements in 10 geographic regions served by the company. These have historically had similar levels of sales and firms paying within the 10-day limit. The firm's controller initiated the experimental change in the firm's traditional 2/10 net 30 policy. The expression "2/10 net 30" means the customer gets a 2 percent discount if he or she pays within 10 days; otherwise the total is due in 30 days. The controller's desire was to determine the effects of reducing the percentage discount offered for early payment. The marketing manager became quite upset with the proposed reduction, feeling that this was an important sales point with some customers. Therefore, a decision was made to run a six-month test on the impact of the discount rate on both prepayment and sales. You have been asked to analyze the results of the experiment and to prepare a report for management. What are your recommendations?

Region	Discount Rate	Sales	Percent of Firms Paying in 10 Days
1	Range from .5/10	98	4
	1.0/10	95	12
	1.5/10	93	45
	2.0/10	92	60
	2.5/10	97	80
2	Range from .5/10	121	6
	1.0/10	130	15
	1.5/10	125	43
	2.0/10	126	58
	2.5/10	123	78
3	Range from .5/10	86	5
	1.0/10	98	13
	1.5/10	101	42
	2.0/10	110	61
	2.5/10	126	82

Region	Discount Rate	Sales	Percent of Firms Paying in 10 Days
4	Range from .5/10	70	8
	1.0/10	81	16
	1.5/10	140	41
	2.0/10	158	60
	2.5/10	162	76
5	Range from 5./10	15	4
	1.0/10	26	11
	1.5/10	28	41
	2.0/10	32	53
	2.5/10	68	69
6	Range from .5/10	38	5
	1.0/10	48	13
	1.5/10	32	42
	2.0/10	62	58
	2.5/10	75	79
7	Range from .5/10	130	9
	1.0/10	128	15
	1.5/10	148	39
	2.0/10	156	72
	2.5/10	162	98
8	Range from .5/10	108	5
	1.0/10	121	16
	1.5/10	138	46
	2.0/10	162	65
	2.5/10	170	82
9	Range from .5/10	105	9
	1.0/10	126	17
	1.5/10	131	48
	2.0/10	170	52
	2.5/10	200	80
10	Range from .5/10	48	10
	1.0/10	59	15
	1.5/10	69	25
	2.0/10	80	40
	2.5/10	102	80

10. The accounting department of Unico Manufacturing is involved in developing an accurate estimate of the direct cost involved in producing 15,000 air-hose couplings. Unico has been involved in the manufacture and sale of couplings for over 20 years. Many of its orders are built to specification. That is, each order is, to some extent, unique.

 Most of Unico's sales are to large purchasers and are based on competitive bids. Competition within the industry has become severe in recent years. The increase in competition has produced a need for very accurate cost estimates. A bid that is higher than the competition will not be accepted, while a bid below direct costs loses money.

Unico is able to estimate the cost of materials per coupling with accuracy. However, the direct-labor component is not as clear. Labor rates are fixed by union contract, so the problem is to predict the unit direct-labor time for a production run of 15,000 couplings.

An examination of company records reveals the information shown below about the 15 most recent production runs of similar couplings.

Contract No.	Direct-Labor Time per Unit	Units Produced
1	11.0	1,000
2	10.3	2,000
3	10.2	4,000
4	9.6	6,000
5	9.3	7,000
6	9.2	8,000
7	9.1	9,000
8	8.5	10,000
9	8.2	11,000
10	8.0	12,000
11	7.8	13,000
12	7.6	14,000
13	7.1	16,000
14	6.8	18,000
15	6.1	20,000

Write a report justifying your estimate of the total direct labor required to produce the 15,000 couplings.

11. U.S. Forest Service

MEMO

TO: *Bob Smith, Statistician*

FROM: *Rus Hotelling, Director*

We are facing the problem of projecting and scheduling personnel requirements on a monthly basis. This is a particularly difficult problem during the summer months since the personnel required for fighting forest fires varies dramatically from region to region and from year to year.

It seems to me that rainfall and forest fires are directly related. However, personnel requirements and schedules must be set by the first of each month and current rainfall is not known until the end of the month. It occurs to me that rainfall from the preceding 30 days might be closely related to the occurrence of forest fires during the next month and thus to personnel requirements. If this proves to be the case, rainfall figures for the preceding 30 days could be used from each of the five districts in the region to make personnel level and allocation decisions for the next month. Accordingly, I am enclosing last summer's rainfall levels and personnel requirements data. I will certainly appreciate your recommendation.

(SIGNED) *Rus*

What recommendation should Bob make to the director?

Rainfall in Inches during Preceding Period		Personnel Requirements in Current Period
District 1		
May	8	100
June	2	120
July	4	108
August	3	108
September	2	105
District 2		
May	10	98
June	5	76
July	6	76
August	7	76
September	8	72
District 3		
May	10	112
June	12	103
July	14	102
August	8	121
September	9	120
District 4		
May	4	146
June	6	145
July	7	147
August	5	146
September	3	150
District 5		
May	2	120
June	4	108
July	6	100
August	8	98
September	3	113

Chapter 11

Multiple Regression

11.1 SITUATION I: WHY?

Suppose you have recently joined the advertising staff of the national magazine Apartment Life (AL). *This magazine is designed to cover a variety of topics which would appeal to apartment dwellers. Special emphasis is given to furnishings, plants, pets, recreation, cooking, and entertaining.*

Ms. Fluharty, director of advertising, has called you into her office and relayed the following information. Total paid circulation of AL *averages around 589,762 copies per month. Of the $1 cover price, only 60 percent is returned to you as the publishers. The remaining 40 percent goes for distribution and retailing. Thus, magazine sales account for approximately $353,857 of the total monthly return to* AL. *The remainder of revenues comes from paid advertisements. One type of paid advertisement is the mail-order section in the back of the magazine. The advertising rates for single-issue, black-and-white mail-order ads are as follows:*

Size	Price
Full page	$6760
2/3 page	$4530
1/2 page	$3380
1/3 page	$2265
1/6 page	$1130
1/12 page	$ 565

The magazine averages around 150 pages per issue and the mail-order section had grown month by month to approximately 20 full pages per issue. Therefore, the mail-order paid advertisements account for $135,200 of monthly revenues to AL.

Ms. Fluharty informs you that she regards $135,200 as material—of sufficient importance. Being in charge of advertising, she is very interested in seeing this level of mail-order advertising develop stability. She wants to be able to count on maintaining this level of revenues from the mail-order section. Ms. Fluharty continues telling you that it is her perception that new accounts for the mail-order section usually start with the small 1/12-page ad and increase the ad size if they experience success. In order to obtain stability in mail-order ad revenues, AL *would like to assist potential and/or new advertisers in any way possible to obtain success. Almost without exception, individuals and mail-order firms placing their first 1/12-page ad with* AL *ask for information on expected returns. What one 1/12-page ad will generate is dependent on many factors, such as the product itself, competition, price, income and buying habits of* AL*'s readers, etc. The only information* AL *has at present is the geographic distribution of the magazine and some crude information on age and income of readers.*

Ms. Fluharty turns to you and asks, "What kind of information could we provide to our prospective mail-order advertisers that could assist them in finding success with their AL *ads? Give it some thought and let's talk again next week."*

Suppose your week has come and gone. You did indeed give the question serious thought, and you ruled out providing a marketing research service for potential ads. Such a service (identifying the market, estimating its size, evaluating competition and the product itself, and ultimately estimating response to a single ad) would be far too expensive. However, you have concluded that 1/12 of a page is small and relatively little information can be conveyed in that size of space. Thus, purchases resulting from such ads may be almost impulse. Consistent with your conclusion are the "presumed" true stories where the following two mail-order ads were placed:

ABSOLUTELY YOUR LAST CHANCE TO
SEND $1 TO (Address)

SEND ONE DOLLAR TO (Address)
AND YOU WILL RECEIVE
ABSOLUTELY NOTHING!!

and people sent their dollars! If sales through small mail-order ads are impulse, then price may actually be more important than the product. Price, independent of the product, may indeed be a strong determinant of sales.

Your phone rings and you learn that it is time for your meeting with Ms. Fluharty. At the meeting you point out that providing the service of marketing research is just not feasible. You express your hypothesis that sales generated by small mail-order ads are largely impulse and that price is a strong *determinant of sales.*

Ms. Fluharty stands up and says, "I think you are right. There may be almost a price barrier. Up to a certain price our readers are willing to gamble. However, as the price continues to increase, our readers feel the need to actually see the product. And, you can't do much showing in a 1/12-page black-and-white ad." You quickly point out that if all this is true, being able to describe the buying habits of your readers would certainly assist potential advertisers. Ms. Fluharty asks you how you would describe the buying habits and what data you need to do so. Your thoughts are as follows. A potential advertiser would like to know how many sales a 1/12-page ad would produce. So, your variable of interest, number of responses, is quantitative. Further, you hypothesize that the number of sales is to some extent related to price, which is also a quantitative variable. Regression! *You recall that regression is a technique for estimating functional relationships between quantitative variables.*

You respond to Fluharty's question by explaining that you think the best approach would be to derive an estimate of the presumed functional relationship between product price and the number of sales generated by a 1/12-page mail-order ad. With the estimated relationship AL *would be able to describe the buying habits of its readers. That is, a potential advertiser could submit a product price to* AL. *In return,* AL *would be able to give the potential advertiser an estimate of the number of sales one 1/12-page ad has generated for products at that same price. That would indeed be a unique, valuable service. The only problem is that the data are or would not be readily available. Yes, we could easily obtain product prices for various 1/12-page ads. However, only the person or firm who placed the ad would know how many sales were generated. For the most part, those people are not going to give you this information—they would regard it as being proprietary. The only way to obtain information on sales would be to buy it.*

Suppose AL *decides to go ahead with your proposal and agrees to buy the required information. The decision is to offer the 35 firms placing 1/12-page ads in the next issue a $50 rebate on the price of their ads for their information on number of sales. Assume that all 35 firms agree, an appropriate period of time has passed, and you now have the set of data shown in Table 11-1.*

Recall from Chapter 10 that the first step in investigating the possible relationship between two quantitative variables is to plot the data on a two-dimensional grid. Again, this plot is called a scatter diagram *and was used in Chapter 10 to determine whether the relationship could be approximated by a straight line.*

Figure 11-1 is the scatter diagram for the data in Table 11-1. Now you have a problem. You know how to fit a straight line to paired data (Chapter 10). The pattern of observations shown in Figure 11-1 indicates that there does appear to be a relationship between product price and number of sales. However, this relationship does not appear to be linear. Rather, it looks like the relationship is curvilinear. You need to be able to fit a curve!

Table 11-1 Number of Sales and Product Price for Thirty-Five 1/12-Page Black-and-White Mail-Order Ads

	Number of Sales y	Product Price x
1	123	$ 1.95
2	137	3.20
3	105	21.30
4	181	12.95
5	160	8.00
6	158	11.00
7	92	1.00
8	20	27.45
9	178	5.95
10	144	17.00
11	143	4.90
12	70	25.00
13	155	6.95
14	130	17.00
15	170	9.95
16	79	23.00
17	150	14.10
18	120	20.00
19	158	4.35
20	164	5.00
21	157	16.00
22	161	10.00
23	0	30.00
24	140	15.00
25	177	11.00
26	155	12.00
27	2	28.00
28	82	22.00
29	155	9.00
30	150	5.00
31	150	16.00
32	163	11.95
33	158	9.50
34	134	4.55
35	144	15.95
	4565	$456.00

Figure 11-1

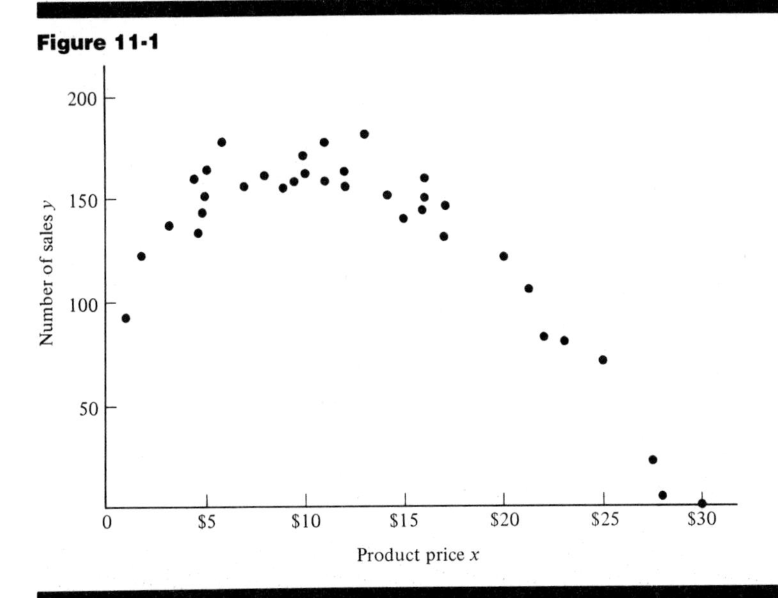

11.2 FITTING A CURVE: HOW?

The shape of the relationship in Figure 11-1 resembles a parabola. Equation (11-1) is a parabola—often referred to as a *quadratic function*—and in regression it is called a *second-order polynomial with one independent variable.*

$$y = \beta_0 + \beta_1 x + \beta_2 x^2 \tag{11-1}$$

The second-order polynomial is capable of the type of curvature shown in Figure 11-1. Thus, your job is to fit an equation of this form to the sample data in Table 11-1.

Consistent with our notation in Chapter 10, we will denote the equation of the *fitted* curve (second-order polynomial) to sample data as

$$\hat{y} = b_0 + b_1 x + b_2 x^2 \tag{11-2}$$

As in Chapter 10, we seek the set (b_0, b_1, b_2) of coefficients that will minimize the sum of the squared deviations of y about the fitted curve. That is, via the method of least squares we will determine the *unique set* of values b_0, b_1, and b_2 for Equation (11.2) that will minimize the quantity

$$\sum_{i=1}^{n} (y_i - \hat{y}_i)^2$$

Recall from Chapter 10 that this quantity is referred to as the *sum of squares error* (SSE) and can, in this case, be expressed as

$$\text{SSE} = \Sigma[y - (b_0 + b_1x + b_2x^2)]^2$$

Taking the partial derivatives of SSE with respect to b_0, b_1, and b_2 and equating these three partial derivatives to zero yields the following three equations:

$$\begin{aligned} &\text{(I)} \quad nb_0 + b_1\Sigma x + b_2\Sigma x^2 = \Sigma y \\ &\text{(II)} \quad b_0\Sigma x + b_1\Sigma x^2 + b_2\Sigma x^3 = \Sigma xy \\ &\text{(III)} \quad b_0\Sigma x^2 + b_1\Sigma x^3 + b_2\Sigma x^4 = \Sigma x^2 y \end{aligned} \tag{11-3}$$

The simultaneous solution of these three equations (known as *normal equations*) for the three unknowns (b_0, b_1, and b_2) gives the coefficients of Equation (11-2) that minimize SSE. That is, of the set of all second-order polynomial equations, the one with the coefficients which are the solution of the normal Equations (11-3) will have the lowest or minimum value for SSE. In our terminology, this second-order polynomial is the "best" fit to our sample data.

11.3 SITUATION I REVISITED

First we must obtain the various sums contained in the three normal equations. The required calculations are shown in Table 11-2. Therefore, the three normal equations to be solved are

$$\begin{aligned} &\text{(I)} \quad 35b_0 + 456b_1 + 8{,}040.52b_2 = 4{,}565 \\ &\text{(II)} \quad 456b_0 + 8{,}040.52b_1 + 167{,}476.32b_2 = 50{,}482.90 \\ &\text{(III)} \quad 8{,}040.52b_0 + 16{,}746.32b_1 + 3{,}859{,}419.75b_2 = 712{,}503.72 \end{aligned}$$

which are merely Equations (11-3) with the appropriate sums inserted.

A set of m equations with m unknowns can be solved (if they have a solution) by a process of elimination. The hand solution of most regression

Table 11-2 Required Calculations, Rounded to Two Decimal Places

Observation	y	x	x^2	x^3	x^4	xy	x^2y
1	123	1.95	3.80	7.41	14.44	239.85	467.40
2	137	3.20	10.24	32.77	104.86	438.40	1,402.88
.							
.							
.							
35	144	15.95	254.40	4,057.68	64,720.63	2,296.80	36,633.60
Sums	4,565	456.00	8,040.52	167,476.32	3,859,419.75	50,482.90	712,503.72

Figure 11-2

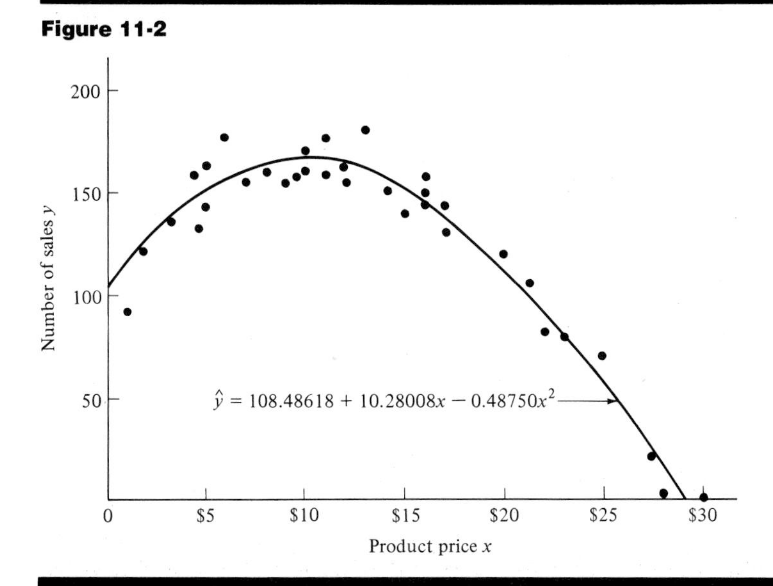

problems that are more complex than the simple linear regression with one independent variable is quite tedious and time-consuming. We will address this issue in Section 11.4. For now let us assume that we have solved the three normal equations and our result is

$$b_0 = 108.48618$$
$$b_1 = 10.28008$$
$$b_2 = -.48750$$

Thus, our second-order polynomial best fitted to the data in Table 11-1 is

$$\hat{y} = 108.48618 + 10.28008x - .48750x^2$$

Figure 11-2 shows our fitted curve superimposed on the scatter diagram. We can see that our fitted model or equation seems to adequately represent the apparent relationship between unit sales and product price.

You now have an equation which could be used to estimate the buying habits of the *Apartment Life* readers relative to 1/12-page mail-order ads. That is, you could estimate the number of sales generated by a 1/12-page mail-order ad in *AL* for a given price. Say, for example, a potential advertiser inquires as to how many orders could be expected from a 1/12-page ad in the mail-order section of *AL* for a product with price $10.85. By applying your fitted equation,

$$\hat{y} = 108.48618 + 10.28008(10.85) - .48750(10.85^2)$$
$$= 162.64$$

Your response would be: "The estimated unit sales generated by the initial $^{1}/_{12}$-page mail-order ad in *AL* for a product with price of \$10.85 per unit is 162.64 units." Providing such a service would certainly be an asset for *Apartment Life*.

A comment seems in order at this point. *We must again caution against extrapolating beyond the range of the data*. This is particularly critical when we are dealing with curvilinear relationships. Note that for any price in excess of \$28.81, the fitted model or equation yields a negative value for $\hat{y}$. Obviously, one could not experience a negative number of units sold.

11.4 SITUATION II: WHY?

Suppose Ms. Fluharty has enthusiastically supported your proposal to use the fitted equation. As an apparent result AL *has experienced both a substantial increase in revenues from the mail-order section and stability in volume from month to month.*

Apartment Life *is published by the Meredith Corporation. While* AL *has a monthly circulation of around 600,000, Meredith also publishes the national magazine* Better Homes and Gardens *with monthly circulation between* nine and ten million! Better Homes and Gardens (BHG) *also has a mail-order section, and because of your work and success at* AL, *you have been temporarily assigned to assist* BHG.

After your meeting with the people at BHG, *you realize that their problem is a little more complex.* Better Homes and Gardens *offers the same size mail-order ads as does* Apartment Life, *which range from the small $^{1}/_{12}$ page to full page. However,* BHG *would like to be able to provide information to potential advertisers for* any size ad. *This differs from your prior analysis where* AL *was interested only in the $^{1}/_{12}$-page ad. Since it is common belief that the physical size of a given ad could have an effect on the unit sales generated by the ad, you must somehow account for variations in ad size.*

An additional problem is that ads of the same size may not have the same circulation. Unlike Apartment Life, Better Homes and Gardens *has regional issues. The only difference among the regional issues is the mail-order section. An advertiser can place his or her ad in all regional issues, any one regional issue, or any combination thereof. The nine regions and associated circulations are shown in Table 11-3.*

Again, unlike your problem with Apartment Life, *the circulation of an ad can vary. One would certainly expect the number of copies of a given ad that are circulated to have some effect on the unit sales generated by the ad. Thus, circulation is a characteristic that should be considered.*

Just what is your problem at this point? You want to provide Better Homes and Gardens *with a means of describing the buying habits of readers in relation to mail-order ads. Specifically,* BHG *would like to be able to estimate the number of units sold for a type of ad. From your prior analysis at* Apartment

Table 11-3 Defined Regions and Circulation for *Better Homes and Gardens* Magazine

Region	Circulation
New England	408,000
Middle Atlantic	1,621,000
East North Central	2,203,000
West North Central	903,000
South Atlantic	1,114,000
East South Central	389,000
West South Central	652,000
Mountain	387,000
Pacific	1,889,000
Total	9,566,000

Life *you are convinced that the number of units sold as a result of a specific ad is related to the product price. In addition to price, you have two characteristics (ad size and circulation) that are variable in the case of* BHG *and also seem to be potentially important determinants of unit sales. Your problem is to somehow estimate the realtionship between a* set *(more than one) of quantitative independent variables and a quantitative dependent variable.*

This is exactly the function of the technique referred to as *multiple regression*. While simple regression refers to the empirical estimation of the relationship between a *single* independent quantitative variable and the dependent variable, multiple regression applies to situations where there are *two or more* independent or predictor quantitative variables being considered.

Equation (11-4) shows the form of the general linear multiple regression model with k independent variables.

$$E(y) = \beta_0 + \beta_1 x_1 + \beta_2 x_2 + \cdots + \beta_k x_k \tag{11-4}$$

where $E(y)$ = the expected value of the dependent variable given the independent variable values

$\beta_0, \beta_1, \beta_2, \ldots, \beta_k$ = the parametric regression coefficients

$x_1, x_2, \ldots, x_k$ = the k independent variables

Before we continue, let us make an interesting observation. In Situation I we fit a quadratic expression to data. The form of the quadratic expression was

$$E(y) = \beta_0 + \beta_1 x + \beta_2 x^2$$

If we define $x_1 = x$ and $x_2 = x^2$, the quadratic expression with one independent variable x can be expressed as a general linear multiple regression model with two independent variables:

$$E(y) = \beta_0 + \beta_1 x_1 + \beta_2 x_2$$

Thus, fitting a curve to data can be viewed as merely one type of application of the general linear multiple regression model.

Now let us return to your problem at *Better Homes and Gardens*. Given your experience at *Apartment Life* and our prior discussion, you might expect the relationship between the number of unit sales generated by an ad in the *BHG* mail-order section and your set of quantitative independent variables to be of the following form:

$$E(y) = \beta_0 + \beta_1 x_1 + \beta_2 x_2 + \beta_3 x_3 + \beta_4 x_4 \qquad (11\text{-}5)$$

where $E(y)$ = expected unit sales
x_1 = circulation of the ad
x_2 = ad size
x_3 = product price
$x_4 = x_3^2$

Recall in your *AL* situation that you found the relationship between unit sales and product price to be quadratic. This is the reason for including the variable x_4. Following your approach at *AL*, suppose *BHG* offers a rebate to 20 new advertisers with varing ad size and circulations. The resulting data are presented in Table 11-4.

Deriving estimates for the set of betas in Equation (11-5) ($\beta_o, \beta_1, \beta_2, \beta_3, \beta_4$) follows the same form as before. That is, the SSE which we seek to minimize is

$$\text{SSE} = \sum_{i=1}^{n} [y_i - (b_0 + b_1 x_1 + b_2 x_2 + b_3 x_3 + b_4 x_4)]^2$$

By taking the partial derivatives of SSE with respect to b_0, b_1, b_2, b_3, and b_4, we obtain five expressions. Each expression is equated to zero and the set of five equations is solved simultaneously for b_0, b_1, b_2, b_3, and b_4.

Most multiple regression analyses with two or more independent variables involve a *great deal of work* in the hand solution. For this reason we ordinarily seek the use of a digital computer. Most computing facilities have one or more "canned" multiple regression programs which require the user to merely input the data. Although the output of such programs varies, certain values are typically shown. Table 11-5 is a reproduction of the computer output for our *BHG* problem using the data in Table 11-4.

First, let's focus on the analysis-of-variance portion of Table 11-5. Given a set of y_i values, there exists a total amount of variation of the y_i's about their mean $\bar{y}$. By definition, the *sum of squares total*, SST, is

$$\text{SST} = \sum_{i=1}^{n} (y_i - \bar{y})^2$$

Table 11-4 *Better Homes and Gardens* Sales Data

Unit Sales y	Ad Circulation* x_1	Ad Size† x_2	Product Price x_3
310	1.621	1	$12.45
480	0.387	6	16.30
1141	9.566	2	5.95
372	1.889	2	10.95
495	0.741	8	23.60
690	3.106	6	14.95
520	2.276	4	9.95
175	0.408	1	3.95
750	5.346	2	5.45
274	1.889	1	18.95
552	2.928	4	2.50
75	1.889	8	45.00
527	2.029	6	21.95
587	3.143	4	7.95
252	0.387	2	12.00
1268	9.566	4	10.95
310	1.114	2	5.95
217	0.903	1	4.50
997	9.566	1	9.90
427	0.389	6	1.95

*Ad circulation in millions.
†One unit of ad size equal 1/12 page.

and for our data set of $n = 20$ observations, SST = 1,922,189.477. As in Chapter 10, algebraically

$$\sum_{i=1}^{n} (y_i - \bar{y})^2$$

can be partitioned into two parts:

$$\sum_{i=1}^{n} (y_i - \bar{y})^2 = \Sigma(\hat{y}_i - \bar{y})^2 + \Sigma(y_i - \hat{y}_i)^2$$

where $\hat{y}_i = b_0 + b_1x_{i1} + b_2x_{i2} + \cdots + b_kx_{ik}$

The first part, $\Sigma(\hat{y}_i - \bar{y})^2$, is the sum of the squared deviations of the estimated y_i values, using the fitted regression equation, about the mean of all y_i observations. This quantity is the sum of the explained deviations squares; it is explained by the regression and is shown as the *sum of squares regression*, SSR. Note the value for SSR is 1,909,448.000. The second or other part of this

Table 11-5 Computer Output for the *Better Homes and Gardens* Data

MULTIPLE R	.9967
STD ERROR OF EST	29.1450

ANALYSIS OF VARIANCE

	DF	SUM OF SQUARES	MEAN SQUARES	F RATIO
REGRESSION	4	1909448.000	477362.00	561.978
ERROR	15	12741.477	849.432	
TOTAL	19	1922189.477		

VARIABLE	COEFFICIENT	STD DEV	COMPUTED T
CONSTANT	48.01199		
CIRCULATION	95.78384	2.21428	43.2573
AD SIZE	55.77458	3.51406	15.8718
PRICE	9.32127	2.06636	4.5110
PRICE SQUARED	−.50463	.04507	−11.1966

partition is $\Sigma(y_i - \hat{y}_i)^2$, which is the sum of the squared deviations of the y_i observation about their predicted values $\hat{y}_i$. This term is referred to as the *sum of squares error*, SSE, and represents that part or portion of SST that is not explained by the regression. In general terms, SSR will have k degrees of freedom (DF), where k is the number of independent variables in the fitted regression equation. Thus, in our case $k = 4$. The degrees of freedom associated with SSE will equal $n - (k + 1)$. A mean square is merely defined to be a sum of squares divided by its degrees of freedom. Thus,

$$\text{MSR} = SSR/k \quad \text{and} \quad \text{MSE} = SSE/[n - (k + 1)]$$

Now that we have the analysis-of-variance portion, *we have it made*!

Strength of the Multiple Relationship

The multiple R is the *coefficient of multiple correlation*. It indicates the strength of the relationship between the dependent variable, y, and the set of independent variables, $x_1, x_2, \ldots, x_k$, and is simply the multiple counterpart of the simple coefficient of correlation given in Equation (10-8).

$$\text{MULTIPLE } R = \sqrt{\frac{\text{SSR}}{\text{SST}}}$$

$$.9967 = \sqrt{\frac{1{,}909{,}448.000}{1{,}922{,}189.477}}$$

If we square R we get R^2, which is the *coefficient of multiple determination*. As $R^2 = \text{SSR/SST}$, we can see that R^2 represents the proportion of the total variation in y that has been or is explained by the fitted regression equation. Since $.9967^2 = .9934$, we have explained 99.34 percent of the variation in unit sales by accounting for variations in ad circulation, ad size, and product price.

The mean square error, MSE, is the variance of the sample observations about the fitted regression function. Its positive square root is then the standard deviation of the sample observation about the fitted regression and is entitled the *standard error of the estimate*. Note that

$$\sqrt{\text{MSE}} = \sqrt{849.432} = 29.1450$$

which is shown as STD ERROR OF EST.

Estimated Regression Coefficients

Finally, we get to the values we are really interested in. Our regression coefficients, $b_0, b_1, \ldots, b_k$, are given in the column headed COEFFICIENT. Recall that we wanted to fit the following equation:

$$\hat{y} = b_0 + b_1x_1 + b_2x_2 + b_3x_3 + b_4x_4$$

where x_1 = ad circulation in millions
x_2 = ad size
x_3 = product price
x_4 = product price squared

Based upon the computer output in Table 11-5, our prediction equation is

$$\hat{y} = 48.01199 + 95.78384x_1 + 55.77458x_2 + 9.32127x_3 - .50463x_4$$

As the b_i's are statistics, they have sampling distributions which in turn have standard deviations. The values listed in the column STD DEV are estimates of the various standard deviations. That is,

$$s(b_1) = 2.21428$$
$$s(b_2) = 3.51406$$
$$s(b_3) = 2.06636$$
$$s(b_4) = .04507$$

where $s(b_i)$ denotes the estimated standard deviation of the sampling distribution of b_i.

Statistical Tests

There are two types of statistical tests typically associated with multiple regression analyses. They are shown in Table 11-5 as the F RATIO and as the COMPUTED T values. In order for these tests to be valid, the following must be true.

$$y_i = \beta_0 + \beta_1 x_{i1} + \beta_2 x_{i2} + \cdots + \beta_k x_{ik} + \epsilon_i \tag{11-6}$$

where ϵ_i are independent, normally distributed, with mean = 0 and variance = σ^2. The first test is

H_0: $\beta_1 = \beta_2 = \cdots = \beta_k = 0$
H_1: not all β_j equal zero

If there is no relationship between the dependent variable y and the set of x variables $(x_1, x_2, \ldots, x_k)$, then β_1 through β_k will each be zero. It can be shown that if this is true (H_0: true), the sampling distribution of the statistic MSR/MSE is the F distribution with $\nu_1 = k$ and $\nu_2 = n - (k + 1)$. Thus, if H_0: $\beta_1 = \beta_2 = \beta_3 = \beta_4 = 0$ is true, the ratio

$$\frac{\text{MSR}}{\text{MSE}} = \frac{477362.00}{849.432} = 561.978$$

which is shown on Table 11-5 as the F RATIO, can be viewed as a random observation from the following distribution:

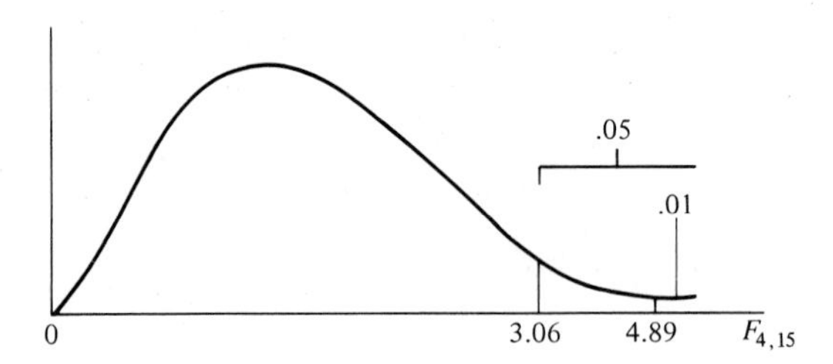

Table A-7 shows that only 5 percent of the possible F values in this distribution exceed 3.06, and according to Table A-8, only 1 percent exceed 4.89. Needless to say, the conditional (given H_0 is true) probability of obtaining a ratio MSR/MSE as large as or larger than 561.978 is minute! Our *BHG* data strongly refute H_0, and we would conclude that at least one of our independent variables is related to unit sales. Note we have performed a one-tailed F test. It can be shown that if any $\beta_j \neq 0$, the $E(\text{MSR}) > E(\text{MSE})$. *Thus, large values for the ratio* MSR/MSE *would refute the null hypothesis and we would place all of α in the right-hand or upper tail of the appropriate distribution.*

The other statistical test which is typically shown on a multiple regression computer output is the COMPUTED T. Actually, there are k tests, one for each independent variable. Given our assumptions, Equation (11-6), the sampling distribution of the statistic

$$\frac{b_i - \beta_i}{s(b_i)}$$

is the student t with $n - (k + 1)$ degrees of freedom. The following test is performed for each independent variable:

H_0: $\beta_i = 0$

H_1: $\beta_i \neq 0$

If $\beta_i = 0$, then

$$\frac{b_i - 0}{s(b_i)} = t_{(n-(k+1))}$$

The COMPUTED T values are the corresponding values for

$$\frac{b_i}{s(b_i)} \quad (i = 1, 2, \ldots, k)$$

In our case, if β_1, for circulation is zero, the value 95.78384/2.21428 = 43.2573 can be viewed as a random observation from the student t distribution with 15 degrees of freedom.

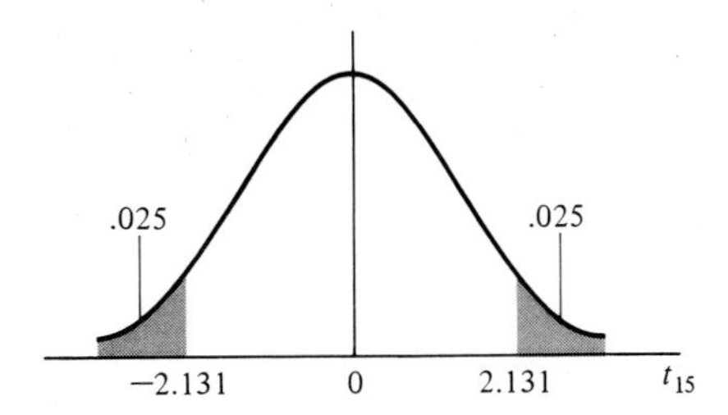

Our sample clearly refutes the hypothesis that $\beta_1 = 0$. Similar arguments can be made for our other three independent variables. Given our COMPUTED T values, we would conclude that each of our independent variables makes a contribution in explaining the dependent variable.

Should it happen that one or more of the COMPUTED T values is so small as to be "not significant," the procedure would be to drop the variable with the smallest absolute value for its COMPUTED T and refit the regression model. This process would be continued until all remaining independent variables had "significant" COMPUTED T values.

Now that we have completed the multiple regression analysis, let's summarize our findings and see where we are. Our statistical description of the buying habits of *BHG* readers in relation to the mail-order ads in *BHG* is

$$\hat{y} = 48.01199 + 95.78384x_1 + 55.77458x_2 + 9.32127x_3 - .50463x_4$$

where $\hat{y}$ = estimated unit sales

x_1 = ad circulation in millions

x_2 = ad size where 1/12 page equals 1

x_3 = product price

x_4 = square of product price

The F RATIO and COMPUTED T values would be "significant" at almost any value of α one might choose. The significant F RATIO leads us to the conclusion that at least one of our independent variables is related to the dependent variable. The fact that each of the COMPUTED T values is significant indicates that all of our independent variables are important predictors or contributors in explaining variation in the dependent variable. Finally, the R^2 value of .9934 suggests that the underlying relationship is rather strong.

As an example of the use of the estimated or fitted equation, consider the following. Suppose a potential advertiser contacts *BHG* and indicates that he or she is considering placing a 1/3-page ad in the New England regional edition for a product that would sell for \$5.95. The potential advertiser has asked if *BHG* could give some indication with regard to what might be expected for the number of unit sales that such an ad might produce. Data for this case would be as follows.

$$x_1 = .408$$

$$x_2 = 4$$

$$x_3 = 5.95$$

$$x_4 = 35.4025$$

$$\begin{aligned}\hat{y} &= 48.01199 + (95.78384)(.408) + (55.77458)(4) \\ &\quad + (9.32127)(5.95) - (.50463)(35.4025) \\ &= 347.7865\end{aligned}$$

Better Homes and Gardens could then respond by saying that the estimate is that the proposed ad would produce about 348 unit sales, and that this estimate is based upon past experience for the relationship between ad circulation, ad size, and product price. As was the case for *Apartment Life*, providing this service for potential advertisers would be unique within the industry and perhaps give *Better Homes and Gardens* a competitive advantage relative to attracting new advertisers.

11.5 IDENTIFICATION OF APPLICATIONS: WHEN?

While simple regression refers to the empirical estimation of the relationship between a *single* independent quantitative variable and the dependent variable, multiple regression allows estimation in situations where there are *two or more* independent or predictor quantitative variables being considered. Therefore, the important characteristics of a situation where multiple regression might be applied are:

1. The dependent variable is *quantitative*.
2. One can identify *two or more* independent *quantitative* variables that are believed to affect or be related to the dependent variable.

Although there are certain variables that are primarily determined or affected by a single variable, most economic variables are affected by several. Therefore, many of the valuable applications of regression lie in the domain of multiple regression.

We have seen how to utilize the information contained in a set of variables to assist in predicting the variable of interest. In addition, we saw in Situation I how the multiple regression model could be used to estimate a curvilinear relationship. The multiple regression model is indeed quite versatile.

PROBLEMS

1. Identify an application of multiple linear regression. What conditions are present in your identified situation that make multiple regression analysis appropriate and of value?

2. Suppose that you have gone to work in the accounting department of a medium-sized furniture manufacturer. Shortly after you begin work, you are called into the general manager's office, where you are greeted with the following statement:

   "I was looking through one of the back issues of the *Harvard Business Review* (March-April 1971) and I came across a formula that a firm in our industry uses to forecast sales. They call it a multiple regression formula. I don't understand what it means. Would you prepare a brief report for me explaining in everyday language exactly what this formula means?"

   The formula is presented below.

$$Y_{t+1} = .02 + .373Y_t + .033H_t + .672\, I_{t+1} - 11.03T$$

where Y_{t+1} = forecast sales for the coming year
Y_t = current year's sales
H_t = housing starts during the current year
I_{t+1} = estimated disposable income for the coming year
T = time trend (first year of data = 1, second year = 2, and so on)

3. Recently you must have seen the American Cancer Society (ACS) TV commercials about cancer, where the announcer tells you in a deep voice, "It is a matter of life and breath." Suppose ACS feels that the level of fear conveyed in the advertisement may not be enough. To test this, ACS prepares 10 antismoking messages and assigns an interval score between 1 and 10 based on the severity of the threat contained in the message. Twenty subjects are exposed to the message and their degree of attitude change is measured on an interval scale of 1 to 100. The results are presented below.

Individual	Severity of Threat Score	Attitude-Change Score
1	1	60
2	1	67
3	2	63
4	2	66

Individual	Severity of Threat Score	Attitude-Change Score
5	3	71
6	3	74
7	4	73
8	4	78
9	5	81
10	5	81
11	6	83
12	6	86
13	7	85
14	7	87
15	8	77
16	8	81
17	9	67
18	9	63
19	10	59
20	10	51

If there is a relationship between the "severity of threat" and "attitude change," ACS would like to choose the level of threat that would have the greatest impact on attitude change. Since you are a new trainee, the chief statistician tells you to proceed as follows:

a. Plot the scatter diagram.

b. Identify the functional form of the relationship.

c. Fit the appropriate equation.

d. Test the hypothesis that threat and attitude-change scores are independent. Let $\alpha = .05$.

e. Give your advice to ACS.

4. After receiving your bachelor's degree in personnel management, you are hired by a small but expanding life insurance firm. Your first assignment is to develop a more efficient technique for the preliminary screening of applicants for sales positions. Since the firm employs only college graduates, you decide to work with information focusing on their performance during college.

   A random sample of 25 from the firm's current sales force is selected and the following information is obtained: (1) last year's performance evaluation score, (2) college grade point average (GPA), (3) percent of total college expenses earned by the individual, and (4) number of social organizations the individual belonged to.

Performance Score	GPA	Percent of Expenses Earned	No. of Social Organizations
43	2.1	50	2
47	2.8	20	5
53	2.6	10	3
56	2.7	60	1
57	3.8	0	0
64	2.6	30	2

Performance Score	GPA	Percent of Expenses Earned	No. of Social Organizations
68	3.2	10	1
68	2.8	30	2
74	2.6	10	2
75	2.9	40	1
77	3.0	30	0
78	3.2	15	1
81	3.4	20	2
83	2.8	40	3
87	2.6	60	5
88	3.1	50	0
89	2.4	80	4
90	3.3	10	2
91	2.9	50	6
92	3.5	40	1
93	3.7	30	2
94	3.1	20	5
95	3.6	70	1
96	3.2	10	4
97	3.4	40	0

a. Identify the response and explanatory variables.

b. Plot each of the explanatory variables against response variables.

c. Identify the functional form between response and explanatory variables.

d. Statistically fit the data.

e. Can you suggest how these data be used in screening potential sales personnel? Do you think that each of the explanatory variables is necessary to answer your questions? Use $\alpha = .05$ whenever necessary.

5. An electric utility firm must consistently forecast demand for electricity on a weekly basis in order to adequately manage inventories of petroleum used to generate the electricity. The firm has used the prior week's consumption of petroleum to forecast next week's demand. Since the utility is located in a Southwestern city, demand for electricity appears to vary directly with the temperature. The reason for this is the increased use of air conditioning as the temperature increases.

   Suppose you have gathered the following types of information: (1) current week's petroleum consumption and (2) preceding week's forecast of current week's average temperature from a random sample of 20 weeks over the past year:

Current Week's Consumption (In Thousands of Gallons)	Preceding Week's Prediction of Current Week's Temperature (In Degrees Farenheit)
101	83
121	86

Current Week's Consumption (In Thousands of Gallons)	Preceding Week's Prediction of Current Week's Temperature (In Degrees Farenheit)
102	82
108	82
112	89
128	93
98	79
96	78
89	76
101	81
121	93
122	92
111	99
120	93
131	101
100	100
98	84
93	82
101	89
121	96

a. Check whether demand for electricity varies directly with the temperature.

b. Identify the model which the utility firm has used to forecast next week's demand.

c. Fit the model.

d. Can you now statistically test your findings of part *a* at the $\alpha = .05$ level?

e. What recommendation would you make to the utility company concerning their forecasting procedure?

6. BrightGlo is the brand name of a paint produced by Southern Products. Every spring for the past 10 years, the wholesale price of BrightGlo has been reduced from 10 to 50 percent. Some or all of the reduction is generally passed on to customers during retailers' spring sales. Bill Johnson, BrightGlo's young marketing manager, wants to have a large discount sale each spring, as this appears to cause a substantial sales increase and consequently a large market share. He argues, "True, we do not make excess profit due to price reduction, but it helps us build our social image. Times are changing, you know." Karen O'Neill, the comptroller of the firm, has expressed concern and commented, "We are a for-profit organization. Our main concern is to maximize profits, and these incremental sales are not sufficient to offset the profit margin lost by the price cut." However, after a little debate, they agree that a profit of $325,000 with a 20 percent market share would be satisfactory to both of them. The percentage discount, market share during the sale period, and the net profit from these sales follow.

Discount, %	Market Share, %	Profit (In Thousands of Dollars)
15	08	250
30	17	400
10	05	200
35	19	370
50	24	280
40	22	320
20	11	300
15	09	240
45	25	300
25	15	360

a. What discount level would you recommend given the desired market share and profit?

b. Do you think that either market share or profits is related to discount rate? Use a significance level of 5 percent to justify your answer.

7. Suppose that GM is concerned about employee turnover among its assembly-line personnel. It takes substantial resources to train these employees, and if they leave the firm too soon, GM cannot recoup this training "investment." Therefore, in addition to consistently attempting to improve working conditions, the firm tries to be as selective as possible in terms of its recruiting programs.

   Your task is to develop guidelines to help in the selective recruiting of personnel for assembly-line positions. You randomly select the personnel files of 25 employees that were hired nine years ago and gather the data reported below. What recruiting guidelines can you suggest?

Months Employed	Age When Hired	Marital Status When Hired*	Years of School Completed	Score on the Firm's Aptitude Test (Range 0–100)
73	24	1	5	55
85	22	0	7	79
95	22	1	8	82
75	24	1	4	38
93	22	1	7	69
83	23	0	6	71
88	22	0	7	88
78	24	1	5	43
71	24	0	4	44
81	22	1	6	88
68	22	1	6	82
80	23	0	7	92
91	22	1	7	89
89	22	0	8	76
90	23	0	7	72
88	22	1	6	77

Months Employed	Age When Hired	Marital Status When Hired*	Years of School Completed	Score on the Firm's Aptitude Test (Range 0–100)
95	24	1	4	48
96	21	0	3	43
88	22	0	8	89
78	22	0	1	22
76	24	1	2	24
72	24	1	8	89
81	25	1	6	62
88	24	1	7	90
92	25	0	2	42

*Marital-status code: Married = 1; Single = 0.

8. Suppose the manager of the First Federal Bank is interested in obtaining a better "feel" for the characteristics of families that write substantial numbers of checks. Improved understanding of the characteristics of these families will assist the manager in advertising decisions and in designing improved features in the checking options offered by the bank. The manager does not want to offer these improved features across the board, i.e., to all customers. Therefore you are asked to "statistically characterize families that write a substantial number of checks." A random sample of 45 joint checking accounts and a phone call to each account holder reveals the information shown below.

Acct. No.	Average No. of Checks per Month	Age of Head of Household	Household Income (In Thousands of Dollars)	Size of Family	Both Spouses Employed Full-time*
1	29	37	16.2	2	1
2	42	34	25.4	3	0
3	21	36	9.9	1	1
4	9	28	12.4	5	1
5	35	60	21.5	1	0
6	12	29	14.1	4	1
7	2	23	8.0	2	1
8	6	29	0.1	3	1
9	29	50	16.9	5	1
10	45	35	27.5	4	1
11	51	32	28.3	3	0
12	10	25	8.3	3	0
13	30	47	18.2	2	1
14	56	38	25.0	4	0
15	22	39	11.1	5	1
16	54	47	16.0	7	1
17	36	37	18.0	5	1
18	4	25	7.9	3	0
19	13	34	9.7	2	1

Acct. No.	Average No. of Checks per Month	Age of Head of Household	Household Income (In Thousands of Dollars)	Size of Family	Both Spouses Employed Full-time*
20	47	39	14.6	6	1
21	48	43	24.2	5	0
22	8	33	11.1	3	0
23	30	62	17.5	6	1
24	3	21	9.5	1	0
25	30	55	19.0	3	0
26	18	51	16.1	4	1
27	27	46	16.8	4	0
28	23	43	18.4	3	1
29	5	36	8.3	2	1
30	58	36	24.4	4	1
31	22	42	13.8	2	0
32	19	39	13.4	3	0
33	25	45	20.2	2	0
34	53	36	26.2	2	1
35	6	42	9.2	4	0
36	29	49	13.4	3	0
37	43	39	19.3	2	0
38	17	36	8.9	5	1
39	37	29	26.0	4	1
40	14	42	10.1	1	0
41	14	37	9.3	3	1
42	20	22	10.2	2	0
43	34	53	22.0	2	0
44	16	43	12.2	2	0
45	7	31	9.6	2	0

*Employment code: Yes = 1; No = 0.

Write a report describing the criteria you have used to statistically characterize these families. The manager does not know any statistical jargon, so your report should be as simple as possible.

9. Hardin Construction Materials, Inc., has a nationwide sales force which it has paid on a salary plus an annual bonus system for the past 15 years. The bonus has been based on the individual salesperson's annual sales throughout this period. Since the total demand for construction materials is not the same in the firm's 82 sales territories, and in fact varies from year to year, it has been suggested that the ratio of the salesperson's sales to the total industry sales, i.e., market share, would be a more equitable basis. Utilization of such a system requires estimates of total industry sales in each territory. You are asked to develop a system for making these estimates. From an extensive series of interviews, you obtain accurate figures for total industry sales in 12 territories. An examination of secondary sources of information reveals the following additional data. What would you recommend?

Territory No.	Total Industry Sales (In Thousands of Dollars)	No. of Housing Starts (In Thousands)	Total Local Government Expenditures (In Millions of Dollars)	Total Retail Sales (In Thousands of Dollars)
1	4888	23	12.8	245
2	3926	26	22.9	193
3	4286	42	40.6	210
4	3898	52	12.8	203
5	4800	28	13.9	254
6	5200	28	28.6	249
7	6839	43	26.2	351
8	4889	49	28.4	244
9	2883	72	49.2	141
10	4293	32	13.6	210
11	6892	42	48.2	376
12	7232	28	41.3	369

10. *Fourteen–Twenty-one Food Stores*

MEMO

TO: *Anne Galloop, Analyst*

FROM: *Gus Draper, Director of Planning*

At the present time we have 25 outlets. In the last meeting it was decided to expand our convenience food store chain at the rate of one outlet per month. I would like you to develop a set of guidelines for selecting locations for future outlets. As you know, executive committee members are not very familiar with modern-day statistical techniques. This does not mean that you should not apply them, but your final report should be as comprehensible as possible to them. If it is of any help to you, I would like to make this suggestion. First identify the dependent (response) and independent (explanatory) variables, then use your multiple regression analysis. Before you set the criteria for required sales, make sure to identify the significant variables, so that we may not use the insignificant ones in future analyses. It would also be helpful to identify the *most* significant ones. You know that we can easily find the weekly auto traffic volumes, etc., from city hall. Therefore, our main interest is, given the explanatory variables, how much sales would be needed to break even? I am enclosing the relevant data pertaining to all of our 25 outlets. One thing more, I would also like your opinion about the model. You know, things like whether it depicts a good fit, etc.

(SIGNED) *Gus*

Encl: Weekly Sales and Other Data for 25 Outlets of Fourteen–Twenty-one Food Stores

Sales (In Thousands of Dollars)	Weekly Auto Traffic Volume (In Thousands)	Ease of Entry/Exit (Scale of 1 to 100)	Average Yearly Household Income of Area (In Thousands of Dollars)	Supermarket within 1 Mile (Yes/No)
532	198	48	26	Yes
1082	199	86	15	No

Sales (In Thousands of Dollars)	Weekly Auto Traffic Volume (In Thousands)	Ease of Entry/Exit (Scale of 1 to 100)	Average Yearly Household Income of Area (In Thousands of Dollars)	Supermarket within 1 Mile (Yes/No)
938	59	29	17	No
637	66	32	15	Yes
1223	256	99	23	No
1032	248	97	16	No
438	45	21	22	Yes
1028	101	42	28	No
1432	121	49	21	No
383	26	13	23	Yes
1222	49	28	15	No
562	50	59	16	Yes
989	32	28	19	No
458	21	19	20	Yes
469	52	14	23	Yes
1022	186	100	24	No
1089	282	100	28	No
1288	389	98	36	No
1312	308	87	32	No
498	131	48	18	Yes
298	45	45	15	Yes
1451	200	64	22	No
1211	285	89	21	No
1288	301	100	19	No
599	208	98	13	Yes

Chapter 12

Analysis of Variance

12.1 SITUATION I: WHY?

Suppose you have recently been hired by a city with a population of around 240,000. Your boss calls you into his office and proceeds to tell you the following "sad story." As property taxes in your locality have continued to increase, the efficiency of every unit within city government is being questioned. One of the functions or responsibilities of your unit is the purchasing of light bulbs for streetlights within the city. Since there are 8642 streetlights in your city that require a 400-watt bulb and the average price of a 400-watt bulb is $11.70, the item "light bulbs (street)" in the city budget is indeed material.

About a year ago a new high-pressure sodium bulb came onto the market. Although it was considerably more expensive, the manufacturer claimed that the bulb's extended life more than compensated for the differential in price. The emergence of the new bulb and the manufacturer's claim prompted your unit to question its past purchase decision. Specifically, had your unit been buying the most economical 400-watt bulb available?

Your predecessor was to some extent a statistical analyst. Her assignment was to reevaluate the purchasing decision in light of the new bulb and make a formal recommendation for the future. Which brand of 400-watt bulb is the most economical? She initially obtained a sample of seven 400-watt bulbs from each of the four brands on the market. These 28 bulbs were then randomly assigned to streetlights within the city and their lives under standard operating conditions were observed and recorded. Brands 1, 2, and 3 are mercury-vapor bulbs; brand 4 is the new high-pressure sodium bulb.

Unfortunately, before your predecessor could finish the analysis, she was transferred to another position and is no longer available. The job is now yours! You are asked to complete the analysis and make your recommendation. The data are given in Table 12-1.

As you leave your boss's office, he points out that the question is not "which brand of bulb has the largest mean life?" but rather "which brand has the largest mean life per dollar cost?"

Given your assignment, the first step would be to divide each observation by its price, thus converting the data to life in thousands of hours per dollar cost. These data are shown in Table 12-2.

Our variable of interest, the hour life per dollar cost, is quantitative. We now want to know what we can statistically conclude about the set of four population means μ_1, μ_2, μ_3 and μ_4. Could all μ_j's, j = 1, 2, 3, 4, be equal? If so, the prices are just right. That is, you get what you pay for and no brand is more economical than any other. On the other hand, if all four μ_j's are not equal, then some brand or brands are more economical than others.

What you need is some way of analyzing the four samples such that you can draw conclusions about the four μ_j's. Yes, we could apply the student t *test (given in Chapter 8) for two populations. This approach would be cumbersome, as it would require several tests, each with its own α. What we need is a test like the student* t *test for two independent populations that is sufficiently general to be used for, say,* r *populations where* $r \geq 2$.

> ANALYSIS OF VARIANCE IS A COLLECTION OF TECHNIQUES DESIGNED TO ANALYZE SAMPLES FROM TWO OR MORE QUANTITATIVE POPULATIONS.

Specifically we desire a test of the following hypotheses:

H_0: $\mu_1 = \mu_2 = \cdots = \mu_r$

H_1: not all μ_j's, $j = 1, 2, \ldots, r$, are equal

Table 12-1 Life of Samples of Four Brands of 400-Watt Light Bulbs
(In Thousands of Hours)

Observation	Brand 1	Brand 2	Brand 3	Brand 4	Brand	Price
1	22	24	22	40	1	$ 9.60
2	24	24	25	36	2	$12.49
3	24	28	24	38	3	$13.01
4	25	24	25	37	4	$18.00
5	21	23	22	39		
6	22	25	21	35		
7	19	27	24	39		

Table 12-2 Life per Dollar Cost for a Sample of Four Brands of 400-Watt Light Bulbs
(In Thousands of Hours)

Observation	Brand 1	Brand 2	Brand 3	Brand 4
1	2.29	1.92	1.69	2.22
2	2.50	1.92	1.92	2.01
3	2.50	2.24	1.84	2.11
4	2.60	1.92	1.92	2.06
5	2.19	1.84	1.69	2.19
6	2.29	2.00	1.61	1.94
7	1.98	2.16	1.84	2.17

12.2 ONE-WAY ANOVA: HOW?

In Chapter 10 we described how the coefficient of determination r^2 related to two sums of squares: the sum of squares of $\hat{y}_i - \bar{y}$ and the sum of squares of $y_i - \bar{y}$. Furthermore, Figure 10-7 showed how the deviations of $y_i - \bar{y}$ could be broken down (partitioned) into two additive components—a deviation of y_i from $\hat{y}_i$ and a deviation of $\hat{y}_i$ from $\bar{y}$. Similar ideas underlie analysis of variance (ANOVA). We again partition the total variation in the entire collection of observations into components. For a *one-way* ANOVA the total variation is partitioned into two components. The first component is the sum of the squared deviations of the sample observations of each population around *its respective sample mean* and is called *sum of squares within* (SSW). The second component is the squared deviations of the sample means around the mean of *all* sample observations, and it is called the *sum of squares among* (SSA). Thus, we obtain the following expression:

$$\text{SST} = \text{SSW} + \text{SSA}$$

The basic idea in analysis of variance is that the SSA component contains information about the differences, if they exist, among the population means. Through a certain logical procedure we are able to infer about the population means by comparing functions of these sums of squares.

Let's develop the basic idea again, this time using symbols rather than words. We know it may not always look nice, but stay with us. We will numerically illustrate every expression we develop. First let us adopt a standard arrangement of the data as shown in Table 12-3. Note a sample observation x_{ij} has two subscripts. The i subscript denotes the row or which sample observation, while the j subscript denotes the column or population from which the observation came. Thus, x_{ij} is the ith observation from the jth population. T_j denotes the sum or total of the n sample observations from the jth population and $T.$ is the total of all T_j's, which is the same as the total of *all* x_{ij} observations. Finally, $\bar{x}_j$ is the mean of the n sample observations from the jth population and equals $(1/n)\, T_j$ while $\bar{x}.$ is the mean of *all* sample observations and equals $(1/N)T.$ where $N = nr$.

Algebraically the total variation in the entire collection of sample observations can be partitioned into two components as shown below.

$$\sum_{j=1}^{r}\sum_{i=1}^{n}(x_{ij} - \bar{x}.)^2 = \sum_{j=1}^{r}\sum_{i=1}^{n}(x_{ij} - \bar{x}_j)^2 + n\sum_{j=1}^{r}(\bar{x}_j - \bar{x}.)^2$$

The first term, $\Sigma\Sigma(x_{ij} - \bar{x}\,.)^2$, is called *sum of squares total* and is denoted as SST. SST is the total variation in the entire collection of sample data. As you can see, it is the sum of the squared deviations of all N observations about their mean. The first component on the right side of the above equation, $\Sigma\Sigma(x_{ij} - \bar{x}_j)^2$, is called *sum of squares within* and is denoted as SSW. SSW is the sum of the squared deviations of the sample observations of each population about its respective sample mean. Thus, for the n sample observations drawn from population 2, their deviations are measured about $\bar{x}_2$ rather than $\bar{x}.$ as in the case of SST. The final components, $n\Sigma\,(\bar{x}_j - \bar{x}.)^2$, is the *sum of squares among* populations and is denoted as SSA. This component is of particular interest to us. Note that it is a weighted (by n) sum of the squared deviations of the sample means $\bar{x}_j$'s about their mean $\bar{x}.$, which is the mean of all N sample observations. *If all* $\bar{x}_j$'s *were equal*, they would equal $\bar{x}.$, since $\bar{x}.$ is the mean of the $\bar{x}_j$'s. In this case each $\bar{x}_j - \bar{x}.$ would equal zero, and SSA $= n\Sigma(\bar{x}_j - \bar{x}.)^2 = 0$. Therefore, if the sample means $\bar{x}_j$'s are nearly equal in magnitude, SSA will be small. Also, the more unequal the sample means, the larger SSA will be. SSA is then to some extent a measure of the similarity among the sample means.

Table 12-3 Data Arrangement for a One-Way ANOVA

	Populations				
Sample Observations	1	2	3 ...	r	
1	x_{11}	x_{12}	x_{13}	x_{1r}	
2	x_{21}	x_{22}	x_{23}	x_{2r}	
3	x_{31}	x_{32}	x_{33}	x_{3r}	
.	.	.	.	.	
.	.	.	.	.	
.	.	.	.	.	
n	x_{n1}	x_{n2}	x_{n3}	x_{nr}	
Totals	T_1	T_2	T_3	T_r	$T.$
Means	$\bar{x}_1$	$\bar{x}_2$	$\bar{x}_3$	$\bar{x}_r$	$\bar{x}.$

Although

$$\text{SST} = \Sigma\Sigma(x_{ij} - \bar{x}.)^2$$
$$\text{SSW} = \Sigma\Sigma(x_{ij} - \bar{x}_j)^2$$
$$\text{SSA} = n\Sigma(\bar{x}_j - \bar{x}.)^2$$

these are not the easiest forms for computing purposes. The following are algebraically equivalent and much easier to compute.

$$\text{SST} = \sum_{j=1}^{r} \sum_{i=1}^{n} x_{ij}^2 - \frac{T.^2}{N} \tag{12-1}$$

$$\text{SSA} = \frac{1}{n}\sum_{j=1}^{r} T_j^2 - \frac{T^2.}{N} \tag{12-2}$$

Since we know that SST = SSW + SSA, we can obtain the value for SSW as

$$\text{SSW} = \text{SST} - \text{SSA}$$

It is customary to present the computed results in an analysis-of-variance table like the one shown in Table 12-4.

There are two new terms shown in Table 12-4. First, in the second column, "df" stands for "degrees of freedom." We see that the SSA has $r - 1$ degrees of freedom while SSW has $N - r$. A number of degrees of freedom is associated with each sum of squares (SS) and can be viewed as the number of independent terms in the particular SS. Therefore, although there are r terms in SSA, only $r - 1$ of the terms are independent, and similarly, only $N - r$ of the N terms in SSW are independent.

Second, the last column in Table 12-4 is headed "MS," which stands for "mean squares." As shown in Table 12-4, a mean square is a sum of squares divided by its degrees of freedom. Thus by definition:

$$\text{MSA} = \text{SSA}/(r - 1)$$
$$\text{MSW} = \text{SSW}/(N - r)$$

To this point we have made *no* assumptions. We have seen only algebra and definitions. Now we are ready for our analysis of variance model and assumptions.

Table 12-4 ANOVA for a One-Way Classification

Source of Variation	df	SS	MS
Among populations	$r - 1$	SSA	$\text{SSA}/(r - 1) = \text{MSA}$
Within populations	$N - r$	SSW	$\text{SSW}/(N - r) = \text{MSW}$
Total	$N - 1$	SST	

12.3 FIXED-EFFECTS ONE-WAY ANOVA MODEL

We will view an x_{ij} sample observation as actually being made up of three components or parts. First let us define μ as the mean of the r population means. That is,

$$\mu = \frac{1}{r}\sum_{j=1}^{r} \mu_j$$

which is merely the mean of the μ_j's, $j = 1, 2, \ldots, r$. With this definition we have the identity,

$$x_{ij} = \mu + \mu_j - \mu + x_{ij} - \mu_j$$

As the right side of the equation contains both a plus and minus μ and a plus and minus μ_j, the identity really implies that

$$x_{ij} = x_{ij}$$

which is certainly true. However, if we group certain terms we can see the three components,

$$x_{ij} = \mu + (\mu_j - \mu) + (x_{ij} - \mu_j)$$

μ represents the general level mean. $\mu_j - \mu$ represents the deviation of the jth population mean from the general level mean. And $x_{ij} - \mu_j$ represents the "error" or deviation of the ith sample observation about the mean of its population, which we denote as ϵ_{ij}. If our sample observations are the result of random sampling from each of the r populations and each population is normally distributed with equal variance, we can symbolically show the one-way ANOVA model as follows:

$$x_{ij} = \mu + (\mu_j - \mu) + \epsilon_{ij}, \qquad \begin{array}{l} i = 1, 2, \ldots, n \\ j = 1, 2, \ldots, r \end{array}$$

$$\sum_{j=1}^{r} (\mu_j - \mu) = 0$$

ϵ_{ij}'s are independent, normally distributed, with mean zero and variance σ^2. In other words, our assumptions are that each of the r populations is normally distributed and have the same variance σ^2. The n observations from each population are the result of random sampling.

Given these assumptions, it can be shown mathematically that the expected value of MSW is σ^2, which is the variance of each population, and the expected value of MSA is

$$\sigma^2 + \frac{n}{r-1}\sum_{j=1}^{r} (\mu_j - \mu)^2$$

That is,

$$E(\text{MSW}) = \sigma^2$$

$$E(\text{MSA}) = \sigma^2 + \frac{n}{r-1}\sum_{j=1}^{r}(\mu_j - \mu)^2$$

Let us focus attention on the last component of $E(\text{MSA})$. Since the $\mu_j - \mu$ terms are squared, this component cannot be negative. If the r population means are quite dissimilar in magnitude, the component $n/(r-1)\,\Sigma(\mu_j - \mu)^2$ will be large and positive. On the other hand, if the population means are nearly equal, this component will be small. Indeed if $\mu_1 = \mu_2 = \cdots = \mu_r$, then the component $n/(r-1)\,\Sigma(\mu_j - \mu)^2$ will equal 0.

Our objective is to devise a test of the null hypothesis

$$H_0: \mu_1 = \mu_2 = \cdots = \mu_r$$

against the alternative

H_1: not all μ_j's are equal

If H_0 is true, both MSA and MSW are independent unbiased estimates of the same variance σ^2. It can be shown that a ratio of two independent unbiased estimates of a common variance follows an F distribution. Thus, *if H_0 is true*,

$$\frac{\text{MSA}}{\text{MSW}} = F_{r-1,N-r}$$

That is, under the condition that H_0 is true, the ratio of MSA to MSW has as its sampling distribution the F distribution with $r - 1$ degrees of freedom in the numerator and $N - r$ degrees of freedom in the denominator. There is our test. We know the conditional probability distribution for MSA/MSW when H_0 is true. Since we always place MSA in the numerator, and large values for MSA correspond to unequal sample means, only large values for MSA/MSW would lead us to reject H_0. Thus, all of α is placed in the upper tail of the appropriate F distribution.

12.4 SITUATION I REVISITED

We can now use the material in Section 12.2 to determine statistically whether there is any difference in the true mean hours per dollar cost of the four brands of 400-watt bulbs in Situation I. For convenience, we have reproduced Table 12.2.

Brand 1	Brand 2	Brand 3	Brand 4	
2.29	1.92	1.69	2.22	
2.50	1.92	1.92	2.01	
2.50	2.24	1.84	2.11	
2.60	1.92	1.92	2.06	
2.19	1.84	1.69	2.19	
2.29	2.00	1.61	1.94	
1.98	2.16	1.84	2.17	
$T_1 = 16.35$	$T_2 = 14.00$	$T_3 = 12.51$	$T_4 = 14.70$	$T. = 57.56$
$\bar{x}_1 = 2.34$	$\bar{x}_2 = 2.00$	$\bar{x}_3 = 1.79$	$\bar{x}_4 = 2.10$	$\bar{x}. = 2.06$

Our first step is to compute SST and SSA. Using Equations (12-1) and (12-2) we find

$$\text{SST} = \Sigma\Sigma x_{ij}^2 - \frac{T.^2}{N} \tag{12-1}$$

$$= (2.29^2 + 2.50^2 + \cdots + 2.17^2) - \frac{57.56^2}{28}$$

$$= 119.97 - 118.33$$

$$= 1.64$$

and

$$\text{SSA} = \frac{1}{n}\Sigma T_j^2 - \frac{T.^2}{N} \tag{12-2}$$

$$= \frac{1}{7}(16.35^2 + 14.00^2 + 12.51^2 + 14.70^2) - \frac{57.56^2}{28}$$

$$= 119.42 - 118.33$$

$$= 1.09$$

Since

$$\text{SST} = \text{SSW} + \text{SSA},$$

$$\text{SSW} = \text{SST} - \text{SSA}$$

$$= 1.64 - 1.09$$

from the above calculations. Thus,

$$\text{SSW} = .55$$

Recall that by definition MSA $=$ SSA/$(r - 1)$ and MSW $=$ SSW/$(N - r)$. Thus

MSA $= 1.09/3 = .363$ and MSW $= .55/24 = .023$

Let us present our results to this point in the accompanying analysis-of-

Table 12-5 ANOVA

Source of Variation	df	SS	MS
Among brands	3	1.09	.363
Within brands	24	.55	.023
Total	27	1.64	

variance table (Table 12-5). If H_0: $\mu_1 = \mu_2 = \mu_3 = \mu_4$ is true, the ratio of MSA/MSW = $F_{3,\,24}$. That is, the sampling distribution of MSA/MSW is the F distribution with 3 degrees of freedom in the numerator and 24 degrees of freedom in the denominator *if and only if* H_0 is true.

MSA/MSW = .363/.023 = 15.78

Suppose we decide to let $\alpha = .01$. From Table A-8 we find the critical F value to be 4.72. As $15.78 > 4.72$, we reject H_0: $\mu_1 = \mu_2 = \mu_3 = \mu_4$ and conclude that the mean hours per dollar cost for the four brands of 400-watt bulbs are not equal. Now what? Does this solve your problem? Are you ready to make your recommendation as to which brand is the most economical? We suspect not.

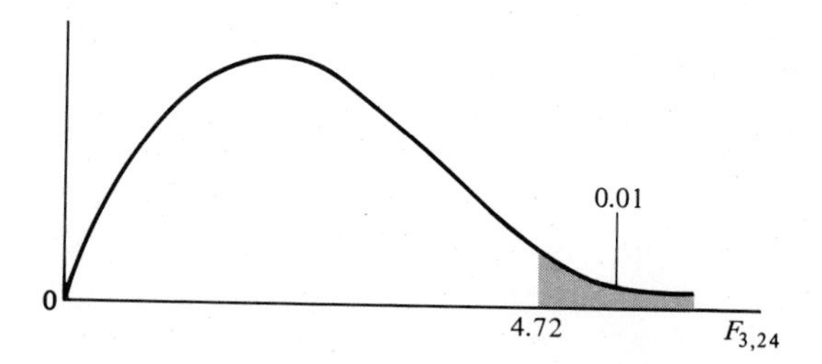

It seems as though the next logical question would be "Which μ_j's are not equal?" We could answer this by testing the following hypotheses:

H_{0_1}: $\mu_1 = \mu_2$

H_{0_2}: $\mu_1 = \mu_3$

H_{0_3}: $\mu_1 = \mu_4$

H_{0_4}: $\mu_2 = \mu_3$

H_{0_5}: $\mu_2 = \mu_4$

H_{0_6}: $\mu_3 = \mu_4$

John Tukey has devised a procedure which does fit our situation. Tukey has shown that any absolute difference between two sample means that exceeds $T\sqrt{\text{MSW}}$ is "significant." That is, any $|\bar{x}_j - \bar{x}_j'| > T\sqrt{\text{MSW}}$ leads to the rejection of the hypothesis H_0: $\mu_j = \mu_j'$. The assumptions for the Tukey method are the same as for our analysis-of-variance test. The value of T is determined by

$$T = \frac{1}{\sqrt{n}} q_{r,N-r}$$

where $q_{r,N-r}$ is known as the studentized range and is given in Table A-9. For our problem we find

$$\begin{aligned} T &= \frac{1}{\sqrt{n}} q_{4,24} \\ &= \frac{1}{\sqrt{7}} 3.90 \\ &= 1.474 \end{aligned}$$

and

$$\begin{aligned} T\sqrt{\text{MSW}} &= 1.474\sqrt{.023} \\ &= .224 \end{aligned}$$

Therefore, any $|\bar{x}_j - \bar{x}_j'|$ which exceeds .224 is "significant" and leads to the rejection of the hypothesis H_0: $\mu_j = \mu_j'$.

A convenient way of displaying the paired differences between sample means is shown in Table 12-6. Note that for the four absolute differences that exceed $T\sqrt{\text{MSW}} = .224$, we reject the corresponding null hypotheses. As a result of the Tukey analysis we can statistically conclude that

$\mu_1 > \mu_2 \qquad \mu_1 > \mu_3 \qquad \mu_1 > \mu_4 \qquad \mu_4 > \mu_3$

Actually the statistical conclusions are that

Table 12-6 Paired Comparisons

Hypothesis	$\lvert\bar{x}_j - \bar{x}_j'\rvert$	Conclusion
H_{0_1}: $\mu_1 = \mu_2$	.34	Reject H_{0_1}
H_{0_2}: $\mu_1 = \mu_3$	.55	Reject H_{0_2}
H_{0_3}: $\mu_1 = \mu_4$	.24	Reject H_{0_3}
H_{0_4}: $\mu_2 = \mu_3$	.21	Accept H_{0_4}
H_{0_5}: $\mu_2 = \mu_4$	.10	Accept H_{0_5}
H_{0_6}: $\mu_3 = \mu_4$	.31	Reject H_{0_6}

$\mu_1 \neq \mu_2$ $\mu_2 = \mu_3$ $\mu_3 \neq \mu_4$

$\mu_1 \neq \mu_3$ $\mu_2 = \mu_4$

$\mu_1 \neq \mu_4$

However, as

$\bar{x}_1 > \bar{x}_2$ $\bar{x}_1 > \bar{x}_3$ $\bar{x}_1 > \bar{x}_4$ $\bar{x}_4 > \bar{x}_3$

we would choose to conclude the above inequalities. The Tukey method and the specific studentized range table used are such that the probability is less than 0.05 of rejecting one or more true hypotheses.

Now you are ready to make your recommendation. Based upon the statistical analyses of the sample data, we can conclude that brand 1 has a mean life per dollar cost that exceeds those of the other three. All other considerations equal, the most economical bulb is brand 1. Note that this is a statistical conclusion and may not in fact be true, as it is based upon sample information. However, we have reached our set of statistical conclusions via objective techniques with known probabilities of error.

One further note: Had the difference between $\bar{x}_1$ and $\bar{x}_4$ been less than $T\sqrt{\text{MSW}} = .224$, we would have been unable to conclude $\mu_1 > \mu_4$. In this case, your recommendation might have been to focus attention on these two brands for further testing. It may be that one is indeed superior to the other but that the true difference is so small that a sample of seven from each is just not enough information.

12.5 SITUATION II: WHY?

Suppose you are the director of marketing for the glue division of Bordon Chemical Co. At present Bordon markets two types of white glue: (1) Water-soluble household, and (2) industrial-strength aliphatic resin. Your counterpart in charge of research and development has just told you that "one of our chemists has come up with a new-formula white glue and believes the new-formula glue to be superior in bond strength to Bordon's existing products." As you can recall the effect that the discovery and introduction of epoxy had on the entire adhesive industry, you certainly do not want to miss any new potential. The research and development director has further told you that he saw a test of the new glue on wood and the bond strength was indeed amazing. The discovery is so new that the research and development unit has not as yet had time to obtain an estimate of production costs. It is not clear at this time whether the new glue would be less expensive, as expensive, or more expensive to produce than your existing products. The research and development director turns to you and asks, "What do you think our next step should be? Should we work up some cost estimates for the new glue?"

Your immediate response is, "Sure, we will certainly need such information before we decide what to do with the new glue." However, as your responsibility is really marketing and sales, your question is, "What is the market for the new glue?" Should the product be marketed as a household glue, or would it be more appropriate for the industrial market? In terms of promotion and sales volume, it does make a difference whether the market is household or industrial. Before you can determine the appropriate market, you really need to know the characteristics of the new-formula glue. How does it compare with your existing glues relative to bond strength? The appropriate marketing strategy would certainly depend upon the production costs and bond strengths for the various glues.

No matter what the market, white glues are used as an adhesive for a variety of materials. The most common usages are with (1) wood, (2) paper, (3) cloth, and (4) plastics. It is your belief that the type of material being glued has an effect on the bond strength. In any event, you do need to know how the three glues compare in bond strength over the four types of materials. You certainly would not want to test your household glue on, say, paper, while the new glue is tested on plastic. Nor would you just want to test the three glues on only one type of material.

You would want to test all three glues on all four types of materials. Suppose you obtain six "identical" pieces of wood, paper, cloth, and plastic. Two pieces of each type of material are randomly assigned one of the three glues. The bond strengths are measured and shown in Table 12-7.

Your primary objective at this point is to determine whether there is a difference in mean bond strength among the three glues averaged over the four types of materials.

12.6 TWO-WAY ANOVA WITHOUT REPLICATIONS: HOW?

This analysis differs from the one-way ANOVA in that we are studying two independent variables, called *factors* (glues and materials), rather than just one (brand of bulbs). For this reason we must use slightly different notation. Again let us adopt a standard arrangement of the data as shown in Table 12-8. In the one-way ANOVA the subscript i merely indicated which sample observation the value was. The observations within a given row were not related. Now,

Table 12-7 Bond Strengths for Three White Glues
(In Pounds per Square Inch)

	Glues		
Materials	Water-Soluble Household	Industrial-Strength Aliphatic Resin	New Formula
Wood	125	140	152
Paper	112	130	124
Cloth	85	105	93
Plastic	85	97	98

they are. The b sample observations within any row are related in that they are observations with the same condition or level of factor A present. In Situation II the rows represent types of material. Thus, the general term x_{ij} stands for the sample observation for the ith level of factor A in combination with the jth level of factor B. Table 12-8 shows factor A with a levels, factor B with b levels, and one sample observation for each cell or combination of factors A and B. The term "without replications" refers to the fact that we have only one sample observation for each combination of factors A and B. A replication is a repeat of an experimental condition. Note that in Table 12-8 we see an extension of our original notation. Totals and sample means now require two subscripts. We need to be able to distinguish between totals and means for rows and for columns. As the i subscript of x_{ij} represents the row, $T_{1\cdot}$ and $\bar{x}_{1\cdot}$ are the total and mean for the b sample observations in row 1. Similarly, $T_{\cdot j}$ and $\bar{x}_{\cdot j}$ would be the total and mean of the a sample observations in column j.

As the rows now have special meaning, the total variation in the data set can be algebraically partitioned into three components:

$$\sum_{i=1}^{a}\sum_{j=1}^{b}(x_{ij} - \bar{x}_{\cdot\cdot})^2 = \sum_{i=1}^{a}\sum_{j=1}^{b}(\bar{x}_{i\cdot} - \bar{x}_{\cdot\cdot})^2 + \sum_{i=1}^{a}\sum_{j=1}^{b}(\bar{x}_{\cdot j} - \bar{x}_{\cdot\cdot})^2 + \sum_{i=1}^{a}\sum_{j=1}^{b}(x_{ij} - \bar{x}_{i\cdot} - \bar{x}_{\cdot j} + \bar{x}_{\cdot\cdot})^2$$

The first term on the right represents the sum of squared deviations among levels of factor A and is denoted as SSA. The second term on the right represents the sum of squared deviations among levels of factor B and is denoted SSB. The remaining term (third component on the right) is that part of the total variation that cannot be accounted for by factors A and B. This term is referred to as sum of squares error and is denoted as SSE. Thus, for the two-way ANOVA without replications, SST = SSA + SSB + SSE. Again we will use computational formulas to obtain the various sums of squares rather than the definitional forms shown above. The ANOVA table for the two-way classification without replications is shown in Table 12-9.

Table 12-8 Data Arrangement for a Two-Way ANOVA without Replications

		Factor B					
		1	2		b	Totals	Means
	1	x_{11}	x_{12}	. . .	x_{1b}	$T_{1\cdot}$	$\bar{x}_{1\cdot}$
	2	x_{21}	x_{22}		x_{2b}	$T_{2\cdot}$	$\bar{x}_{2\cdot}$
Factor A							
	a	x_{a1}	x_{a2}		x_{ab}	$T_{a\cdot}$	$\bar{x}_{a\cdot}$
	Totals	$T_{\cdot 1}$	$T_{\cdot 2}$		$T_{\cdot b}$	$T_{\cdot\cdot}$	
	Means	$\bar{x}_{\cdot 1}$	$\bar{x}_{\cdot 2}$		$\bar{x}_{\cdot b}$		$\bar{x}_{\cdot\cdot}$

Table 12-9 ANOVA for a Two-Way Classification without Replications

Source of Variation	df	SS	MS
Factor A	$a - 1$	SSA	$\text{SSA}/(a-1) = \text{MSA}$
Factor B	$b - 1$	SSB	$\text{SSB}/(b-1) = \text{MSB}$
Error	$(a-1)(b-1)$	SSE	$\text{SSE}/(a-1)(b-1) = \text{MSE}$
Total	$ab - 1$	SST	

$$\text{SST} = \sum_{i=1}^{a}\sum_{j=1}^{b} x_{ij}^2 - \frac{T..^2}{ab} \tag{12-3}$$

$$\text{SSA} = \frac{1}{b}\sum_{i=1}^{a} T_{i\cdot}^2 - \frac{T..^2}{ab} \tag{12-4}$$

$$\text{SSB} = \frac{1}{a}\sum_{j=1}^{b} T_{\cdot j}^2 - \frac{T..^2}{ab} \tag{12-5}$$

and

$$\text{SSE} = \text{SST} - \text{SSA} - \text{SSB}$$

12.7 FIXED-EFFECTS TWO-WAY ANOVA MODEL

We now have ab populations. Each population represents, at least in theory, an infinite collection of measurements for experimental units that received a specific combination of levels of factors A and B. Therefore μ_{ij} is the mean of the population that received the ith level of factor A and the jth level of factor B.

We now view an x_{ij} observation as being made up of four components. If we define the following

$$\mu_{i\cdot} = \frac{1}{b}\sum_{j=1}^{b} \mu_{ij}$$

$$\mu_{\cdot j} = \frac{1}{a}\sum_{i=1}^{a} \mu_{ij}$$

$$\mu = \frac{1}{ab}\sum_{i=1}^{a}\sum_{j=1}^{b} \mu_{ij}$$

and then let

$$\alpha_i = \mu_{i\cdot} - \mu \qquad \text{and} \qquad \beta_j = \mu_{\cdot j} - \mu$$

we can express x_{ij} as

$$x_{ij} = \mu + \alpha_i + \beta_j + \epsilon_{ij}$$

where ϵ_{ij} is the error term $(x_{ij} - \mu_{ij})$.

The assumptions of this model are that a random sample of size 1 is drawn from each of the ab populations, each of the ab populations is normally distributed with equal variances, and the factor effects are additive. As μ is the mean of the $\mu_{i.}$'s and the $\mu_{.j}$'s, and we have defined $\alpha_i = \mu_{i.} - \mu$ and $\beta_j = \mu_{.j} - \mu$, it follows that

$$\sum_{i=1}^{a} \alpha_i = \sum_{j=1}^{b} \beta_j = 0$$

Therefore we can express our model and assumptions in symbolic form as

$$x_{ij} = \mu + \alpha_i + \beta_j + \epsilon_{ij} \qquad i = 1, 2, \ldots a$$
$$j = 1, 2, \ldots b$$

$$\sum_{i=1}^{a} \alpha_i = \sum_{j=1}^{b} \beta_j = 0$$

where ϵ_{ij}'s are independently, normally distributed with mean zero and variance σ^2.

Given these assumptions, it can be shown mathematically that the expected values for the mean squares are

$$E(\text{MSE}) = \sigma^2$$

$$E(\text{MSA}) = \sigma^2 + \frac{b}{a-1}\sum_{i=1}^{a} \alpha_i^2$$

$$E(\text{MSB}) = \sigma^2 + \frac{a}{b-1}\sum_{j=1}^{b} \beta_j^2$$

The last component in E(MSA) cannot be negative as it is a sum of squared numbers, namely, α_i^2.

$$\frac{b}{a-1}\sum_{i=1}^{a} \alpha_i^2 \geq 0$$

And, it will equal 0 only if all α_i's = 0, which is when all $\mu_{i.}$'s are equal. A similar argument holds for the last term in E(MSB). That is,

$$\frac{a}{b-1}\sum_{j=1}^{b} \beta_j^2 \geq 0$$

and it equals 0 only when all $\mu_{.j}$'s are equal. As we know a ratio of two independent unbiased estimates of a common variance follows an F distribution, we have our two tests:

Test I (No difference among levels of factor A)

H_0: $\alpha_i = 0$ for all $i = 1, 2, \ldots a$

H_1: not all $\alpha_i = 0$

and

Test II (No difference among levels of factor B)

H_0: $\beta_j = 0$ for all $j = 1, 2, \ldots b$

H_1: not all $\beta_j = 0$

That is, if there are no differences among the mean effects of the levels of factor A, all α_i's will equal 0 and MSA and MSE will be independent unbiased estimates of σ^2. Thus

$$\text{MSA/MSE} = F_{a-1,(a-1)(b-1)}$$

Similarly, if there are no differences among the mean effects of the levels of factor B, all β_j's will equal 0 and

$$\text{MSB/MSE} = F_{b-1,(a-1)(b-1)}$$

12.8 SITUATION II REVISITED

We are now in a position to statistically compare the three glues applied to the four types of materials in Situation II. Again, for convenience, Table 12-10 is a reproduction of Table 12-7.

Our first step is to compute the various sums of squares. Using Equations (12-3), (12-4), and (12-5), we obtain the following.

$$\text{SST} = \Sigma\Sigma x_{ij}^2 - \frac{T..^2}{ab} \qquad (12\text{-}3)$$

$$= (125^2 + 112^2 + \cdots + 98^2) - \frac{1346^2}{12}$$

$$= 5309.67$$

$$\text{SSA} = \sum_{i=1}^{a} \frac{T_{i.}^2}{b} - \frac{T..^2}{ab} \qquad (12\text{-}4)$$

$$= (417^2 + 366^2 + 283^2 + 280^2)/3 - \frac{1346^2}{12}$$

$$= 4468.33$$

$$\text{SSB} = \sum_{j=1}^{b} \frac{T_{.j}^2}{a} - \frac{T..^2}{ab} \qquad (12\text{-}5)$$

$$= (407^2 + 472^2 + 467^2)/4 - \frac{1346^2}{12}$$

$$= 654.17$$

Table 12-10 Bond Strengths for Three White Glues
(In Pounds per Square Inch)

		Factor *B*				
		Water-Soluble Household	Industrial-Strength Aliphatic Resin	New Formula	Totals	Means
Factor *A*	Wood	125	140	152	$T_{1\cdot} = 417$	$\bar{x}_{1\cdot} = 139.00$
	Paper	112	130	124	$T_{2\cdot} = 366$	$\bar{x}_{2\cdot} = 122.00$
	Cloth	85	105	93	$T_{3\cdot} = 283$	$\bar{x}_{3\cdot} = 94.33$
	Plastic	85	97	98	$T_{4\cdot} = 280$	$\bar{x}_{4\cdot} = 93.33$
	Totals	$T_{\cdot 1} = 407$	$T_{\cdot 2} = 472$	$T_{\cdot 3} = 467$	$T_{\cdot\cdot} = 1346$	
	Means	$\bar{x}_{\cdot 1} = 101.75$	$\bar{x}_{\cdot 2} = 118.00$	$\bar{x}_{\cdot 3} = 116.75$		

As SST = SSA + SSB + SSE,

SSE = SST − SSA − SSB

SSE = 5309.67 − 4468.33 − 654.17

SSE = 187.17

Summarizing our results we have

ANOVA

Source of Variation	df	SS	MS
Factor *A*	3	4468.33	1489.44
Factor *B*	2	654.17	327.08
Error	6	187.17	31.20
Total	11	5309.67	

Test I

H_0: $\alpha_i = 0$ for $i = 1, 2, 3, 4$

H_1: not all $\alpha_i = 0$

Let $\alpha = .05$.

If H_0: $\alpha_i = 0$, for all i, is true,

$\text{MSA/MSE} = F_{3,6}$

That is, given the assumptions of the two-way ANOVA model (that each of the *ab* populations is normally distributed with equal variances and that the sample observations are the result of random sampling), the conditional (given H_0 is true) probability distribution of MSA/MSE is the F distribution with 3 degrees of freedom in the numerator and 6 in the denominator.

1489.44/31.20 = 47.74

As $47.74 > 4.76$, we reject H_0: $\alpha_i = 0$, for all i.

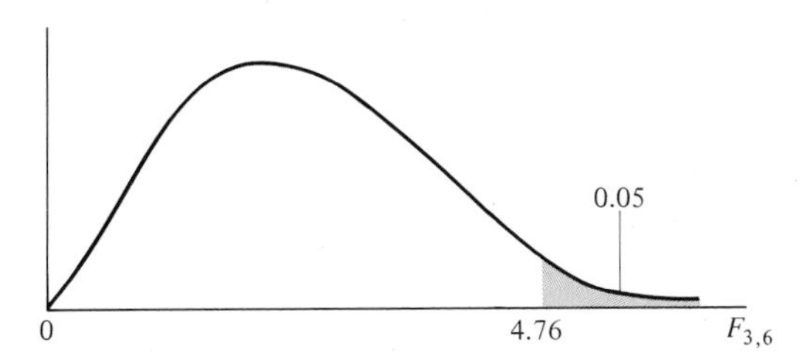

This result certainly supports our belief that type of material has an effect on bond strength.

The hypotheses for the second test are

Test II

H_0: $\beta_j = 0$ for $j = 1, 2, 3$

H_1: not all $\beta_j = 0$

Again, let $\alpha = .05$.

If H_0: $\beta_j = 0$, for all j, is true,

$$\text{MSB/MSE} = F_{2,6}$$

$$327.09/31.20 = 10.48$$

As $10.48 > 5.14$, we reject H_0: $\beta_j = 0$, for all j, and conclude there is a difference in the mean bond strengths of the three glues applied to the four types of material.

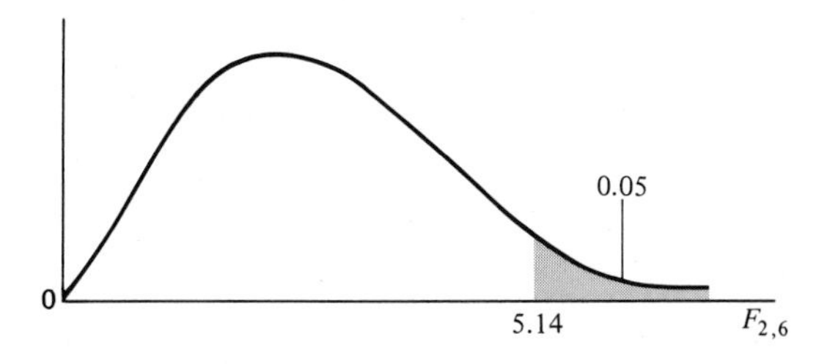

Now that we have decided that the three glues do not produce the same mean bond strength, our question is, "How do they compare?" What can we conclude about the glues other than they are not equal? Fortunately the Tukey method is again applicable. For comparing differences between pairs of sample means for factor B,

$$T = \frac{1}{\sqrt{a}} q_{b,\ (a-1)(b-1)}$$

and any

$$|\bar{x}_{\cdot j} - \bar{x}_{\cdot j'}| > T\sqrt{\text{MSE}}$$

leads to the rejection that $\mu_{\cdot j} = \mu_{\cdot j}'$. Therefore, if we choose $\alpha = .05$, our critical difference is

$$\frac{1}{\sqrt{4}} 4.34\sqrt{31.20} = 12.12$$

As

$$\bar{x}_{\cdot 2} - \bar{x}_{\cdot 1} = 16.25 > 12.12$$

$$\bar{x}_{\cdot 3} - \bar{x}_{\cdot 1} = 15.00 > 12.12$$

and

$$\bar{x}_{\cdot 2} - \bar{x}_{\cdot 3} = 1.25 < 12.12$$

we conclude that

$$\mu_{\cdot 2} > \mu_{\cdot 1}$$

$$\mu_{\cdot 3} > \mu_{\cdot 1}$$

and are unable to reject H_0: $\mu_{\cdot 2} = \mu_{\cdot 3}$.

Based upon the sample observations we can statistically conclude that both glues 2 and 3 (industrial-strength and new-formula, respectively) have mean bond strengths that are greater than that for glue 1 (household water soluble) averaged over the four types of material (wood, paper, cloth, and plastic). However, the difference between $\bar{x}_{\cdot 2}$ and $\bar{x}_{\cdot 3}$ was not large enough to conclude any real difference between glues 2 and 3.

Returning now for your recommendation, if the production cost for glue 3 is less than or equal to that for glue 1, you would probably want to drop glue 1 and replace it with the new formula glue 3. If on the other hand the production cost for glue 3 is less than that for glue 2, your existing industrial product, you may want to replace glue 2 with glue 3. By this move we would not expect to be marketing a "stronger" product, but rather one of basically equal strength but lower production cost.

12.9 IDENTIFICATION OF APPLICATIONS: WHEN?

You have seen two of the basic models of analysis of variance, one-way ANOVA and two-way ANOVA without replications. Although the two models differ, all analysis-of-variance models have the following in common:

1. The dependent variable is *quantitative*.
2. The independent variable or variables are *qualitative*.

In Situation I the dependent variable, sometimes referred to as the "response variable," was life in thousands of hours per dollar cost for 400-watt light bulbs—certainly a quantitative variable; the independent variable (brand of light bulb) was qualitative. In Situation II the dependent variable was bond strength measured in pounds per square inch (quantitative), and the two independent variables (type of glue and type of material) were qualitative. The distinction between the two models shown in this chapter relate to the number of independent variables.

There are basically two types of situations where the one-way ANOVA model would be appropriate.

1. We desire to test the hypothesis that means of $r \geq 2$ *existing* populations are equal.
2. We desire to test the hypothesis that the mean effects for $r \geq 2$ *treatments* applied to homogeneous experimental units are equal.

In the first case the populations exist, and we draw random samples from each. Alternatively, the one-way ANOVA model would be applicable in a situation where we are interested in two or more treatments. Treatments are to be assigned to homogeneous experimental units, and we are interested in the mean effects of the treatments. The experimental units would be randomly assigned to the various treatments.

The two-way ANOVA model would be applicable in situations where we are interested in analyzing the effects that combinations of two qualitative variables (factors) produce on a quantitative variable. In Situation II, we were interested in combinations of type of glue and type of material on the quantitative variable bond strength.

PROBLEMS

1. Identify an application of either one-way or two-way ANOVA in your major field of study. What conditions are present in your identified situation that make ANOVA both appropriate and of value?
2. Assume that Safeway recently conducted an experiment focused on the effects of store layout on purchase size per customer. The motivation for the experiment was the belief by certain individuals in the Safeway Corporation that the layout of a store may actually have an effect on the size of purchase. Given the size of the Safeway chain, an increase of only a few cents in the mean purchase amount per customer could have a substantial impact on corporate profits. A random sample of five stores was assigned to each of the four different layouts considered. The mean purchase amount per customer over the first six months the new layouts were in operation was computed for each store as presented below.

Layout 1	Layout 2	Layout 3	Layout 4
$20.78	$23.98	$19.45	$17.54
21.59	23.45	20.06	20.37
21.78	21.90	18.56	15.40
20.34	24.93	20.15	17.70
19.73	24.50	20.16	20.98

a. Test the hypothesis that the mean purchase amount per customer is the same for all four store layouts. Let $\alpha = .05$.

b. What assumptions did you find necessary in order to answer part *a*?

3. Use of analysis of variance in asset acquisition: Dynco Inc., a major manufacturer of ballbearings, was facing tough competition from Japanese manufacturers. Part of the reason was the aging machines which Dynco was using. The production manager sent a request to the comptroller for replacement of the existing (brand A) machines. Three different brands (B, C, D) of new machines were being considered. From the experiences of two of the firm's small subsidiary operations and information supplied by a personal friend of the production supervisor, the output per hour for four machines of each type was obtained and is presented below. Since each type of machine has similar acquisition and operating costs, the output per hour is seen as the critical variable in the purchase decision.

	Output per Hour			
Machine Type	A	B	C	D
1	502	896	726	989
2	451	900	700	950
3	631	897	826	1001
4	529	915	750	1120

In order to make a decision, the comptroller asked you to analyze the data as follows:

a. Test H_0: $\mu_A = \mu_B = \mu_C = \mu_D$. (Use $\alpha = .05$.)

b. If you reject the null hypothesis in part *a*, do the paired comparisons using Tukey's method.

c. On the basis of results obtained in part *b*, can you now determine which machine or machines are better than machine A? Which brand would you recommend that Dynco purchase?

4. As the firm's newly appointed finance manager, you were surprised to find that Wheelwright Wholesaling has always had a policy requiring full payment of accounts receivable within 30 days of billing with no discount for early payment. Had Wheelwright commanded a monopolistic (one seller and many buyers) market, this policy would have made sense. No wonder the director was grumbling during lunch about losing market share, tight credit, and consistent cash problems. You asked and received the approval of the director for a new financial management policy. However, the director told you that the effects of your new policy would be reviewed in six months.

You thought that an early payment discount would provide cash at a lower rate than the prevailing credit rate. Therefore, to determine the impact of early payment discounts, you decided to offer four different discount levels to a total of 52 firms. Thirteen firms received each discount level. After five months the average number of days between billing and receipt of payment was calculated for each firm.

Average Number of Days between Billing and Receipt of Payment

	1 No Discount, Net 30	2 1/10, Net 30	3 2/10, Net 30	4 2/10, 1/20, Net 30	5 3/10, Net 30
1	28	21	19	17	8
2	27	20	18	15	9
3	30	19	20	16	10
4	26	22	19	14	7
5	27	18	17	12	8
6	28	19	20	13	7
7	26	20	16	9	4
8	25	22	18	10	7
9	26	23	19	12	8
10	28	26	21	12	9
11	27	27	23	11	7
12	29	20	20	10	6
13	29	19	18	8	8

You asked your data analyst to analyze the data as follows:

a. Test: H_0: $\mu_1 = \mu_2 = \mu_3 = \mu_4 = \mu_5$. (Use $\alpha = .05$.)

b. Using Tukey's method, provide all the paired comparisons.

After receiving the analysis from the data analyst, you made your recommendations to the director. The director was very pleased with your recommendations but could not understand your "statistical" language. You are asked to rewrite the recommendation so that everybody can understand it. Do that for the director. Do not forget to qualify your recommendations with all the assumptions etc. you have made.

5. Suppose the personnel department of Fuller Brush Inc. was spending a substantial amount of money on different sales training techniques. The personnel manager was always full of enthusiasm and lost no time in describing how her experiment was going to find the best method to adopt for subsequent training programs. The director asked the personnel manager, "Will you be kind enough to substantiate your findings with some figures?" The personnel manager replied, "Here are the average weekly sales of seven randomly selected salespersons from each of the five equivalent sale territories. They were trained by five different methods. The performance of salespersons trained by method E is consistently superior. The director was impressed. He showed these figures to his statistician and asked him, "Do you also agree with the personnel manager's findings?" A couple of hours later, the statistician gave the director his findings. Do you think the statistician's findings were the

same as the personnel manager's? What assumptions, if any, should the statistician have made in his analysis and why?

Average Weekly Sales
(In Thousand of Dollars)

Method A	Method B	Method C	Method D	Method E
20.00	24.89	16.00	17.50	25.15
16.85	21.33	20.12	18.21	26.23
17.89	22.65	17.33	20.22	26.88
21.22	30.22	20.86	17.02	29.33
23.86	29.87	22.00	19.11	30.36
26.77	21.99	26.79	18.17	29.88
21.98	20.68	20.07	16.02	28.23

6. Use of analysis of variance in exploratory research: Suppose Smith and Wilson is a relatively small West Coast advertising agency. Assume that they have hired you to help in their newly established advertising research department. Your first task is to evaluate the results of a copy test for a new consumer product. Three different versions of the advertisement were prepared and each was shown to a random sample of 10 people. After the commercial was shown, each consumer's attitude toward the product was measured on a scale of 1 to 100 (the higher the score, the more favorable the attitude). What would you recommend?

Commercial A	Commercial B	Commercial C
49	88	31
53	87	23
50	92	26
51	98	28
48	79	24
47	68	11
49	99	13
53	83	40
52	84	28
51	78	39

7. ToughGuy Truck Leasing is a local truck-leasing agency that services the West Texas oil field. Its fleet of over 200 trucks includes pickups and vans, larger highway trucks, and massive off-road vehicles. ToughGuy is responsible for all vehicle maintenance. Gerry Hopkins recently concluded an experiment in which one of each category of vehicle utilized either a standard, premium, or super premium grade of oil for two years. The motivation for conducting this experiment was to see whether different kinds of oils have any effect on maintenance cost. Total costs of engine maintenance, including oil changes, per 100 hours of operation, are presented below for each of the nine vehicles involved in the test.

Maintenance Cost per 100 Hours of Operation

	Pickup	Highway Truck	Off-Road Vehicle
Standard oil	$102	$110	$63
Premium oil	68	88	41
Super premium oil	49	76	35

a. Give the type of design employed for this experiment and justify your answer.

b. Perform an analysis of variance.

c. Using a 5 percent level of significance, test the hypothesis that there is no difference in mean maintenance cost among the three oils averaged over the three type of vehicles.

d. If you decide that the three types of oil really affect the maintenance cost, i.e., do not produce the same mean maintenance cost, what else can you conclude about the costs other than that they are not equal?

8. Suppose the manager of Ford Motor Company recently read an article in *Business Week* as to how the democratic style of supervision of plant workers in Volvo plants in Sweden has reduced the cost of production. Unlike their American counterparts, Volvo workers are divided into teams and each team is required to finish the allotted work within the prescribed time. According to *Business Week*, not only has this resulted in reduction of costs at all levels, but the workers suffer less from one of the most dreaded diseases—boredom. In order to see whether the same style of work schedule can be implemented in a Ford plant, the plant manager ordered a feasibility study.

   Six of the firms' plants were expanded, reorganized, and assigned to assemble components for passenger cars. With the reorganizations and expansion, the production manager implemented three different salary plans in the six plants: a straight hourly wage, a straight piece rate, and a piece rate that increased as output increased. In addition, the manager instituted a very close, authoritarian supervising style in three of the plants and a more democratic style in the other three. The labor cost per unit produced from each plant for the first two months of the experiment is shown below.

	Authoritarian Supervision	Democratic Supervision
Hourly rate	6.89	4.83
Constant price rate	3.86	2.75
Increasing price rate	4.25	3.28

a. Identify the design the manager used for this experiment and justify your diagnosis.

b. Perform an analysis of variance on the data.

c. Do the data provide sufficient evidence that mean labor costs among the two styles of supervision averaged over the three types of rates are different? (Use $\alpha = .05$.)

d. Do the data provide sufficient evidence that there are differences among the levels of labor rates averaged over styles of supervision? (Use = .05.)

e. On the basis of the analysis performed above, do you think that the proposed plan is feasible for the Ford Motor Company?

9. Suppose Del Monte Foods is considering the introduction of a new type of frozen mixed vegetables with a cream sauce. The company is not sure whether the price should be below, the same as, or above the existing competition. Likewise, it is uncertain about the effectiveness of television advertising versus newspaper coupons as a device for inducing initial sales. Therefore, six similar markets are selected and used as test markets such that each possible combination of price level and advertising technique occurs once. The case sales at the end of the month in each area are shown below. What would you recommend?

	Case Sales per Advertising Medium	
	Coupons	Television
Price below existing competition	101	106
Price same as existing competition	94	100
Price above existing competition	83	87

10. Assume you have just been hired as a statistician by the Federal Trade Commission and assigned to the section working on deceptive advertising. A recent advertisement for a major detergent contained the following claim: ". . . cleans better than the two best selling brands in hot, cold, or warm water." The agency decided to conduct a test of this claim to determine if action should be taken against the firm involved.

A box of each of the three brands of detergent involved was purchased from a local supermarket. Nine identical piles of laundry were carefully soiled to the same extent. Each detergent was used to wash three piles of laundry, one at a hot setting, one at a warm setting, and one at a cold setting. Each pile was then assigned a score of 1 to 100 based on the amount of soil removed (the higher the score, the cleaner the laundry). The results are presented below.

	Hot Water	Warm Water	Cold Water
Advertised brand	98	86	79
Competitor A	93	84	63
Competitor B	46	44	40

Should the FTC take action against the advertised brand?

11. A major distillery for alcoholic beverages is interested in gaining a large share of the market composed of people aged 18–23 (in those states where 18 is the legal drinking age). Beer and wine are the traditional favorites of this age group. However, Bill Schuessler, the firm's marketing manager, and Carol Pringle, manager of research

and development, feel that a new version of distilled apple cider will appeal to this younger market. While they are sure that the alcohol content should remain below 5 percent, they are not sure exactly what it should be. Therefore, five versions of the cider have been prepared and each version has been tasted by a random sample of 25 young consumers. Each consumer has evaluated the drink on a 100-point scale, with the higher scores representing a greater liking for the product. The results of the test appear below. What should Schuessler and Pringle conclude?

1% Version	2% Version	3% Version	4% Version	5% Version
16	40	81	89	53
18	62	81	95	20
72	11	35	10	98
57	49	16	70	67
30	40	83	94	92
16	63	35	06	08
96	77	20	20	59
39	93	64	20	49
31	62	76	38	03
78	91	19	47	32
03	93	33	30	47
74	42	33	28	72
09	06	09	90	95
42	23	12	42	25
16	15	08	02	04
21	45	40	74	20
21	58	57	80	50
56	61	78	95	97
45	65	66	99	24
91	34	84	98	49
91	22	21	20	35
50	07	38	14	21
65	39	05	22	58
27	50	96	13	44
37	16	94	85	17

12. After you return from the annual SOSBA (Supplies of the Street Barricades Association) Conference in New York City, you find the following memo on your desk:

MEMO National Barricade Co.

TO: *Dick Hawkins, Purchaser*

FROM: *Shirley Barton, Director*

Hope you have some nice things to say about the conference during lunch today. Meanwhile, you will be happy to note that our bid to supply barricades to the city of Eugene has been approved. Chief Sight Overseer, Bob Pitwins, informs me that we will need 50,000 batteries. We cannot put away this fact as trivial because all our barricades have battery-powered flashing lights. The way you have analyzed this kind of problem in the past, I took the liberty to ask three main suppliers to supply five batteries each. Bob Pitwins made a special effort to test the life of these three brands of batteries. The data (life in thousands of hours and cost per battery) are enclosed.

I will really appreciate it if you could send me your recommendation as soon as possible. See you in the cafeteria at noon.

(SIGNED) *Shirley*

Encl.

Brand A	Brand B	Brand C
1.0	1.9	1.3
0.7	1.8	1.6
1.1	2.0	1.3
0.9	1.6	1.5
1.0	2.4	1.4

Cost per battery:

Brand A	$1
Brand B	$2
Brand C	$1

Write your recommendations to Shirley Barton.

Chapter 13

Analysis of Frequencies

13.1 SITUATION I: WHY?

The insurance business isn't what it used to be. Ten to fifteen years ago, rates could be established differently for various age groups, sexes, and individuals with different marital status. Now, however, with the advent of a number of laws barring price discrimination on these bases alone, insurance companies are required to provide substantial documentation justifying any price differential established for any of these groups.

Suppose you are employed at the main office for State Farm Insurance in Bloomington, Illinois, and you are called upon to deal with a problem created by recent legislation in the state of Oregon. The new law forbids insurance companies to charge different rates on the basis of sex and age unless sufficient statistical evidence can be presented which indicates a true difference among claim rates for the various classes. Nationwide experience in automobile insurance underwriting indicates that the differential rates are justifiable. Your task is to develop statistical evidence which can be used in the first presentation to the insurance commissioner in Oregon.

If the necessary documentation cannot be provided, one of two alternatives remain. First, State Farm can refuse to alter its rate schedule. This would result in State Farm's losing its permit to sell insurance in Oregon. Currently, State Farm insures approximately 130,000 motorists in Oregon with an average premium of $240 per year. Losing its permit would mean State Farm would lose *revenues of $31,200,000 in the coming year. With costs increasing as they are, revenue losses in subsequent years would be even greater.*

The second alternative, if evidence cannot be provided, is to establish a uniform rate. This would result in a 25 to 30 percent increase in rates charged the lowest-risk motorists. The majority of these would likely cancel their policies or at least become somewhat disenchanted with State Farm. Further, there would be a decrease in rates charged higher-risk drivers. This would result in State Farm's insuring a larger percentage of these motorists. Not only would revenues decrease but claims would likely increase as well. The increased risk would put any insurance company in a precarious position.

Your documentation must include evidence that the different premiums are justified and evidence identifying the degree of differentiation justified. The first you expect to accomplish with hypothesis testing, the second with estimation.

You begin your analysis with the hypothesis-testing phase. In a preliminary investigation, you gather claims data from a random sample of 100 policies. Currently, State Farm has different rates for male and female drivers and different rates for four age groups. These groups are identified with the data in Tables 13-1 and 13-2.

Analysis of the first table provides no difficulties for you at this time. Using the normal approximation to the binomial and the procedures in Chapter 8, you can test the following hypothesis:

$$H_0: \pi_1 = \pi_2$$

where π_i denotes the claim rate for the ith class (male, female). If you were to choose $\alpha = .10$ as the level of significance, you would reject the null hypothesis, since

$$z = \frac{\frac{24}{58} - \frac{10}{42}}{\sqrt{\frac{34}{100}\left(\frac{66}{100}\right)\left(\frac{1}{58} + \frac{1}{42}\right)}} = 1.831$$

which exceeds 1.645. This result would justify your belief in different claim rates for the two sexes.

Table 13-1 Number of Occurrences by Sex

		Claims	No Claims	Total
Sex of Primary Operator	Male	24	34	58
	Female	10	32	42
	Total	34	66	100

Table 13-2 Number of Occurrences by Age

		Claims	No Claims	Total
Age of	16–17	14	18	32
Youngest	18–25	9	24	33
Driver	26–55	4	16	20
Insured	Over 55	7	8	15
	Total	34	66	100

Analysis of the second table is a different matter, however. The test employed above can only analyze two samples. Furthermore, while the analysis of variance can be used to test the equality of the means of more than two populations, the assumptions for that method include the presence of continuous response variables which are normally distributed. Since your data are frequencies *or* counts, *you are not at all comfortable with making those assumptions. How can you document evidence supporting price differentials for the different age groups? Specifically, you need to be able to test the hypothesis that all four age classes have identical claim rates, that is,*

H_0: $\pi_1 = \pi_2 = \pi_3 = \pi_4$

13.2 *r* DICHOTOMOUS POPULATIONS: HOW?

An observation from a dichotomous population identifies one of two possible values. Most often, our interest lies in the number observed of one of the values. These numbers are called *counts* or *frequencies*, as we saw in Chapter 2, and they are indicative of what are called *enumerative data*.

The outcome of random sampling from a large dichotomous population can be described by a binomial probability distribution. In Situation I you face two problems involving such samples. In the first, you chose to approximate the two binomial distributions by normal distributions. Those approximations allow you to test the difference in the proportions of claims in the two samples. Note, however, that the binomial distributions being approximated had sample sizes of $n = 58$ and $n = 42$, respectively. These sample sizes can reasonably be considered large enough to allow the approximation.

In every hypothesis-testing problem we have encountered, we have noticed the same general analytical format. If the null hypothesis is true, we expect to observe a certain outcome. For instance, if the null hypothesis is

H_0: $\mu = 70$

we expect the sample mean, say $\bar{x}$, to be close to 70. To determine if $\bar{x}$ is close to 70, we examine the difference between $\bar{x}$ and 70, namely, $\bar{x} - 70$. The sampling distribution of $\bar{x}$, or equivalently $\bar{x} - 70$, then gives us a measure for judging whether $\bar{x}$ is different enough from 70 to warrant rejection of the null hypothesis.

In the analysis of frequencies, we analyze the data using the same principle. The frequencies are arranged in a table as was done in Situation I. Each position in the table is called a *cell*. The analysis is performed on a cell-by-cell basis. That is, we compare what we observe in each cell with what we would expect to observe *if the null hypothesis were true*.

To facilitate the discussion, we present the notation used in the analysis in the table of observed frequencies.

Observed Frequencies

	Outcome 1	Outcome 2	Total
Sample 1	O_{11}	O_{12}	R_1
Sample 2	O_{21}	O_{22}	R_2
.	.	.	.
.	.	.	.
.	.	.	.
Sample r	O_{r1}	O_{r2}	R_r
Total	C_1	C_2	n

Note in this table that R_i is the sum of the observed frequencies in row i, that is,

$$R_i = O_{i1} + O_{i2}$$

Note that the column sums for the table are denoted by C_j's. That is,

$$C_j = O_{1j} + O_{2j} + \cdots + O_{rj}$$

Finally, we point out that the sample size n satisfies

$$\begin{aligned} n &= C_1 + C_2 \\ &= R_1 + R_2 + \cdots + R_r \end{aligned}$$

We let π_i denote the proportion of population i which belongs to the class labeled "Outcome 1". Equivalently, π_i can be thought of as the probability of selecting one item from population i which belongs to the class identified as "Outcome 1". With this added notation, we can express the null hypothesis of interest in this statistical analysis. Namely, the null hypothesis is

$$H_0: \pi_1 = \pi_2 = \cdots = \pi_r$$

As in other hypothesis-testing problems, we will compare the observed cell frequency, O_{ij}, with the cell frequency *we expect to observe if the null hypothesis is true*. We label the expected cell frequency as E_{ij}. Thus, E_{ij} is the value that the null hypothesis suggests we will observe in cell i, j. For each cell, we form the difference $O_{ij} - E_{ij}$ to provide the means for comparison. The mathematics which is used to derive a sampling distribution dictates that we sum the terms

$$\frac{(O_{ij} - E_{ij})^2}{E_{ij}}$$

for all the cells. When we do that, if certain conditions hold, we can treat the sum as an observation from a chi-square random variable. The appropriate chi-square probability distribution can then be used to determine if the observed cell frequency O_{ij} differs enough from the hypothesized expected cell frequency E_{ij} to warrant rejection of the null hypothesis.

All this can be summarized by stating that, if H_0 is true,

$$\chi^2 = \sum_{i=1}^{r} \sum_{j=1}^{2} \frac{(O_{ij} - E_{ij})^2}{E_{ij}} \tag{13-1}$$

is approximately distributed as a chi-square random variable with $r - 1$ degrees of freedom. You should note that the sum includes terms from every cell.

An important observation should be made at this point. If the null hypothesis is true, each observed cell frequency O_{ij} will be close to its associated expected cell frequency E_{ij}. Consequently, the value of the expression in Equation (13-1) will be small. Because of this, if the value of the test statistic from Equation (13-1) is small, we should conclude that the data support the null hypothesis. Conversely, if the test statistic value computed by Equation (13-1) is very large, then the observed cell frequencies are quite different from the hypothesized expected cell frequencies, and the data are evidence contrary to the null hypothesis. Thus, we would tend to reject the null hypothesis when the test statistic of Equation (13-1) is large. This says that the rejection region includes only large values, that is, values in the upper tail of the appropriate chi-square distribution.

We've made a couple of vague statements about the sampling distribution of the expression in Equation (13-1). First, we said that "if certain conditions hold," we could use the chi-square distribution. Then we said that χ^2 was "approximately" distributed as a chi-square random variable. The important property of the data which affects the accuracy of the approximating chi-square distribution is the size of each expected cell frequency E_{ij}.

The difficulty with small expected cell frequencies becomes clear by examining Equation (13-1). The value E_{ij} appears in the denominator of the term associated with it. The smaller the value of the expected cell frequency, the smaller is the denominator; therefore the more sensitive is the value

$$\frac{(O_{ij} - E_{ij})^2}{E_{ij}}$$

to small changes in $O_{ij} - E_{ij}$. This distorts the behavior of the random variable χ^2 so that, as an expected cell frequency decreases, the chi-square distribution becomes a less accurate approximation to the sampling distribution of χ^2. In practice, the approximation is quite accurate if *all* expected cell frequencies are *at least as large as* 5.

A question quite naturally arises concerning the appropriate methods for dealing with problems in which one or more of the expected cell frequencies is less than 5. This problem is more easily discussed in Section 13.4 where we extend the present analysis to contingency tables.

Let us return to the problem of testing the equality of several probabilities. The appropriate null hypothesis is

$$H_0: \pi_1 = \pi_2 = \cdots = \pi_r$$

where π_i is the probability of success for the ith binomial distribution. When the binomial distribution describes random sampling from a dichotomous population, this probability can also be viewed as the proportion of the population with the particular characteristic of interest (outcome 1 or outcome 2). Consequently, these hypothesis-testing problems are spoken of as tests for differences in probabilities or for differences in proportions.

If the null hypothesis is true, all populations have the same proportion of items with the property "outcome 1". Therefore, each of the samples can be combined to provide an estimate of the probability of observing an outcome 1. Consequently, we would get our best estimate of this probability by using the entire collection of n observations. Therefore, our best estimate of the probability for outcome 1 is given by the ratio C_1/n, and our best guess at the probability for outcome 2 is C_2/n. To be more concise, our estimate of the probability of outcome j is $C_j/n, j = 1, 2$. Since our ith sample contains R_i observations, the number of outcomes j we expect under the null hypothesis is

$$\frac{C_j}{n} R_i$$

So, the expected frequency for the cell corresponding to the ith population and outcome j is

$$E_{ij} = \frac{C_j}{n} R_i = \frac{R_i C_j}{n} \qquad \text{for } i = 1, 2, \ldots, r \quad j = 1, 2 \qquad (13\text{-}2)$$

For example, for the expected number of observations in sample 1 of outcome 1, we have

$$E_{11} = \frac{R_1 C_1}{n}$$

The expected number of observations in sample 2 of outcome 1 is

$$E_{21} = \frac{R_2 C_1}{n}$$

Table 13-3 Expected Cell Frequencies

	Outcome 1	Outcome 2	Total
Sample 1	$\frac{R_1C_1}{n}$	$\frac{R_1C_2}{n}$	R_1
Sample 2	$\frac{R_2C_1}{n}$	$\frac{R_2C_2}{n}$	R_2
.	.	.	
.	.	.	
.	.	.	
Sample r	$\frac{R_rC_1}{n}$	$\frac{R_rC_2}{n}$	R_r
Total	C_1	C_2	n

We present these expressions in Table 13-3. Note that each expected cell frequency is nothing more than the appropriate *row sum multiplied by the associated column sum and divided by the total number of observations*.

Table 13-1 (Restated) Number of Occurrences

	Claims	No Claims	Total
Male	24	34	58
Female	10	32	42
Total	34	66	100

To illustrate the analysis, we return to Table 13-1 in Situation I. The analysis of that problem has already been performed but could have been accomplished by the methods we've just described.

We calculate the expected cell frequencies as

$$E_{11} = \frac{R_1C_1}{n} = \frac{58(34)}{100} = 19.72$$

$$E_{12} = \frac{R_1C_2}{n} = \frac{58(66)}{100} = 38.28$$

$$E_{21} = \frac{R_2C_1}{n} = \frac{42(34)}{100} = 14.28$$

$$E_{22} = \frac{R_2C_2}{n} = \frac{42(66)}{100} = 27.72$$

These expected cell frequencies are given in Table 13-4.

Table 13-4 Expected Cell Frequencies

	Claims	No Claims	Total
Male	19.72	38.28	58
Female	14.28	27.72	42
Total	34	66	100

The analysis now proceeds as

Cell	O	E	$O - E$	$(O - E)^2$	$(O - E)^2/E$
1, 1	24	19.72	4.28	18.32	.929
1, 2	34	38.28	−4.28	18.32	.479
2, 1	10	14.28	−4.28	18.32	1.283
2, 2	32	27.72	4.28	18.32	.661
					$\chi^2 = 3.352$

We have arranged the analysis in tabular form to facilitate the understanding of the calculations that take place. We have used one row for each cell of the table. In the row corresponding to any cell, we include:

1. The observed cell frequency O for that cell.
2. The expected cell frequency E for that cell.
3. The difference between the observed cell frequency and the expected cell frequency, $O - E$.
4. The square of the difference, $(O - E)^2$.
5. The value of the term in Equation (13-1) which corresponds to that particular cell.

Now, if the null hypothesis of equal probabilities is true, the value $\chi^2 = 3.352$ can be viewed as an observation from a chi-square random variable with 1 degree of freedom $(r - 1)$. For $\alpha = .10$, the value from Table A-6 in the Appendix is 2.70554. Since $3.352 > 2.70554$, the null hypothesis is rejected as it was in Situation I. This test will always agree with the test used in Situation I.

13.3 SITUATION I REVISITED

We want to test the null hypothesis that the probabilities of claims from the various age groups are all equal. That is, we wish to test the null hypothesis

$$H_0: \pi_1 = \pi_2 = \pi_3 = \pi_4$$

For $\alpha = .10$, we have already rejected the null hypothesis that different sexes have the same claims rate. Now let's turn our attention to Table 13-2, which gives observed frequencies for claims as broken down by age groups. The row sums and column sums are included in the table. They are:

$R_1 = 32 \quad R_2 = 33 \quad R_3 = 20 \quad R_4 = 15 \quad C_1 = 34 \quad C_2 = 66$

The total number of observations is $n = 100$.

Table 13-2 Number of Occurrences by Age

		Claims	No Claims	Total
Age of	16–17	14	18	32
Youngest	18–25	9	24	33
Driver	26–55	4	16	20
Insured	Over 55	7	8	15
	Total	34	66	100

Let us calculate the expected cell frequencies and arrange them in Table 13-5. All these values are greater than 5, so we approximate the sampling distribution of χ^2 by the chi-square distribution with $r - 1 = 4 - 1 = 3$ degrees of freedom. The critical value for $\alpha = .10$ is found in Table A-6 in the Appendix to be 6.25139.

The calculation of χ^2 proceeds as

Cell	O	E	$O - E$	$(O - E)^2$	$(O - E)^2/E$
1, 1	14	10.88	3.12	9.73	.894
1, 2	18	21.12	−3.12	9.73	.461
2, 1	9	11.22	−2.22	4.93	.439
2, 2	24	21.78	2.22	4.93	.226
3, 1	4	6.80	−2.80	7.84	1.153
3, 2	16	13.20	2.80	7.84	.594
4, 1	7	5.10	1.90	3.61	.708
4, 2	8	9.90	−1.90	3.61	.365
					$\chi^2 = 4.840$

Since $\chi^2 = 4.840 < 6.25139$, our test statistic value is in the acceptance

Table 13-5 Expected Cell Frequencies

		Claims	No Claims
	16–17	10.88	21.12
Age	18–25	11.22	21.78
Group	26–55	6.80	13.20
	Over 55	5.10	9.90

Figure 13-1

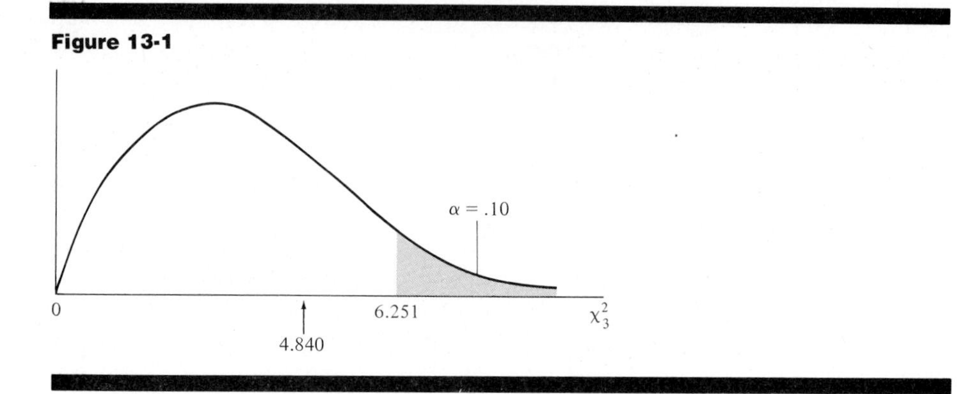

region as diagrammed in Figure 13-1. The random sample of 100 policies does not lead to the rejection of the null hypothesis of equal probabilities. Consequently, these data do not support differential rates for the various age groups. Therefore, your data and analyses would support differential rates on sex, but not on age.

13.4 CONTINGENCY TABLES

You are rather disturbed by the results of your analysis, and you should be. The policy of price discrimination on the basis of the age of the youngest driver insured was instituted by State Farm in response to experiences over many years. Yet your data do not support the policy. You reason that perhaps the difficulty lies in the structure you've given the problem. If costs are higher for some age groups, it may not be a result of *more claims*. Rather it may be that the *amounts claimed* are different for the different age groups. In other words, one age group may not file claims more often but may simply file claims of greater value. So you arrange the data you already have into Table 13-6.

Table 13-6 is an example of a 4-by-4 *contingency table*. In general, we can have r-by-c contingency tables, that is, tables with r rows and c columns. A contingency table represents the classification of each item in a sample according to two variables. The classification usually represents various levels of qualitative characteristics. This points to one advantage of this analysis, that is, that it is applicable to qualitative as well as quantitative variables.

In Table 13-6, each insurance policy sampled is classified on the basis of the age of the youngest driver insured by it and on the basis of the amount claimed under that policy. The rows identify r different classes of one of the variables, and the columns represent the c different categories of the other. Each item in the sample is associated with exactly one row and exactly one column. Again the cell entries are frequencies. For instance, the observed value in cell 2, 3 is the number of items in the sample which belong to class 2 of the row variable and at the same time class 3 of the column variable.

The choice of classes for each variable is made by the decision maker. Because more classes allow better discrimination, more classes mean more information is contained in the table. However, there is a limit to the number of classes possible for a given sample size, as we will see shortly. Sometimes the decision maker is presented with only one possible classification scheme for a variable. This is the case with the column variable in the problems discussed in Section 13.2. Of course, the classification scheme is dictated in these cases.

One use of contingency tables is the analysis of differences in probabilities or proportions. Problems of r dichotomous populations are presented in r-by-2 contingency tables. In this usage, analysis of the r-by-c contingency table is an extension of the procedure presented in Section 13.2.

More often, however, contingency tables are used for making judgments of the *independence* or *dependence* of the two variables represented by the rows and columns. *The only difference is one of interpretation*: Does the analyst view the rows as different populations or simply as different values or classes of a variable? Table 13-6 will be analyzed to test the independence of the two variables. So, while in Section 13.3 the age groups were treated as populations, here the age of the youngest driver insured is treated as a variable. This should illustrate that the essence and the analysis of the problem remain the same whether the analyst is testing for differences in probabilities or testing independence.

The analysis of contingency tables is accomplished by a straightforward extension of what was presented in Section 13.2. For each cell, we form

$$\frac{(O - E)^2}{E}$$

where O is the observed cell frequency and E is the hypothesized expected cell

Table 13-6 Number of Occurrences

		Amount of Claim				
		<$50	$50–$100	$100–$200	>$200	Total
Age of Youngest Driver Insured	16–17	19	1	4	8	32
	18–25	24	6	1	2	33
	26–55	16	3	0	1	20
	Over 55	8	3	3	1	15
	Total	67	13	8	12	100

frequency. Now, if H_0 is true,

$$\chi^2 = \sum_{i=1}^{r} \sum_{j=1}^{c} \frac{(O_{ij} - E_{ij})^2}{E_{ij}} \tag{13-3}$$

is approximately distributed as a chi-square random variable if all the E_{ij}'s are at least as large as 5. The appropriate degrees of freedom for the approximating chi-square can be calculated as $(r - 1)(c - 1)$. We should point out that this value, $(r - 1)(c - 1)$, is a more general formula for calculating the degrees of freedom which can be applied to the test statistic of Section 13.2. There we had an r-by-2 contingency table. The degrees of freedom for the approximating chi-square random variable would be

$$(r - 1)(2 - 1) = r - 1$$

the same value given in Section 13.2.

To test the independence of the row and column variables, we set

H_0: the two variables are independent

Now the task is to determine the expected cell frequencies *if H_0 is true*. Again we let R_i represent the sum of the frequencies in row i, C_j denote the sum of the frequencies in column j, and n be the total number of items in the sample. As in Section 13.2,

$$n = R_1 + R_2 + \cdots + R_r \qquad \text{and} \qquad n = C_1 + C_2 + \cdots + C_c$$

If the two variables are independent, then the probability π_{ij} that any item in the sample will fall into cell i, j is the product of two probabilities. Namely, for any item in the sample,

$$\pi_{ij} = Pr(\text{item in cell } i, j)$$

$$= Pr(\text{item in class } i \text{ of row variable})Pr(\text{item in class } j \text{ of column variable})$$

This expression represents an application of the special multiplication rule developed in Chapter 4. Recall that this rule is employed whenever we compute the probability of a joint occurrence of events which are *independent*. Now, R_i/n is our best estimate of the probability that the item falls into class i of the row variable; and C_j/n is our best guess at the probability that the item falls into class j of the column variable. So, our best estimate for π_{ij} is

$$\left(\frac{R_i}{n}\right)\left(\frac{C_j}{n}\right)$$

Given this probability and a sample size of n we expect to observe

$$E_{ij} = \left(\frac{R_i}{n}\right)\left(\frac{C_j}{n}\right) n = \frac{R_i C_j}{n} \tag{13-4}$$

items in cell i, j. That is, *if the row and column variables are independent*, the expected cell frequency is the product of the row sum and the column sum

Table 13-7 Expected Cell Frequencies

		Amount of Claim			
		<$50	$50–$100	$100–$200	>$200
Age of Youngest Driver Insured	16–17	21.44	4.16	2.56	3.84
	18–25	22.11	4.29	2.64	3.96
	26–55	13.40	2.60	1.60	2.40
	Over 55	10.05	1.95	1.20	1.80

divided by the sample size. Note that Equation (13-4) gives the same formula for the calculation of the expected cell frequencies as Equation (13-2).

OK, now let's use Equation (13-4) to calculate the expected cell frequencies for our test of the independence of the two variables, the age of the youngest driver insured by a policy and the total amount of claims filed. Table 13-7 gives these values.

At this point, the analysis can proceed no farther without some modification. Some of the expected cell frequencies are less than 5. One of two things can be done to rectify the situation. First, we could gather more data. This may not be practical. Further, it is not always clear how many more observations are needed.

The second possibility is to collapse the table by combining and/or redefining classes. For instance, in Tables 13-6 and 13-7, we might combine the age groups 26–55 and over 55. We might also redefine the amount-claimed classes as <$50, $50–$150, and >$150. If we do collapse Table 13-6 in those ways, we obtain Table 13-8. With these data, the expected cell frequencies are as given in Table 13-9.

The form of the collapsed table is, for the most part, left to the discretion of the analyst just as the original classifications were. However, it must remain capable of providing the information necessary for decision making. At any rate, collapsing the table reduces the information content. For instance, Table

Table 13-8 Number of Occurrences

		Amount Claimed			
		<$50	$50–$150	>$150	Total
Age of Youngest Driver Insured	16–17	19	1	12	32
	18–25	24	7	2	33
	Over 25	24	8	3	35
	Total	67	16	17	100

13-8 can no longer tell you anything about the group of policies with the youngest insured motorist in the over-55 category. For that reason, collapsing should be kept to a minimum.

The analysis of the 3-by-3 contingency table of Table 13-8 proceeds as

Cell	O	E	$O - E$	$(O - E)^2$	$(O - E)^2/E$
1, 1	19	21.44	−2.44	5.95	.278
1, 2	1	5.12	−4.12	16.97	3.314
1, 3	12	5.44	6.56	43.03	7.910
2, 1	24	22.11	1.89	3.57	.161
2, 2	7	5.28	1.72	2.96	.561
2, 3	2	5.61	−3.61	13.03	2.323
3, 1	24	23.45	.55	.30	.013
3, 2	8	5.60	2.40	5.76	1.029
3, 3	3	5.95	−2.95	8.70	1.462
					$\chi^2 =$ 17.051

For the $\alpha = .10$ level of significance, the critical value for a chi-square with $(r - 1)(c - 1) = (3 - 1)(3 - 1) = 4$ degrees of freedom is 7.77944. The calculated χ^2 value clearly exceeds the critical value of 7.77944. Thus, the indication is that the conditional probability of observing a sample outcome as extreme as that observed is less than .10 (see Figure 13-2).

Consequently, we reject the null hypothesis of independence and conclude that there is a relationship between the age groups specified and the amount claimed. Now the hypothesis-testing phase of your study for State Farm is complete. You will be able to justify differential rates based upon the sex of the primary operator and the age of the youngest driver insured. What remains is the estimation phase, which can be accomplished by methods similar to those presented in Chapter 7.

13.5 GOODNESS-OF-FIT TEST

We have seen in this chapter how to test the hypothesis that r population proportions are equal and how to test the hypothesis that two characteristics

Table 13-9 Expected Cell Frequencies

		Amount Claimed		
		<$50	$50–$150	>$150
Age of Youngest Driver Insured	16–17	21.44	5.12	5.44
	18–25	22.11	5.28	5.61
	Over 25	23.45	5.60	5.95

Figure 13-2

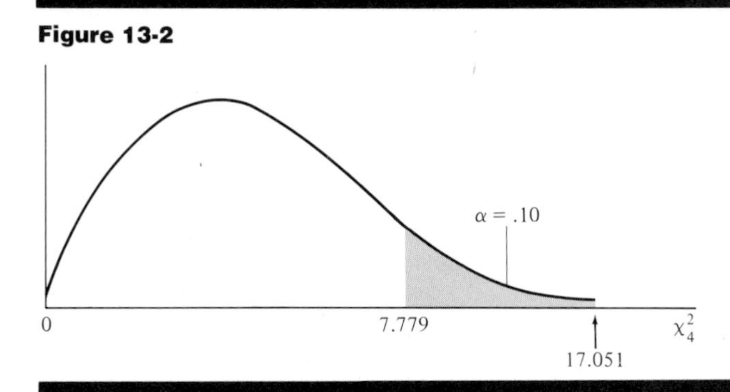

(qualitative or quantitative) are independent. The common ingredient in these situations is the analysis of frequencies. Given random sampling from each population, we observe frequencies in the various classes or categories. We are also able to determine the expected frequencies per class or category when the null hypothesis is true. The statistic $\Sigma \frac{(O - E)^2}{E}$ allows us to statistically compare the observed and hypothesized expected frequencies to determine the conditional (given the null hypotheses is true) probability of obtaining a sample result as extreme or more so than that observed.

Another interesting application of our statistic is known as the *chi-square goodness-of-fit test*. Many of the techniques we have studied have the assumption that either the population or the sampling distribution of the statistic is approximately normally distributed. The chi-square goodness-of-fit test provides us with a way of investigating such an assumption. Let's illustrate this test by returning to one of our earlier data sets.

Recall in Situation I, Chapter 7, you were estimating the mean miles per tire for a particular brand of truck tire. In order to develop a confidence interval estimate for μ we had to assume that either the population was normally distributed or our sample size was large enough so that the sampling distribution of $\bar{x}$ was approximately normal via the central limit theorem. Let's test the hypothesis that the population of x_i values is normal. For convenience we have reproduced the sample data below. Our null hypothesis is the following:

H_0: the variable x (miles to first recap) is normally distributed

The first step, once we have a random sample of n observations on x, is to establish a number of nonoverlapping intervals. As you will see, we must have at least four intervals, and the more intervals we establish, the better is the chi-square approximation. However, we require that all expected frequencies equal or exceed 5. So the number of intervals is limited by n. An examination of Table 7-1 reveals that $49.8 \leq x_i \leq 82.2$. Say we define the following set of intervals.

$$x_i \leq 50.0$$
$$50.1 \leq x_i \leq 55.0$$
$$55.1 \leq x_i \leq 60.0$$
$$60.1 \leq x_i \leq 65.0$$
$$65.1 \leq x_i \leq 70.0$$
$$70.1 \leq x_i \leq 75.0$$
$$75.1 \leq x_i \leq 80.0$$
$$80.1 \leq x_i$$

Table 7-1 Miles to First Recap for Test Tires
(Miles in Thousands)

Tire	Miles	Tire	Miles	Tire	Miles	Tire	Mile	Tire	Mile
1	80.1	11	73.4	21	58.8	31	69.9	41	75.5
2	62.4	12	60.1	22	69.4	32	57.4	42	62.1
3	79.3	13	58.3	23	71.2	33	66.6	43	74.6
4	77.2	14	73.7	24	64.4	34	71.1	44	82.2
5	54.1	15	79.2	25	65.7	35	55.3	45	64.1
6	68.2	16	60.8	26	65.6	36	58.2	46	58.3
7	70.0	17	62.2	27	59.8	37	62.1	47	65.1
8	49.8	18	58.1	28	64.7	38	59.4	48	70.0
9	52.7	19	73.0	29	78.2	39	70.5	49	63.2
10	68.7	20	62.4	30	71.1	40	68.7	50	58.4

Table 13-10 gives the intervals for x_i and their corresponding observed frequencies from our data set in Table 7-1. Now we need to determine what the expected interval frequencies are if H_0 is true. That is, what are the expected frequencies for our intervals if the population of x is normally distributed? To calculate the expected frequencies, we need to determine the probabilities for each of the intervals. Let E_i be the expected frequency for interval i, π_i be the

Table 13-10 Observed Frequencies

Interval	O
$x_i \leq 50.0$	1
$50.1 \leq x_i \leq 55.0$	2
$55.1 \leq x_i \leq 60.0$	10
$60.1 \leq x_i \leq 65.0$	11
$65.1 \leq x_i \leq 70.0$	11
$70.1 \leq x_i \leq 75.0$	8
$75.1 \leq x_i \leq 80.0$	5
$80.1 \leq x_i$	2

probability that an observation of a normal random variable falls in interval i, and n be the sample size. Then,

$$E_i = \pi_i n$$

But each π_i depends upon the mean and variance of the normal distribution. These parameters must be estimated from the data. For these estimation problems, we must use what are called the *maximum-likelihood estimates*. The maximum-likelihood estimator for the mean is the same as the estimator for the mean given in Chapter 7; namely,

$$\bar{x} = \frac{1}{n}\sum_{i=1}^{n} x_i$$

The maximum-likelihood estimator for the variance, however, is computed slightly differently. Instead of dividing the sum of the squared deviations by $n - 1$, we divide by n. So, the maximum-likelihood estimator for the variance is

$$v^2 = \frac{1}{n}\sum_{i=1}^{n} (x_i - \bar{x})^2$$

Using these estimators, all the π_i's can be determined.

For our data set (Table 7-1), we obtain

$$\bar{x} = 66.106$$
$$v^2 = 58.87$$

So we compute the areas of the normal distribution with mean = 66.106 and variance = 58.87 that correspond to our intervals for x_i. Figure 13-3 shows the defined intervals of the hypothesized normal.

Through our transformation

$$z = \frac{x_i - \mu}{\sigma}$$

Figure 13-3

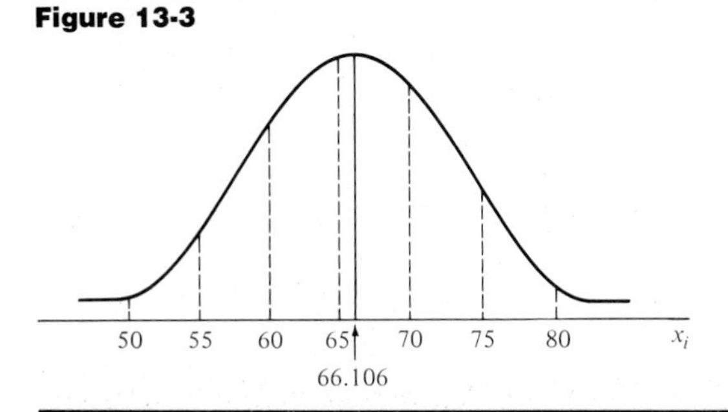

we have obtained the following probabilities π_i's.

$\pi_1 = Pr(x_i \leq 50) = .018$

$\pi_2 = Pr(50 \leq x_i \leq 55) = .056$

$\pi_3 = Pr(55 \leq x_i \leq 60) = .138$

$\pi_4 = Pr(60 \leq x_i \leq 65) = .232$

$\pi_5 = Pr(65 \leq x_i \leq 70) = .251$

$\pi_6 = Pr(70 \leq x_i \leq 75) = .182$

$\pi_7 = Pr(75 \leq x_i \leq 80) = .088$

$\pi_8 = Pr(80 \leq x_i) = .035$

Since $n = 50$, the π_i for each interval must be greater than or equal to .1 in order for the expected frequency to equal or exceed 5. Thus, we must compress intervals 1, 2, and 3 into a single interval and interval 8 must be combined with interval 7.

Table 13-11 shows our resulting intervals for x_i and the corresponding observed and expected frequencies. Next we compute our statistic

$$\chi^2 = \Sigma \frac{(O - E)^2}{E}$$

as shown below.

Class	O	E	O − E	(O − E)²	(O − E)²/E
1	13	10.60	2.40	5.76	.543
2	11	11.60	−0.60	0.36	.031
3	11	12.55	−1.55	2.40	.191
4	8	9.10	−1.10	1.21	.133
5	7	6.15	0.85	0.72	.117
					χ^2 = 1.015

The statistic

$$\chi^2 = \Sigma \frac{(O - E)^2}{E}$$

Table 13-11

Interval	Observed Frequency, O	Expected Frequency, E
$x_i \leq 60.0$	13	10.60
$60.1 \leq x_i \leq 65.0$	11	11.60
$65.1 \leq x_i \leq 70.0$	11	12.55
$70.1 \leq x_i \leq 75.0$	8	9.10
$75.1 \leq x_i$	7	6.15

Figure 13-4 Chi-Square Distribution with 2 Degrees of Freedom

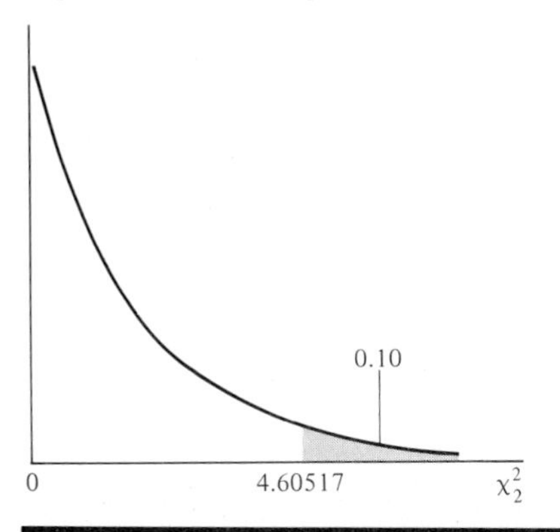

is approximately a chi-square variable with $k - 3$ degrees of freedom, where k equals the number of intervals or classes. Now you see why the number of intervals must be at least 4. Figure 13-4 shows the chi-square distribution for $5 - 3 = 2$ degrees of freedom. Clearly our computed χ^2 value is not significant at any reasonable α level and we accept the null hypothesis that x is normally distributed. Had our χ^2 value exceeded the tabulated chi-square value for a chosen value of α, we would have rejected the hypothesis of normality.

13.6 EXAMPLES

(1) A book publisher is deciding which of two proposed statistics textbooks to contract to publish. One is technique-oriented, the other stresses decision-making applications. The publisher contacts 50 institutions of higher learning to determine if the three types of institutions, two-year, four-year, and graduate, have different preferences regarding statistics texts. The data are summarized in the table.

Number of Occurrences

		Text Preferred		
		Technique-oriented	Applications-oriented	Total
Type of School	Two-year	1	11	12
	Four-year	9	11	20
	Graduate	11	7	18
	Total	21	29	50

The analysis proceeds as

Cell	O	E	$O-E$	$(O-E)^2$	$(O-E)^2/E$
1, 1	1	5.04	−4.04	16.32	3.238
1, 2	11	6.96	4.04	16.32	2.345
2, 1	9	8.40	.60	.36	.043
2, 2	11	11.60	−.60	.36	.031
3, 1	11	7.56	3.44	11.83	1.565
3, 2	7	10.44	−3.44	11.83	1.133
					$\chi^2 = 8.355$

For $\alpha = .05$, the critical value from Table A-6 in the Appendix for a chi-square random variable with 2 degrees of freedom is 5.99147. Since $8.355 > 5.99147$, the publisher rejects the null hypothesis of identical preferences at the three types of institutions.

(2) Suppose one of the executives at State Farm questions whether the two variables, sex of the primary operator and age of the youngest driver insured, are really independent. If they are not independent, price discrimination is justified for only one of the two. Dependence between the two characteristics may mean that the characteristic age is related to the amount of claims *only* because it is related to the characteristic sex. You set up the following contingency table to test the independence of the two variables.

Number of Occurrences

		Sex of Primary Operator		
		Male	Female	Total
Age of Youngest Driver Insured	16–17	19	13	32
	18–25	19	14	33
	26–55	14	6	20
	Over 55	6	9	15
	Total	58	42	100

Cell	O	E	$O-E$	$(O-E)^2$	$(O-E)^2/E$
1, 1	19	18.56	.44	.19	.010
1, 2	13	13.44	−.44	.19	.014
2, 1	19	19.14	−.14	.02	.001
2, 2	14	13.86	.14	.02	.001
3, 1	14	11.60	2.40	5.76	.497
3, 2	6	8.40	−2.40	5.76	.686
4, 1	6	8.70	−2.70	7.29	.838
4, 2	9	6.30	2.70	7.29	1.157
					$\chi^2 = 3.204$

For $\alpha = .10$, the critical value for a chi-square random variable with 3 degrees of freedom is 6.25139. Since $3.204 < 6.25139$, you cannot reject the null hypothesis of independence. Consequently, you conclude that price discrimination is justified on both bases.

(3) A manufacturer of refrigerators is planning a marketing strategy for its three models. It is believed that the different models appeal to people of different income brackets. But before it acts on this belief, the manufacturer would like statistical evidence. The following table is set up.

Number of Each Model Sold to Each Family Group

		Family Income (in $1000's)				
		<8	8–12	12–20	>20	Total
Model	#4302	19	44	41	0	104
	#4402	10	17	26	7	60
	#4502	3	3	9	21	36
	Total	32	64	76	28	200

To test the independence of the two variables, the manufacturer analyzes as follows:

Cell	O	E	$O - E$	$(O - E)^2$	$(O - E)^2/E$
1, 1	19	16.64	2.36	5.57	.335
1, 2	44	33.28	10.72	114.92	3.453
1, 3	41	39.52	1.48	2.19	.055
1, 4	0	14.56	−14.56	211.99	14.560
2, 1	10	9.60	.40	.16	.017
2, 2	17	19.20	−2.20	4.84	.252
2, 3	26	22.80	3.20	10.24	.449
2, 4	7	8.40	−1.40	1.96	.233
3, 1	3	5.76	−2.76	7.62	1.323
3, 2	3	11.52	−8.52	72.59	6.301
3, 3	9	13.68	−4.68	21.90	1.601
3, 4	21	5.04	15.96	254.72	50.540
					$\chi^2 =$ 79.119

The value $\chi^2 = 79.119$ can be viewed as an observation on a chi-square random variable with $(3 - 1)(4 - 1) = 6$ degrees of freedom. For $\alpha = .01$, the critical value is 16.8119. Since $79.119 > 16.8119$, the manufacturer rejects the independence of the two variables.

13.7 IDENTIFICATION OF APPLICATIONS: WHEN?

In this chapter we have extended the analysis of dichotomous populations to problems in which more than two populations were involved. This led us to the analysis of data configurations with r rows and two columns. We extended that analysis to contingency tables with r rows and c columns.

In identifying the possible applications of these methods, two things are important. *The first is that the numbers to be analyzed are counts or frequencies*. Frequencies, of course, are not normally distributed; however, if the sample size is large enough, normal approximations can be used. When the normal approximations cannot be used, we can apply the analysis described in this chapter.

The second important property is the purpose of the analysis. We can use the analysis of frequencies *to make judgments about independence*. Indeed, contingency analysis is unique in its ability to investigate independence of qualitative characteristics. In our application of contingency table analysis, we investigated independence between two quantitative variables. Recall that we categorized the quantitative variables into intervals or levels.

Finally, we discussed goodness-of-fit tests. Because normally distributed variables are required for so many of the techniques presented, it is essential to be able to test the validity of a normal probability distribution as a model for an experiment or process.

PROBLEMS

1. Identify an application of the analysis of frequencies from your major area of study. Point out the features which make this application appropriate and of value.

2. Meier and Frank is a chain of department stores located in the Northwest. The company has issued its own credit cards for a number of years. As a new employee in the finance department, imagine that you have been assigned to a major investigation of the firm's credit policies designed to reduce bad debt losses. You are requested to investigate the relationship between marital status at the time the card is issued and the subsequent payment record of the individual.

   You randomly select 100 credit cards issued three years previously. Of the 70 individuals that were married at the time the cards were issued, 11 have defaulted on a payment. Of the 30 individuals who were single when the cards were issued, 7 have defaulted. Do these data present sufficient evidence to indicate that marital status affects credit-worthiness?

   a. Test using the χ^2 statistic. (Use $\alpha = .05$.)

   b. Test using the z test. (Use $\alpha = .05$.)

   c. Find the square root of the χ^2 calculated in part *a* and compare this to z.

3. A data analyst of Dunford Corporation, manufacturers of a wide variety of industrial lighting equipment, recommended to the management that further orders for the supply of rheostats from Abbot Corporation be canceled. Her recommendation was based on the fact that out of 30 rheostats supplied by Abbot Corporation, 7 were found defective. At the same time, the numbers of defective rheostats from Liberty Electric and Mason Electric were found to be 1 and 2, respectively. The total number of rheostats supplied by these two manufacturers was also 30. When this fact was brought to the attention of Abbot's representative, he dismissed the findings by saying: "A sample of 30 is way too small to allow any conclusion. These results are undoubtedly due to chance." Who is right?

4. Suppose Carrier Air Conditioning has decided to try to increase sales of window units during the month of February. The basic motivation behind the idea is to provide a much-needed boost to the firm's retail dealers at a time when both heating and air-conditioning sales are at a minimum. In addition, if the effort is successful, it will reduce dealers' stocks of the previous year's models and leave them in good shape for spring sales efforts.

   Rudy Lyda, assistant to the vice president of marketing, was asked to design a method to increase sales. He decided to use a combination price reduction–advertising campaign. Carrier dealers were to be provided with a 20 percent rebate on the cost of all units sold during February. In addition, Carrier agreed to pay 75 percent of the cost of any local advertising done by a dealer which stressed a price reduction on Carrier units.

   When Lyda presented his plan, he was told that it looked promising but, given the high cost of the plan, management wanted to be certain it would work even if that meant delaying its initiation for a full year. On receiving this reaction, Lyda made arrangements to test his program. He had the previous February's retail sales figures for both Fort Worth and San Antonio, Texas—110 and 190 units, respectively. The promotional program was implemented for the second week in February in the Fort Worth area but not in San Antonio. The sales were 50 and 55 units, respectively.

   How should Lyda respond to management's desire to be certain that the plan will work?

5. With the compulsory draft no longer in existence, the U.S. Army as well as the other branches of the service have had to devote particular attention to recruitment. Suppose that you have been assigned to a staff group investigating Army recruiting policies. You decide to focus on individuals that talked with recruiters but decided not to enlist. A random sample of 150 such individuals is selected. Past research has shown a number of reasons for not enlisting. However, all previous studies have treated all potential recruits as a homogeneous group. It is your feeling that the reasons for deciding against enlisting vary by educational categories.

   Previous research has shown that more pay, more liberal regulations, and a more "he-man" orientation are frequently mentioned as major improvements needed in the Army. Therefore, you ask the members of your sample to indicate which one of these three improvements would have the most influence on their decision to enlist. The results are presented below.

Education Level	More Pay	Liberalized Regulations	"He-man" Orientation
High school graduate	27	20	13
Some college	24	18	8
College graduate	14	24	2

   Do these data indicate that the preferred improvement and education level are related? (Use $\alpha = .10$.)

6. Due to a rapid expansion of demand, Fuelco, Inc., a carburetor-rebuilding firm, has had to move from one 8-hour shift per day to three. The plant manager is concerned

that the midnight to 8:00 shift will produce a higher number of defective carburetors. Therefore, he wants to institute additional quality control measures during this shift. Prior to incurring the expense associated with additional controls, he tests 125 carburetors randomly selected from each of the three shifts. The results are presented below. What should the plant manager conclude?

Shifts	Defective	Not Defective
Morning	7	118
Afternoon	3	122
Night	12	113

7. Hewlett-Packard, Inc., is planning to build a calculator assembly plant in Corvallis, Oregon. The primary product would be a statistical version of a pocket calculator. The function of this plant would be the assembly of chips (printed circuits designed to perform a specific operation). The statistical calculator is to compute square roots. Thus, one of its components would be a square root chip.

   Hewlett-Packard purchases these chips from independent contractors. Suppose you are hired as purchasing agent for the Corvallis plant. You have identified four potential suppliers of the square root chips. The four brands, or supplier products, have been shown to be equivalent in terms of speed, availability, and cost. However, not all chips function properly. Some chips can be identified as being defective. There is no question but what assembling a calculator with a defective square root chip is expensive. Therefore you would want to purchase these chips from the supplier with the smallest proportion of defective chips. In order to make your selection, you have obtained samples of 100 square root chips from each of the suppliers. Your testing of the chips reveals the following.

	International Electronics	Sansino	Ohio Components	Printed Circuits
Number of defective chips	10	3	8	11
Number of nondefective chips	90	97	92	89

   Assume that these four samples constitute random samples from the respective suppliers.

   a. Is there a difference among the proportions defective for these suppliers? (Use $\alpha = .05$.)

   b. State your assumptions and justify the use of your test statistic.

8. Central Manufacturing operates a major assembly plant near Louisville, Kentucky. In the absence of major, unexpected orders, the plant operates one shift for five days per week and the labor force maintains a constant size. When major orders are received, Central hires additional personnel through the labor union. Carol Perkins, personnel manager, has become concerned that the additional personnel are unreli-

able and poorly trained. However, before approaching union representatives with her concerns, she wants to be quite confident that she is correct. Consequently, she has run a simple linear regression using productivity as the dependent variable and size of the work force as the independent variable. Before making the appropriate test of hypothesis, she must be confident that the normal distribution describes the behavior of the error terms. From the regression analysis output, she has the following data.

Average of the error values $= 0$

Standard deviation v $= .12$

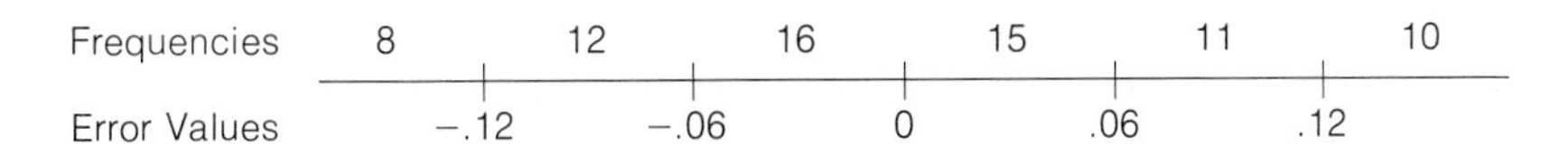

Can Ms. Perkins proceed with her analysis?

9. Myro Pharmaceutical recently developed a drug which the firm believes will be effective in alleviating acne. After securing appropriate approval from the Food and Drug Administration, Myro asked 400 individuals affected with acne to use the drug. An additional 400 acne sufferers received a placebo (an application of medicine that is identical to the drug being tested in appearance, instructions, and so forth but that consists solely of inert ingredients). After one month the results of the study were collected:

	Medicine	Placebo
Improvement noted	208	152
No improvement noted	192	248

Prepare a brief report on the effectiveness of the drug for submission to the Food and Drug Administration.

10. As a staff officer in the Internal Revenue Service, assume that you are working on a revision of income tax forms and instructions for individuals. As a preliminary step you decide to analyze the types of mistakes that individuals made in filling out last year's forms. You randomly select 100 forms which had been audited and found to contain errors. Previous investigation has led you to believe that there are three major categories of mistakes.

Mechanical mistakes involve mistakes in addition or subtraction, recording figures in the wrong space, and so forth. Judgment mistakes involve errors in applying the rules such as treating nonprescription drugs as deductible expenses. Finally, deliberate distortions involve what appear to be conscious attempts to defraud, such as a failure to report a major source of income.

Your immediate concern is whether the type of error varies between individuals who complete the forms themselves and individuals who use tax services. An analysis of the 100 forms which contained an error revealed the information presented below. Is the type of error independent of the preparer of the return?

	Preparer	
Type of Mistake	Individual	Tax Service
Mechanical	22	11
Judgment	35	18
Deliberate	7	7

11. Suppose Armstrong Tile is investigating consumer preferences for floor coverings for kitchens. The marketing vice president feels that varying climates and the life-styles have led to different preferences for tiles, linoleum, and indoor-outdoor carpet among the West Coast, the Midwest, the South, and the Northeast. If this is true, she plans to minimize the firm's current nationwide marketing programs and emphasize regional programs.

    You have been asked to analyze the results of a recent consumer survey conducted by the firm. The survey involved 1000 consumers and, among other things, measured consumer preferences for kitchen floor coverings. The results are presented below. Advise the marketing vice president.

	West Coast	Midwest	South	Northeast
Tile	95	78	85	66
Linoleum	62	100	110	106
Carpet	93	72	55	78

Chapter 14

Nonparametric Methods

14.1 SITUATION I: WHY?

Cessna Aircraft Company has its major production facility in Wichita, Kansas. Suppose the vice president, Paul Marshall, is concerned about a recently observed problem in production. Production rates are down for no apparent reason: there have been no major shortages in supplies, no unusual turnovers in the work force, and no interruptions in the work schedule. Since many studies have shown relationships between morale and productivity, Marshall questions whether the problem may be low morale.

Currently, each crew is responsible for cleaning up the work area before it leaves for the day. Marshall suspects the technical personnel may regard this phase as drudgery. In an effort to deal with this, he has installed a new policy with one of the production crews. Marshall has employed a janitorial service to perform the cleaning function for this crew. Because of the unusual circumstances of the cleaning (for instance, there is a severe time limitation on the length of time allowed for the entire cleaning job), the cost of the janitorial service is considerable. Consequently, before Marshall installs this new policy for all the work crews, he would like to be fairly certain that the new policy is effective.

Suppose you were recently hired by Cessna Aircraft, and Marshall has asked you to help him determine whether the policy change has been effective. Two things are clear to you at the start. First, you do not want to focus your entire analysis on just the one production crew with the cleaning service. There may be external factors influencing all the crews, and these may lead you to conclude that the policy change was effective when, in fact, it was not. So, you decide to compare two work crews—one being the crew with the janitorial service and the other without—to determine if there is a difference in morale.

Second, you cannot simply measure the output of the production crews. Aircraft construction is a time-consuming process, and it may be several weeks before any *production figures are available, let alone a sufficiently large sample. The time lag is too costly.*

Your only hope is to observe the two production crews for a short period of time and attempt to judge whether there is any difference in morale. On 10 different days, the two work crews are observed, and you make a judgment as to which crew has the higher morale. Crew A is used to denote the work crew which has been relieved of the clean-up responsibility; the other crew is labeled crew B. The data are as follows:

Day	Crew A	Crew B	Sign*
1	Higher		+
2	Higher		+
3	No difference		0
4	Higher		+
5	No difference		0
6	Higher		+
7		Higher	−
8	Higher		+
9	Higher		+
10	No difference		0

*A positive sign is assigned to any day in which crew A was judged to have the higher morale. A negative sign is recorded if crew B was judged to have the higher morale. A zero is assigned whenever no difference was observed.

Now what do you do? Can you make a recommendation to Marshall? You need to be able to determine whether your sample result is consistent with the contention that there is no difference in morale between the two work crews. You realize that the data are not very precise. All you are willing to claim is that on a given day you observed a difference in morale, but you are not willing to assign a magnitude to the difference. Note that the data are categorical in nature.

1. Crew A has higher morale than crew B.
2. Crew B has higher morale than crew A.
3. No difference in morale.

These observations are examples of what are called nominal data.

NOMINAL DATA ARE OBSERVATIONS ON A QUALITATIVE VARIABLE.

14.2 ANALYZING NOMINAL DATA: HOW?

As the definition states, nominal data are associated with qualities. The word "data" usually suggests *numbers* which are recorded as observations from an experiment. In the case of Situation I, we have observations of a qualitative nature. These data are typically expressed in symbols or words which represent a class or level of a qualitative characteristic. Other examples of nominal data are "success," "comfortable," "no preferences," and "high morale."

In the preceding chapter, we also encountered nominal data. Observations on a process were identified in terms of two characteristics. But the methods presented in the preceding chapter required two qualitative variables so that the frequencies, the number of times particular levels of the qualitative characteristic were observed, could be put into the contingency table format. The probability distribution used to analyze these data depended upon that format.

For dealing with nominal data naming only one characteristic, we might try to employ the binomial distribution. We have seen that it is not appropriate to use the normal distribution unless we have a large number of observations. Remember that to use the central-limit theorem, we need a sample size large enough so that

$$n\pi \geq 5 \quad \text{and} \quad n(1 - \pi) \geq 5$$

However, in the problem of Situation I and others like it, less than a large amount of data must suffice. Consequently, we need to examine the experiment in question in an attempt to derive a probability distribution which can be used for the analysis.

Let's examine more closely the possibility of using the binomial distribution. Recall that the binomial distribution applies to sampling in which:

1. Each trial results in one of two possible outcomes.
2. The trials are independent.
3. The probability of success remains constant from trial to trial.

In the problem at hand, we want to compare two work crews. The observations will be in terms of three possible outcomes. One crew will be judged as having higher morale—that's two possible outcomes, one for each crew. The third possibility is that the judgment will be that the two production crews have equally high morale. We call such judgments *ties*, indicating that we cannot tell the difference between the two crews. We might reasonably argue that the ties do not tell us anything; they merely identify times at which the observer could not make a judgment. If the ties are discarded, we now have a binomial experiment with each trial producing one of two possible outcomes—crew A has higher morale, or crew B has higher morale.

To make use of the binomial probability distribution, we need to determine a probability for each of the two possible outcomes in any trial. In hypothesis-testing situations, remember, we use the null hypothesis to identify a sampling distribution for the test statistic. The null hypothesis in this situation would be that the two crews have equally high morale. In that case, we should be equally likely to judge crew A as having higher morale as we are to judge crew B as having higher morale. So, the appropriate probability to use for our binomial distribution is .5. This transforms the null hypothesis into

H_0: $\pi = .5$

where

$\pi = Pr(\text{crew A has higher morale}) = Pr(\text{``+''})$

The test we have just described is called the *sign test* because of the means of recording the comparison of two samples. If one is judged "superior" by some standard, then a "+" is recorded; if the other is judged "superior," a "−" is recorded. For the sign test, the null hypothesis is always that the probability $\pi = .5$. That is, the null hypothesis takes one of the following three forms:

(I) H_0: $\pi = .5$
(II) H_0: $\pi \leq .5$
(III) H_0: $\pi \geq .5$

The probability $\pi = .5$ can now be used to determine the probability of our sample outcome given that the null hypothesis is true. This conditional probability, of course, is the basis for our rejecting or accepting the null hypothesis. This probability can be calculated using the binomial probability function given in Chapter 6, or it can be found in Table A-3 in the Appendix.

14.3 SITUATION I REVISITED

To use the binomial distribution, we need to have independent trials, that is, a random sample of judgments. In particular, had you observed the two production crews more than once on a given day, you would have run the risk of having judgments which were not independent of each other. However, since observations were recorded on 10 different work days, we will assume that the trials were independent and therefore constitute a random sample.

Let's recall the data from Situation I.

Day	Crew A	Crew B	Sign
1	Higher		+
2	Higher		+
3	No difference		0
4	Higher		+
5	No difference		0
6	Higher		+
7		Higher	−
8	Higher		+
9	Higher		+
10	No difference		0

If the ties or 0 observations are removed, only those observations which indicate a difference in morale will remain. If no change in the level of morale is produced by the new policy, we would expect that the number of + observations and − observations would be about equal. That is, we would expect the two outcomes to be equally likely. If π denotes the probability of observing +, then the null hypothesis for this decision would be

H_0: $\pi = .5$

This null hypothesis actually represents a broader null hypothesis in this problem; namely,

H_0: $\pi \leq .5$

For, if $\pi = .5$ or if $\pi < .5$, you will recommend that Marshall terminate employment of the janitorial service. In the one case, $\pi = .5$, no improvement will have been discernible from the new policy. In the other, $\pi < .5$, the use of the janitorial service will have apparently reduced the level of morale. So, the appropriate alternate hypothesis is

H_1: $\pi > .5$

If the null hypothesis is rejected in favor of this alternate hypothesis, you might recommend that Marshall expand the use of the janitorial service.

The ties are discarded from the original data, leaving

+, +, +, +, −, +, +

There are seven observations remaining from the original list. Six of these are +. We have, then, seven trials and six of them are "successful." Under the null hypothesis, the probability of crew A's being judged as having higher morale is .5. We are interested in determining whether six successes in seven trials is consistent with a probability of .5 of success. For this hypothesis-testing problem, we choose $\alpha = .10$ as the level of significance. The probability of observing as many as six successes in seven trials when $\pi = .5$ is

$$Pr(x \geq 6) = \binom{7}{6}(.5^6)(.5^1) + \binom{7}{7}(.5^7)(.5^0)$$

where x is the number of successes. Table A-3 can be used to determine this probability.

$$Pr(x \geq 6) = 1 - Pr(x \leq 5)$$
$$= 1 - .9375 = .0625$$

Since $.0625 < .10$, the observation of six successes out of seven trials represents a significant departure from what the null hypothesis claims. Thus, it may be more reasonable to conclude that employment of the janitorial service has improved the morale of crew A.

14.4 EXAMPLES

(1) Ms. Jane Worley is completing plans for the construction of her new family restaurant. She is taking an active role in the design of the interior of the building. She is deciding whether booths holding four adults each should be used along the walls. The alternative is to use tables and chairs. The booths would allow greater capacity, but they may be less comfortable than the tables and chairs. She selects eight families who consent to express their preferences in terms of comfort. She considers these as eight independent trials. Her observations are as follows:

Family	Booths	Chairs	Sign
1		Preferred	+
2	Preferred		−
3	Preferred		−
4		Preferred	+
5		Preferred	+
6		Preferred	+
7		Preferred	+
8		Preferred	+

So, of the eight families questioned, six prefer the tables and chairs. If π is used to denote the probability that the customer will prefer the table and chairs, and since the booths allow greater capacity, the hypotheses are

$$H_0: \pi \leq .5$$
$$H_1: \pi > .5$$

Now, when $\pi = .5$,

$$Pr(x \geq 6) = 1 - Pr(x \leq 5)$$
$$= 1 - .8555 = .1445$$

Using $\alpha = .05$ as the level of significance, the data do not provide enough evidence to support rejection of the null hypothesis. So, Ms. Worley might choose to use the booths.

(2) A second problem that Ms. Worley faces concerns the selection of cushion material for the booths. Under consideration are the standard spring-and-foam-rubber cushion seats and backs and a new, primarily plastic, cushion which the manufacturer claims to be more comfortable than, and equally durable as, the standard construction. Sixteen adults are selected as subjects to help Ms. Worley determine whether the manufacturer's claim of greater comfort is justified. Each person judges the two cushions in the following terms:

1. The standard cushion is more comfortable.
2. The new cushion is more comfortable.
3. The two cushions are equally comfortable.

These judgments are viewed as independent of each other. Since the claim is one of greater comfort, we have a one-tailed test. Let π be the probability that the *standard* cushion will be judged as more comfortable. Then the hypothesis-testing problem is to test

H_0: $\pi \geq .5$

against

H_1: $\pi < .5$

The data are

Adult	Standard Cushion		New Cushion	Sign
1	More comfortable			+
2			More comfortable	−
3		No difference		0
4		No difference		0
5			More comfortable	−
6			More comfortable	−
7	More comfortable			+
8		No difference		0
9		No difference		0
10		No difference		0
11			More comfortable	−
12		No difference		0
13			More comfortable	−
14			More comfortable	−
15		No difference		0
16	More comfortable			+

Only a small number of pluses would lead to the rejection of the H_0. Thus, the appropriate probability to calculate for this test is, when $\pi = .5$, and the number of trials is nine,

$Pr(x \leq 3) = .2539$

Again using $\alpha = .05$ as the level of significance, Ms. Worley concludes that there is no difference in the comfort afforded by the two cushions.

(3) Art Johnson is currently negotiating a new lease for the building which houses his supermarket. The company which owns the building is firm in its demand for a substantially higher rental charge. If the supermarket is no more profitable than it has been through most of its years of operation, Johnson will have no choice but to refuse to sign the new lease agreement. Over the past 12 months, however, Johnson believes that there has been an increase in profits on a monthly basis and that this increase represents a trend rather than random fluctuations from the usual performance. If he is correct, he will be able to sign the lease knowing that the new rent figure can be covered. To determine whether there is a trend in monthly profit, he will compare the last 12 months with the preceding 12 months. He will match each month with the same month one year ago and record a + if the later month showed greater profit, a − if the earlier month showed a greater profit, and 0 if the two months had equal profit figures. (Note that Johnson has formed nominal data out of data which are quantitative.) If the past 12 months have been only random fluctuations from the usual performance, he would expect to observe an equal number of +'s and −'s. If π is the probability of observing a +, then Johnson's hypotheses are

H_0: $\pi \leq .5$

H_1: $\pi > .5$

The monthly profit figures for the past 24 months are

Month	Last Year	This Year	Sign
January	$1600	$1700	+
February	1500	1900	+
March	2000	1800	−
April	1100	1300	+
May	1500	2300	+
June	1200	2100	+
July	1800	1800	0
August	1600	1600	0
September	1900	2200	+
October	1400	1600	+
November	1500	1800	+
December	1700	1700	0

Of the twelve trials, three are ties and these are discarded. Of the remaining nine observations, eight are +'s. The appropriate probability for Johnson's hypothesis-testing problem is $Pr(x \geq 8)$. Using Table A-3, with $n = 9$ and $\pi = .5$, he obtains

$$Pr(x \geq 8) = 1 - Pr(x \leq 7)$$
$$= 1 - .9805 = .0195$$

So, the smallest level of significance for which Johnson can reject his null hypothesis is

$\alpha = .0195$

This small probability leads Johnson to conclude that there has indeed been an increase in monthly profit, and he signs the lease.

14.5 SITUATION II: WHY?

In October of 1976, John Backe was named president of Columbia Broadcasting System, Inc. (CBS). According to Time, *October 25, 1976, "For the first time since the 1950s, the company (CBS) is behind ABC and NBC in audience ratings." The plunge in ratings seemed to coincide with a 6-point drop in CBS stock. Clearly, advertising rates and competition for both shows and sponsors are related to network ratings. Suppose the change in presidents was to some extent related to this situation.*

Let's further suppose that you work for CBS in the program-planning division and the mid-season changes are now being formulated. The decision has been made to add one new police/mystery show at mid-season. However, you have two possibilities. Pilot episodes for each of the two potential police/mystery series have been filmed. Your job is to choose which of the two series should be added. As the series you recommend will have an effect on the network's ratings, you would like to make the wiser choice.

Clearly, the important characteristic is audience preference. A fairly standard way of measuring or obtaining potential audience preference is by free previews. A sample of individuals is given a free preview of a show and asked for their evaluations. The evaluations are on a scale from 0 to 100 with the higher evaluation representing greater preference.

As the free previews are somewhat expensive, each pilot episode is shown to only a few audiences. Suppose pilot A was shown to five audiences and pilot B was shown to eight other audiences. For each audience the average evaluations were computed and the resulting data were

Pilot A	*Pilot B*
83	75
72	64
77	70
91	77
88	81
	65
	80
	78

Are the two pilot shows equal in terms of audience preference? Or is one

pilot more preferred than the other? If we felt comfortable with the assumptions that both populations of evaluation are normally distributed with equal variances, we could use the student t *test for the difference between population means as presented in Chapter 8. If we could increase our sample sizes, these assumptions would not be so critical. However, sample sizes of five and eight are not very large and we may not be willing to make such assumptions.*

We need some way to decide whether there is a difference in locations of two populations based on small samples without assuming normality. *We will accomplish this by replacing the numeric preferences with their ranks or rank position. The resulting analysis is performed on the ranks rather than the numeric preferences themselves. If we group all 13 observations into one group and assign rank 1 to the smallest numeric preference rating, rank 2 to the next smallest, and so on through rank 13, which is assigned to the most preferred rating, we obtain the following:*

Pilot A		**Pilot B**	
Evaluation	*Rank*	*Evaluation*	*Rank*
83	11	75	5
72	4	64	1
77	6.5	70	3
91	13	77	6.5
88	12	81	10
		65	2
		80	9
		78	8

Two of these observations are identical, namely, 77. In this case, the tied observations are assigned the average of the rankings. As 77 is the sixth and seventh smallest observation, it is assigned the rank $6.5 = (6 + 7)/2$. *If three observations were tied for the seven, eight, and nine positions, the three would be assigned the rank* $8 = (7 + 8 + 9)/3$.

As yet, we haven't seen a technique designed to analyze data in the above format. Our observations are rankings or rank-order data which are called ordinal data. *Ordinal data are data in which the observations or classes have a natural ordering. We can say that one observation is "bigger," "better," or "more preferred" than another, but we do not try to attach a numeric value to the difference between observations.*

ORDINAL DATA ARE OBSERVATIONS THAT ARE RANK-ORDERED BASED UPON A NATURAL ORDERING.

14.6 ANALYZING ORDINAL DATA: HOW?

Ordinal data can always be reduced to nominal data. However, ordinal data contain more information than nominal data and such a reduction in information would be wasteful. Analysis of ordinal data is accomplished by examination of the ranks. As in all statistical analyses, judgments will be made using probabilities. So the task, here again, is to establish the appropriateness of a particular probability distribution. The procedure we will use is known as the *Mann-Whitney test*. We have two random samples—each from a different population. The two samples are combined, and the items are ordered from smallest to largest. The smallest item in the two samples receives the rank 1, the second smallest the rank 2, and so on. We select one sample and label it sample 1. The ranks of all the items from sample 1 are summed. This sum we call S. If there are n items in sample 1 and m items in sample 2, the smallest value possible for S would be realized when the items from sample 1 are all assigned smaller ranks than any item in sample 2. In this case,

$$S = 1 + 2 + \cdots + n = \frac{n(n + 1)}{2}$$

So, the minimum value possible for S is $n(n + 1)/2$. The test statistic for the Mann-Whitney test is

$$U = S - \frac{n(n + 1)}{2}$$

which is the difference between S and *its smallest possible value*.

The test statistic U can range from a minimum of 0 to a maximum of nm. The statistic U will equal 0 when all n items from sample 1 assume the n *smallest* ranks and U will equal nm when the n items from sample 1 assume the n *largest* ranks. Either of these situations would suggest a difference in location of the two populations.

Technically, the null hypothesis to be tested is that both populations are identically distributed. However, as the Mann-Whitney test is particularly sensitive to differences in location, we sometimes use it to test hypotheses about differences in population means. Therefore, we will express the null hypothesis as

$$H_0\colon \mu_1 = \mu_2$$

which is appropriate if we can assume that, if there is a difference between the two population distribution functions, it is a difference in location.

If both populations are identically distributed, which includes equal means, we would expect the observed distribution of ranks to be random. That is, every possible arrangement of the nm ranks, n to sample 1 and m to sample 2, would be equally likely. However, if $\mu_1 < \mu_2$, we would expect the n items sampled from population 1 to be smaller in magnitude than the m items from population 2. This would yield a small value for the test statistic U. If, on the other hand, $\mu_1 > \mu_2$, U would tend to be large. What values are too large and what values are too small, of course, depend upon the distribution of U under the null hypothesis of no difference in means. Table A-10 in the Appendix identifies the critical values for U for various levels of significance. Of course, the distribution of U depends upon both n and m. Let's look at a particular entry in Table A-10. Suppose we choose the sampling distribution of U for $n = 4$ and $m = 8$. As Table A-10 gives lower-tail U values and corresponding areas, we see that

$$Pr(U < 0) \leq .001$$
$$Pr(U < 2) \leq .005$$
$$Pr(U < 3) \leq .01$$
$$Pr(U < 5) \leq .025$$
$$Pr(U < 6) \leq .05$$
$$Pr(U < 8) \leq .10$$

when both populations are identically distributed. Since the distribution of U is symmetric when the null hypothesis is true, we can obtain the upper-tail values for U by subtracting the tabulated U_p value from its upper limit of nm. That is,

$$U_{1-p} = nm - U_p$$

Thus,

$$Pr(U > 27) \leq .025$$

as $(4)(8) - 5 = 27$.

Thus, the probability of observing a U value less than 5 or greater than 27 where $n = 4$ and $m = 8$ and H_0 is true is less than or equal to .05.

14.7 SITUATION II REVISITED

Recall the evaluations and corresponding ranks for the 13 audiences were as follows:

Pilot A		Pilot B	
Evaluation	Rank	Evaluation	Rank
83	11	75	5
72	4	64	1
77	6.5	70	3
91	13	77	6.5
88	12	81	10
		65	2
		80	9
		78	8

Suppose we define pilot A as population 1. In this case $n = 5$, $m = 8$, and S, the sum of the five ranks associated with population 1, is

$$S = 11 + 4 + 6.5 + 13 + 12 = 46.5$$

Therefore the test statistic U is

$$\begin{aligned} U &= S - \frac{n(n + 1)}{2} \\ &= 46.5 - \frac{(5)(6)}{2} \\ &= 31.5 \end{aligned}$$

If there is no difference in audience preference between the two pilots, $\mu_1 = \mu_2$. Therefore, our hypotheses are as follows:

H_0: $\mu_1 = \mu_2$

H_1: $\mu_1 \neq \mu_2$

If we set $\alpha = .10$, we would reject H_0 for any $U < 9$ or $U > (5)(8) - 9$, which is 31. As our observed U value is 31.5 and it is greater than 31, we reject the null hypothesis and conclude that the audience preferences are not equal. Given this result you would want to recommend the police/mystery series for pilot A, since the ranks associated with pilot A are *significantly* larger than those for pilot B.

14.8 EXAMPLES

(1) Recall that in Situation II, the decision had already been made to add one police/mystery show to the schedule. We now suppose that this decision is under review, that is, you want to determine whether the general popularity of police/mystery shows is the same as the other regular series currently on the schedules of the major networks.

As samples, you compare the Nielsen ratings for the first month of the season. Suppose there are 14 police/mystery shows and 53 other regularly scheduled series currently offered. The Nielsen ratings are obtained and the 67 (14 + 53) shows are ranked from least popular (rank 1) to most popular (rank 67).

Let population 1 be the ratings of police/mystery shows and population 2 be the ratings for the other shows. Our hypotheses are

H_0: $\mu_1 = \mu_2$

H_1: $\mu_1 \neq \mu_2$

Suppose the ranks of the police/mystery shows were

2, 4, 5, 10, 15, 25, 32, 36, 38, 48, 50, 51, 54, 60

Then

$S = 430$

and the test statistic value is

$$U = 430 - \frac{14(15)}{2} = 325$$

The two samples have 14 and 53 items, respectively, which exceeds the bounds of the table. When one of the samples contains more than 20 items, we can use the following normal approximation. Letting n and m be the two sample sizes,

$$U = \frac{nm}{2} + z\sqrt{\frac{nm(n + m + 1)}{12}}$$

or

$$z = \frac{U - \frac{nm}{2}}{\sqrt{\frac{nm(n + m + 1)}{12}}}$$

where z is the standard normal variable. In this case, $n = 14$ and $m = 53$, and

$$z = \frac{325 - \frac{14(53)}{2}}{\sqrt{\frac{14(53)(68)}{12}}} = -.709$$

If we use $\alpha = .10$, this value is not significant. Consequently, we conclude that there is no evidence that the popularity of police/mystery shows is different from that of the other shows.

(2) A company is considering whether to send its production supervisors to sensitivity-training sessions. The sessions are expensive, and therefore the company would not want to send the supervisors unless there was solid evidence of the training's effectiveness. The company employs fourteen supervisors, and six have been sent, on a trial basis, to the sessions. After they completed the training and were back on the job for two weeks, short questionnaires were distributed to the employees asking them to rate their supervisor. The six who completed the training received the following percentages of favorable ratings:

70%, 40%, 56%, 52%, 67%, 80%

Let us call these data sample 1. The other eight supervisors received the following percentages:

67%, 50%, 55%, 33%, 60%, 30%, 45%, 35%

We are testing the hypothesis

H_0: $\mu_1 \leq \mu_2$

against

H_1: $\mu_1 > \mu_2$

Consequently, we would reject H_0 *only when the test statistic U is too large.*

These data contain one tie. Each of the tied items is assigned the average rank. The data are arranged from smallest to largest as before.

30%, 33%, 35%, **40%**, 45%, 50%, **52%**, 55%, **56%**, 60%, **67%**, 67%, **70%**, **80%**

The percentages for the supervisors who were given the sensitivity training are shown in bold type. The ranks assigned to the percentages for these supervisors are

4, 7, 9, 11.5, 13, 14

Since there are two observations of 67 percent, and they occupy positions 11 and 12, each is assigned the rank

$$11.5 = \frac{11 + 12}{2}$$

So,

$$S = 58.5$$

and

$$U = 58.5 - \frac{6(7)}{2} = 37.5$$

We are interested only in determining if this value is too large. For $\alpha = .05$, $n = 6$, and $m = 8$, the value from Table A-10 is 11. So, the critical value for U is 37.

$$U_{1-p} = (6)(8) - 11 = 37$$

Since $U = 37.5 > 37$, we would recommend that the company send their production supervisors to the sensitivity-training sessions.

14.9 SITUATION III: WHY?

K-Tel Records specializes in producing and marketing record albums and tapes which combine several hit records in the same musical field. All songs included on the albums and tapes are reproductions of the original recordings by the original artists. Each album package is aimed at a specific segment of the market and is comprised of songs which are similar in rhythm and style. Often, the package is based upon a theme or focused on a particular style of dancing.

Suppose you are the marketing manager for K-Tel Records, and the prob-

lem you face at the moment concerns the decrease in sales which most of your fellow executives at K-Tel attribute to a general economic slump.

Currently, all the K-Tel album packages are two-record sets including between 20 and 28 songs. Although some packages are more expensive, the selling price is generally $5.99 per set for the albums and $6.99 for tapes. Your proposal to deal with the slumping sales is to offer one-record albums with 10 to 14 songs per set. The selling price you propose is $3.39 for records and $3.99 for tapes.

This approach would substantially increase the cost of production, for it requires paying higher royalties, more careful selection of songs, and greater packaging expenses. Also, while the product is somewhat more attractive because of the lower total price, it is somewhat less attractive because of the higher price per song. Surely you do not want to take this action without its having a reasonably good chance of success.

Information for this decision will be expensive to acquire, however. First, some one-record sets must be made available. Second, an extensive advertising campaign must be undertaken to inform customers that the one-record sets are available. A one-minute ad on nationwide television during the morning hours costs over $40,000, which is far too expensive. Other advertising media have proved ineffective for K-Tel in the past. The only alternative is to purchase advertising time on local stations. Because of the expense, very few observations will be available for your analysis.

You decide on an amount to allocate to the data-collection process. Based on that, you select six cities: Atlanta, Boston, Los Angeles, Minneapolis, St. Louis, and Seattle. You run television ads promoting both a two-record set and a one-record set consisting of half of the same songs. You observe the following profit figures from each city based on sales and orders placed during a one-week period.

City	Profit from Two-Record Sets	Profit from One-Record Sets
Atlanta	$2213	$3006
Boston	4478	4799
Los Angeles	5396	5404
Minneapolis	1166	1318
St. Louis	2339	3551
Seattle	985	850

The question you need to answer is whether there is any difference between one- and two-record sets relative to profit.

The difference between this problem and the one described in Situation II is that our observations are *paired*. The two profits $2213 and $3006 occurred in

Atlanta and are therefore not independent of one another. We refer to related observations such as these as *paired observations*.

We could test the hypothesis that there is no difference between one- and two-record sets relative to profit by forming the algebraic difference between pairs. If there is no difference, the mean difference between pairs is zero. If we could assume that the distribution of differences between pairs is normal, we could use the student t test given in Chapter 8 to test the hypothesis that the mean difference is zero. As we know, the normality assumption becomes less critical as the sample size increases. However, six observations is not large and we may not be willing to assume normality. Thus, we need a technique for testing the hypothesis of no difference between paired observations without assuming normality.

Incidentally, the above data on profits are known as *interval data*. With interval data, not only can the observations be rank-ordered but the algebraic difference between observations has meaning.

INTERVAL DATA ARE NUMBERS FOR WHICH DIFFERENCES PROVIDE MEANINGFUL INFORMATION.

14.10 SMALL SAMPLE ANALYSIS OF PAIRED DATA: HOW?

It seems somewhat awkward to introduce a concept in Chapter 14 which we have been using throughout the book. However, we are not actually introducing the concept, only the terminology. The phrase "interval data" is used to differentiate this type of data from nominal data and ordinal data. Let's briefly focus on the concept that differences provide meaningful information. With nominal data, the difference between "class 1" and "class 2" has no algebraic meaning whatsoever. For ordinal data, differences do not provide consistent information. The difference between a rank of 2 and a rank of 4 is not necessarily the same as the difference between a rank of 12 and a rank of 14—despite the fact that arithmetically the same number is associated with each difference. With interval data, not only is the difference between two values defined but its meaning is consistently understood. It should be clear that interval data are the most informative data.

In the first situation of this chapter, the problem facing the decision maker was embodied in the data. The outcomes generated by the process were not expressible in numbers which could be analyzed by the methods of previous chapters. In Situations II and III the numbers are not the problem. The figures coming from K-Tel's promotion will be expressed in profit, not ranks or names. However, the problem here, like that in Situation II, is in the quantity of data. We are reluctant to assume that the differences between profits are normally distributed. If we had a large number of observations, the sampling distribution of sample mean differences would approximate normality via the central-limit theorem and we could use techniques presented in Chapter 8. But without a sufficient number of observations, we cannot arbitrarily use normality.

Small amounts of interval data can be analyzed by a statistical test credited to Wilcoxon. The procedure involves ranking the differences and summing some of the ranks as we did in the Mann-Whitney test in Situation II.

Let's be more specific about the procedures of the Wilcoxon test. For convenience, call the pairs (x, y). For each pair (x_i, y_i), we form the difference

$$D_i = y_i - x_i$$

Any pairs for which $D_i = 0$ are discarded from further analysis. We then arrange the absolute values of the D_i's in increasing order of magnitude and assign ranks to the $|D_i|$'s. The rank 1 is assigned to the smallest $|D_i|$, 2 to the second smallest, etc. Tied values, such as $|D_i| = |D_j|$, are again each given the average of the ranks to be assigned. Now, we sum only those ranks which are assigned to the positive D_i's. This sum is the Wilcoxon test statistic W.

A brief illustration of the procedure is in order. Suppose the pairs are (1.0, 3.5); (2.6, 2.5); (3.2, 2.9); and (1.3, 1.4). Then we form the differences

$$D_1 = 3.5 - 1.0 = 2.5$$
$$D_2 = 2.5 - 2.6 = -0.1$$
$$D_3 = 2.9 - 3.2 = -0.3$$
$$D_4 = 1.4 - 1.3 = 0.1$$

The ordering of the absolute values is, from smallest to largest,

$$|D_2| = 0.1 \qquad |D_4| = 0.1 \qquad |D_3| = 0.3 \qquad |D_1| = 2.5$$

Ranks are assigned to these values as 1.5, 1.5, 3, and 4, respectively. Since D_1 and D_4 are the only positive differences, we sum the rank of $|D_1|$, namely 4, and the rank of $|D_4|$, namely 1.5. Our test statistic value, then, is

$$W = 4 + 1.5 = 5.5$$

To use the Wilcoxon test we need to satisfy certain conditions. The first two prerequisites for this test involve the points we made earlier. First, there must be a natural pairing of the observations; and second, the numbers must constitute interval data. To derive a sampling distribution of the test statistic under the null hypothesis, we need the pairs to be independent. In the K-Tel situation, the observations from one city must be independent of the observations from the other cities. Finally, the processes producing the x's and y's should be such that tied values of the $|D_i|$'s are not too likely.

With these conditions and under the null hypothesis that the means of the variables x and y are equal, the sampling distribution of the Wilcoxon test statistic can be calculated. Table A-11 in the Appendix gives the critical values for this test statistic for various levels of significance.

For each n, $n \leq 20$, the size of each sample, Table A-11 gives the quantiles for the distribution of W. That is, for a given value of n, the body of the table contains w_p where

$$Pr(W < w_p) \leq p$$

The various values for n are identified in the left-hand column. The top row identifies the value of p for selected values up to $p = .50$. To find w_{1-p}, where $p \leq .50$ (so $1 - p \geq .50$), we simply compute

$$w_{1-p} = \frac{n(n+1)}{2} - w_p$$

For example, Table A-11 shows that the probability of observing a W value less than 6 when you have 10 pairs and there is no difference between the two populations is less than or equal to .01. As the distribution of W is symmetric when H_0 is true, the probability is no more than .01 that W will exceed

$$\frac{(10)(11)}{2} - 6 = 49.$$

If $E(x) < E(y)$, we would expect $y_i - x_i > 0$, which would lead to large values of W. Conversely, if $E(x) > E(y)$, we would expect $y_i - x_i < 0$ and thus small values of W. Now let's return to the K-Tel Records situation.

14.11 SITUATION III REVISITED

Let x stand for the profit from sales and orders of the two-record sets and y symbolize the profit from sales and orders for one-record albums. With this notation, we now have

$$\mu_1 = E(x) \qquad \text{and} \qquad \mu_2 = E(y)$$

We formulate the following hypotheses:

H_0: $E(x) = E(y)$

H_1: $E(x) \neq E(y)$

Recall that you had observed the values:

City	Profit from Two-Record Sets	Profit from One-Record Sets
Atlanta	$2213	$3006
Boston	4478	4799
Los Angeles	5396	5404
Minneapolis	1166	1318
St. Louis	2339	3551
Seattle	985	850

The differences D_i and the ranks assigned to $|D_i|$ are

City	y_i		x_i		D_i	$\lvert D_i \rvert$	Rank
Atlanta	\$3006	−	\$2213	=	\$ 793	\$ 793	5
Boston	4799	−	4478	=	321	321	4
Los Angeles	5404	−	5396	=	8	8	1
Minneapolis	1318	−	1166	=	152	152	3
St. Louis	3551	−	2339	=	1212	1212	6
Seattle	850	−	985	=	−135	135	2

Only Seattle produced a negative difference, D_6. So, summing the ranks assigned to all the $|D_i|$'s except $|D_6|$ gives

$$W = 5 + 4 + 1 + 3 + 6 = 19$$

From Table A-11 for $n = 6$, we find

$$w_{.05} = 3$$

This tells us that the lower .05 tail of the sampling distribution of W if the null hypothesis is true is given by

$$W < 3$$

To determine the upper .05 tail of the sampling distribution, we compute

$$w_{.95} = \frac{6(7)}{2} - w_{.05}$$
$$= 21 - 3 = 18$$

Thus, $W > 18$ identifies the upper .05 tail of the sampling distribution of W if H_0 is true. Since $19 > 18$ and this is a two-tailed test, we would reject H_0 for $\alpha = .10$, which gives a rejection region of $W < 3$ and $W > 18$. As our observed W value was greater than 18 and large values for W suggest that $E(x) < E(y)$, we might conclude that one-record sets are more profitable than two-record sets and decide to market the one-record sets nationwide.

14.12 EXAMPLES

(1) Suppose Dow Chemical is developing a new herbicide for ridding pastures of certain plant life. This plant life can contaminate the milk given by cows which have been feeding on the plants. Currently, dairy farmers must use a herbicide which also contaminates the milk. Consequently, after being sprayed with the herbicide, the pasture must not be used for a full year.

The question facing Dow executives is whether the new herbicide also contaminates the milk. If it does, development will be discontinued. Experiments will be run to determine if the new herbicide is a contaminant. Dairy cows will graze on treated areas for a one-week period; milk given by these cows before and after the test period will be examined for the presence of a particular chemical. If the herbicide is judged a contaminant, all the cows used will be slaughtered. Consequently, only twelve are purchased for the tests.

Let the variables x and y represent the before and after levels, respectively, of the chemical found in the milk. Identify

H_0: $E(x) \geq E(y)$

H_1: $E(x) < E(y)$

The data are

Animal	Before, x	After, y
1	.011	.011
2	.013	.012
3	.011	.013
4	.006	.012
5	.012	.011
6	.011	.015
7	.010	.019
8	.013	.016
9	.014	.012
10	.012	.007
11	.012	.019
12	.009	.017

The first pair of observations is discarded since the difference between the x and y is zero. The differences, their absolute values, and the assigned ranks are

Animal	D_i	$\|D_i\|$	Rank	
2	.012 − .013 = −.001	.001	1.5	
3	.013 − .011 = .002	.002	3.5	*
4	.012 − .006 = .006	.006	8	*
5	.011 − .012 = −.001	.001	1.5	
6	.015 − .011 = .004	.004	6	*
7	.019 − .010 = .009	.009	11	*
8	.016 − .013 = .003	.003	5	*
9	.012 − .014 = −.002	.002	3.5	
10	.007 − .012 = −.005	.005	7	
11	.019 − .012 = .007	.007	9	*
12	.017 − .009 = .008	.008	10	*

We sum the ranks which are marked by an asterisk, for these are associated with the positive differences. Therefore,

$$W = 3.5 + 8 + 6 + 11 + 5 + 9 + 10 = 52.5$$

For $n = 11$ (the first pair was discarded),

$$w_{.05} = 14$$

So,

$$w_{.95} = \frac{11(12)}{2} - 14 = 52$$

The value of W ($W = 52.5$) in this application exceeds 52. This means that 52.5 lies in the upper .05 tail of the sampling distribution of W as long as H_0 is true. Consequently, we would reject H_0 for $\alpha = .05$ and conclude that the new herbicide does contaminate the milk. Dow Chemical should not pursue development of the new herbicide.

(2) Suppose an auto repair shop assesses the same labor charges for a minor tune-up on a 4-cylinder engine as it does for a minor tune-up on a 6-cylinder engine. The owner questions whether this practice is wise and argues that the 6-cylinder engines require more labor. The chief mechanic disagrees and examines the latest tune-up work orders for evidence to support his argument. On each work order, the mechanic performing the service records his or her time in minutes. The chief mechanic gathers the latest 4-cylinder and 6-cylinder tune-ups for each of his nine mechanics and records the work time reported. He comes to you, his friend, for help in analyzing the data.

Letting x represent the work time for a 4-cylinder tune-up and y represent the work-time for a 6-cylinder tune-up, you set up the hypotheses as

$$H_0\colon E(x) \geq E(y)$$
$$H_1\colon E(x) < E(y)$$

The data the chief mechanic gives you are

	Labor Time (in Minutes)	
Mechanic	4-Cylinder	6-Cylinder
1	35	50
2	40	45
3	60	50
4	75	50
5	45	45
6	55	45
7	50	70
8	60	90
9	80	45

In performing the Wilcoxon test, you discard the times for the fifth mechanic since the difference is zero. The other differences are

Mechanic		D_i	$\|D_i\|$	Rank	
1	50 − 35 =	15	15	4	*
2	45 − 40 =	5	5	1	*
3	50 − 60 =	−10	10	2.5	
4	50 − 75 =	−25	25	6	
6	45 − 55 =	−10	10	2.5	
7	70 − 50 =	20	20	5	*
8	90 − 60 =	30	30	7	*
9	45 − 80 =	−35	35	8	

Summing the ranks associated with the positive differences gives

$W = 4 + 1 + 5 + 7 = 17$

For $n = 8$, this value does not lie in the upper tail of the sampling distribution of the test statistic. Consequently, the data do not allow you to reject the null hypothesis, and, for the time being at least, the owner has no evidence for his argument.

14.13 IDENTIFICATION OF APPLICATIONS: WHEN?

The title of this chapter is, to some extent, misleading. The techniques presented here *are* nonparametric methods. However, so are the methods presented in the preceding chapter.

The phrase "nonparametric methods" differentiates these procedures from the parametric methods presented in Chapters 7, 8, 10, 11, and 12. Nonparametric methods are typically applied when parametric techniques cannot be applied. It is usually the case that parametric statistical procedures are the preferred means of analysis *if they are appropriate*. Consequently, to identify when nonparametric methods should be applied, it is necessary to identify those cases in which parametric methods cannot be applied.

There are two necessary assumptions for the application of the presented parametric techniques:

1. The variables are normally distributed, or the sampling distribution of the statistic is normal or approximately normal.
2. The inferences concern parameters.

If either of these two is not reasonable in a particular decision problem, the application of parametric methods is not justified.

There are a great many nonparametric techniques available. We have only scratched the surface, because our objective was to introduce the area of nonparametric statistical techniques and to explain the value of such analyses.

If we look back at our three situations, we see that the sign test was used to analyze paired nominal data, the Mann-Whitney test was used to analyze two independent samples where normality could not be assumed, and the Wilcoxon test was used to analyze paired interval data again without the normality assumption.

If you have the need to analyze either nominal or ordinal data, you have a situation where nonparametric techniques would be appropriate. Even if you have interval data, if the sample size is small you may not consider the assumption of normality reasonable. This would also lead to the application of nonparametric techniques.

PROBLEMS

1. For each of the nonparametric methods presented in this chapter, identify an application from your major area of study. Point out the features of each which make these applications appropriate and of value.

2. The president of a firm has recently been approached by the computer services company which maintains its information system. The staff at the computer services company has developed new software packages which provide more extensive data to management. The costs involved in a system change include the inconvenience of the break-in period when the "bugs" are worked out. Still, the president sees the advantages of the more extensive data.

   But since such a change is a very expensive undertaking, the president wants to be sure that a majority of the executives who will be affected by the change are in favor of it. A sample of 20 executives is selected, and these executives are asked to state their preferences. Fifteen prefer the new method. Making use of an appropriate statistical test, make a recommendation to the president. (Use $\alpha = .10$.)

3. Suppose you are the editor of a Las Vegas newspaper. One of your investigative reporters comes to you with a story charging one casino with illegal operation of a particular game. You know that a justifiable charge printed in your newspaper means added notoriety and prestige, and as a result, increased circulation. On the other hand, a charge which cannot be documented by evidence will bring a law suit. You ask to see the reporter's evidence. The reporter played the game 11 times and recorded these outcomes:

   lost, lost, won, lost, lost, tied, won, lost, lost, lost, lost

   Suppose Nevada law requires that players of this game have an equal chance of winning or losing on a single play. Should you print the story? (Use $\alpha = .01$.)

4. The treasurer of a company has the responsibility for managing the firm's pension fund. The proceeds from the investments must provide payments to the retired personnel as well as maintain a reserve for future retirements. Joan Maisle is the treasurer of a growing firm in a rapidly growing industry. Because of this growth, it is particularly important that the pension fund investments do well. In addition to maintaining the fund, the investments must actually increase the size of the fund.

   Ms. Maisle is examining two mutual funds for this investment. Historical data for the two funds are given below. Make a recommendation to Ms. Maisle (use all the tests from this chapter that apply.)

Percent Rate of Return

	Fund A	Fund B
Year 1	7.28	No information
Year 2	6.77	No information
Year 3	9.34	8.65
Year 4	11.44	9.81
Year 5	10.12	10.43
Year 6	5.79	6.91
Year 7	−1.12	0.38
Year 8	−2.36	−1.76
Year 9	0.21	2.02
Year 10	4.57	7.50

5. Inland Electronics recently developed a rheostat-type switch that would allow industrial lathes to operate at variable speeds. There are currently four firms marketing similar switches. The Inland switch, while more expensive than its competitors, allows a substantially finer adjustment of speed. As part of the initial market research, 20 lathe operators and 15 purchasing agents were invited to test the switch and were given information on its performance and price characteristics.

   The marketing manager for Inland is concerned that while lathe operators will prefer the new switch, the purchasing agents will not because of its higher price. If this is the case, the marketing program will have to be designed to overcome this problem. To test the marketing manager's idea, you have asked each individual to state a preference rating for the new switch on a scale of 1 (very undesirable) to 100 (very desirable). The ratings assigned to Inland's switch are presented below.

Lathe Operators				Purchasing Agents		
53	70	100	91	51	59	55
72	90	78	74	43	70	100
81	77	25	92	62	75	15
87	39	76	84	73	38	27
82	88	100	67	94	42	19

   Using the Mann-Whitney test, would you conclude that the marketing manager's suggestion was correct?

6. Recently the Federal Trade Commission (FTC) received many complaints from consumers against a television manufacturing company which has claimed in its media advertising that its new Infra-Electo-Ronic YP110-20Z system produces a clearer and better picture than the current best-selling brand X. FTC is the federal agency authorized to enforce the Truth in Advertising Act. You are the newly appointed data analyst. Since you know that the performance of individual sets manufactured by the same firm varies slightly, you select 5 sets from each firm. Each set is assigned a score of 1 (poor) through 7 (excellent). The results are

   Infra-Electo-Ronic YP110-20Z: 7, 2, 6, 5, 1

   Brand X: 3, 4, 3, 6, 6

Using the Mann-Whitney test, would you conclude that the claim by the manufacturer is justified?

7. As a new marketing employee, you feel that one of your firm's products might be better promoted by market segmentation. In particular, you feel the product appeals more to men than women, and the ads should be designed so as to be more slanted toward male consumers. However, this change in advertising would require a costly revision in the entire marketing plan and consequently should not be recommended without supporting data. An inexpensive procedure for gathering evidence would be to ask the 12 male and 10 female employees in your office for their opinions. You do this and code their responses ranging from 1 (extremely unfavorable) to 10 (extremely favorable). The resulting scores for males and females are presented below.

Males			Females		
9	5	6	4	5	2
4	8	8	7	6	2
10	10	7	3	4	
7	3	5	5	1	

Do these data confirm your suspicions?

8. Suppose Western Mutual Life hires approximately 100 agents per year. These agents are given six months training in Denver, Colorado, before going into the field. At the end of the six-month training period, all the agents are rank-ordered in terms of their performance during the previous six months. The personnel manager wants to start a second, one-month managerial training program for those agents who are ranked highest at the end of the training session. However, the president will rule out such a program unless rank in the training session can be shown to be related to promotion at the end of three years in the field (the earliest that promotion is allowed). Of the 103 agents that completed training 3 years ago, 63 are still with the firm and 20 of these were promoted. Their ranks in training class are presented below.

Not Promoted						Promoted		
1	24	37	56	70	90	2	30	62
5	25	42	59	72	91	6	31	73
7	28	46	61	74	93	10	35	85
12	29	49	63	75	96	13	38	85
15	32	50	65	76	97	17	41	
18	34	52	67	77	99	21	45	
20	36	55	68	79	101	23	48	
22						26	53	

On the basis of this evidence, can the personnel manager recommend the managerial training program?

9. Suppose RCA is considering shifting from their standard work group format to one in which the employees themselves completely determine how the product will be assembled. The Volvo plant in Sweden has experienced considerable success since

instituting a similar plan, and this success is well known to RCA management. However, the changeover would be costly, so before instituting the change on a widespread basis, management wants to determine the impact of the shift on output. Therefore, seven work groups are selected and their average daily output is recorded. They are trained in the new procedure and given time to adjust to the shift, and their average daily output is recorded again. The results are presented below.

Group	Prior Output	Current Output
1	60	64
2	50	62
3	59	58
4	50	57
5	60	61
6	58	63
7	63	66

a. Using the Wilcoxon test, what would you conclude concerning the new method?

b. What assumptions, if any, are necessary to apply this test?

10. Suppose Ford Motor Corporation executives were planning a major effort to increase the firm's market share. The strategy included advanced training for dealers. The head of the training department suggested a new approach which could help dealers boost their sales. After much thought and discussion, executives decided to pursue a special two-week training program for the dealers. The program would be operated entirely at Ford's expense and would cost approximately $1500 per dealer. Therefore, a preliminary study was proposed. Prior to initiating the program on a large-scale basis, 15 dealers were selected at random and invited to attend. In addition, their most recent efficiency rating (on an interval scale from 1 to 100) was obtained. Ford executives want to compare these scores with the scores obtained by the same dealers in the reporting period immediately after the training session. The results are presented below:

Individual	Score Prior to Session	Score after Session
1	77	87
2	61	79
3	94	94
4	83	81
5	44	53
6	62	66
7	70	67
8	84	89
9	17	27
10	90	84
11	72	77
12	85	88
13	64	71
14	84	84
15	98	96

a. Using the sign test described in this chapter, what recommendation would you make to Ford? (Use $\alpha = .05$.)

b. If you use the Wilcoxon test, will your recommendation be the same? (Use $\alpha = .05$.)

c. Regardless of the decisions reached in parts *a* and *b* above, which test is more appropriate in this situation and why?

11. As an employee of a large public accounting firm, you have been assigned to a major investigation of the impact replacement-cost accounting will have on the earnings reported by industrial firms. Since converting from standard cost to replacement cost is an expensive process, you decide to estimate the impact on a sample of six medium-sized firms. You need to determine if the shift would lower reported income by more than 10 percent. The results of your investigation are shown below.

Firm	Profit Using Standard Cost	Profit Using Replacement Cost
A	$2,250,000	$1,920,000
B	3,700,000	$2,615,000
C	3,620,000	3,550,000
D	1,214,000	1,100,000
E	5,718,000	4,300,000
F	615,000	−217,000

What would you conclude? What assumptions did you make?

12. General Tire Company

MEMO

TO: *Mary Culbhusan, Chief Accountant*

FROM: *Richard Douglas, Marketing Manager*

We have run into a problem with the new tires which we have developed. I have received complaints from dealers all over the nation that sales are not what we expected because they are 5 percent more expensive than the existing standard tires. You know we cannot reduce the prices. Therefore, we have to change our marketing strategy and try to convince consumers that our tires, though more costly, have considerably longer tread life than the competing brand. But, because of the Federal Trade Commission, we'd better be sure before campaigning on this basis. I took twelve different trucks and mounted the new tire and standard tire side by side. After two months of usage, tread wear was measured and the estimated mileage of each tire computed. The data are given below.

Truck Number	Standard Tire	New Tire
1	63,500	64,000
2	66,250	66,000
3	61,000	63,500
4	64,500	66,000
5	65,000	66,250
6	60,250	64,500
7	63,000	63,000

Truck Number	Standard Tire	New Tire
8	65,500	66,000
9	67,250	68,250
10	63,750	64,500
11	64,250	66,000
12	64,750	66,750

I shall appreciate your recommendation.

(SIGNED) *Dick*

What should Culbhusan's recommendation be?

Chapter 15

Time Series and Forecasting

15.1 SITUATION I: WHY?

Suppose you are vice president for finance for Southern Airways and are considering the purchase of two new airplanes. This is a critical time for Southern. The larger airlines are experiencing more and more traffic, and most of these new customers are former customers of the smaller airlines. Part of the reason for this change in passenger preferences seems to be the desirability of flying on newer and more comfortable airplanes.

There are at this stage three alternatives available to Southern. First, Southern can elect to purchase two new airplanes. This action would likely result in a gradual increase in the market share for Southern with noticeable increases in revenues beginning in the second year after the purchase.

The second alternative is to purchase only one new airplane. This course of action is expected to allow Southern to hold its own in the marketplace for the next three or four years. It is hoped that, in that time, new revenues and other sources of funds would become available, allowing Southern to expand its fleet and thereby gain a greater share of the airline traffic in subsequent years.

The third alternative is to divert funds which might be used to purchase the airplanes to other investments. Southern's market share would deteriorate, making it very difficult for Southern to regain its current position in the market. The end result might well be that Southern Airways would cease to operate as a passenger airline.

Of the three alternatives, Southern would much prefer to purchase the two airplanes. The second and third alternatives might be pursued only in the event that the first alternative is not workable. If it is determined that the purchase of the two new airplanes is not wise, then the two remaining alternatives will be subjected to careful scrutiny with regard to the possible consequences of each.

If the first alternative is not wise, it will be because the company will have difficulty meeting the obligations of the financing of the purchase. The funds for the purchase of the airplanes will come from three sources. First, common stock will be issued. This source will be utilized first, because Southern incurs no real obligations. The second source utilized will be the sale of corporate bonds. Each bond will require Southern to make a lump-sum payment every six months. However, this length of time should be sufficient to allow Southern to accumulate the funds necessary to meet these obligations.

These two sources of funds should produce enough capital to allow the purchase of one airplane. Further, these sources do not present serious problems as far as meeting the financial obligations imposed is concerned.

If the company is to purchase the second airplane, however, additional capital will have to be raised by securing a loan. The obligations imposed by the loan include monthly payments. It is the question of servicing the loan which has you concerned at this time. If Southern defaults on the loan, future attempts to raise capital by any of the three methods mentioned will be extremely difficult. Investors will be very reluctant to purchase either Southern common stock or its bonds.

The decision hinges on Southern's ability to service the loan which will be necessary to accomplish the purchase of two planes. Because of the substantial increase in revenues anticipated in the second year following the purchase of two planes, the critical period for Southern would be the first year after purchase. During that year, Southern will experience the most difficulty in meeting the monthly obligations of the loan.

It is your responsibility as vice president for finance to determine Southern's ability to service the loan in the first year after purchase of the two planes. It is now January 19X9. The two planes, if ordered, would be put into service in July of this year, at which time the loan payments would begin.

The costs involved in the operation of Southern are closely associated with the inflation rate. In fact, very accurate projections of cost can be developed by taking the total cost for December 19X8, namely, $4,320,000, and calculating an increase of 1/2 percent per month. In July, if the two planes are purchased, monthly payments of $210,000 would be added.

Revenues, on the other hand, present a more difficult problem; no simple formula can be used to project revenues. At times revenues have grown, and at times they have fallen. They seem to be subject to more fluctuation than costs are. What you need is a means of projecting or forecasting revenues.

THE PRIMARY PURPOSE OF TIME-SERIES ANALYSIS IS FORECASTING FUTURE VALUES OF A RANDOM PROCESS.

15.2 COMPONENTS OF TIME SERIES

It is often useful to analyze sequences of observations of random variables over time. Such a sequence is called a *time series*. In virtually all of our preceding analyses, however, we have at least implicitly assumed that the observations were independent and identically distributed, thus constituting a random sample. Sometimes this assumption is appropriate in analyzing time series. In this chapter, however, we operate under the assumption that the observations are *dependent*. Essentially, what we're assuming here is that time, or some other variable associated with it, influences the observed values of the random variable.

As we've seen over and over again, the purpose of statistical analysis is the interpretation of data. The process of interpretation has two end results. The first is the *description* of the underlying process in terms of a model—the normal probability distribution, the linear regression model, or one of the others. The second is the *inference* of underlying relationships. For instance, we might test the mean productivity before and after a training session and draw conclusions about the degree of effectiveness of the training session. In time-series analysis, we interpret the influence of time on the random variable we're studying. From that, we infer the future influence of time-related factors on the random variable, and we thereby forecast future observations.

Because economic conditions change over time, the behavior of random variables of interest in business decisions have been seen to change in character over time. Means may change; variances may change. It is perhaps inappropriate to speak in terms of time as the influencing variable, when it is actually the changing economic conditions which lead to the variations in the distribution of the random variable. But time is the common factor, and it has proved productive to link the changes in economic conditions and business variables through the variable *time*.

We would like to isolate and identify the main influences which are observable in time-series data. These main influences will form the components of our time-series models. Once these are isolated, we can analyze them to better interpret economic reasons for the past observations. More important to us in the setting of Situation I, these components can be incorporated into a model for forecasting future occurrences for decision making.

There are four main influences observed in time-series data. These are identified as

1. Trend, T
2. Cycle, C
3. Season, S
4. Irregular, I

We organize our discussion in line with these components.

Trend

Trend is a long-term influence, typically causing slow-moving variations. The phrase "long-term" here means that a trend influence lasts several years. Some general economic conditions producing trend effects are changes in the size of the population, gradual improvements in technology, variations in standards of living, and inflation/deflation. Other factors include changes in consumer preferences and the degree to which efficiency and effectiveness (in the production process, say) are enhanced by learning. Figure 15-1 shows a possible trend influence in time-series data.

All these factors clearly identify important aspects for planning purposes. All suggest potential changes in profit, for instance. In some cases, sales are changing because the market is expanding or contracting. In others, costs are rising or falling because of variations in quality or efficiency in the production processes. Other important performance variables can be seen to be related to these factors as well.

We should spend a moment discussing the inflation/deflation influence. Typically, before we begin the time-series analysis, the inflation/deflation influence in the data is removed via an appropriate price index. Inflation creates the illusion of increased unit sales because revenues are climbing. This might lead executives to increase production, expand the work force, or take some

Figure 15-1

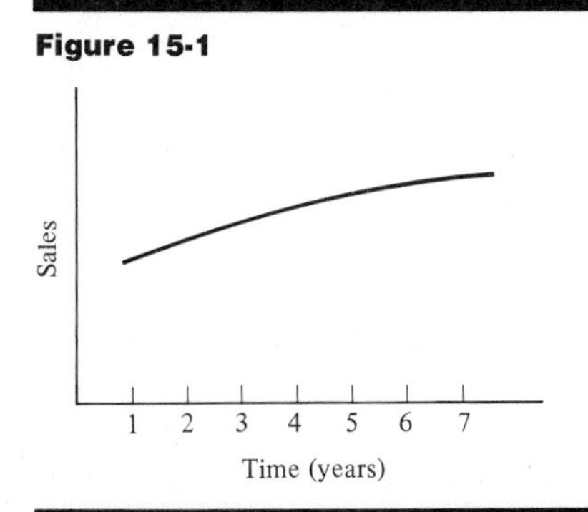

other costly action to meet the erroneously anticipated increase in demand. To avoid this misinterpretation, we must first adjust the data to remove the inflation influence. There are times, however, when we are analyzing time-series data in which the inflation/deflation influence might be appropriately considered a part of the trend component. For instance, with profit figures, inflation/deflation on the revenue side and on the cost side are already incorporated into the figures. The difference between these two inflation/deflation rates might be considered appropriate trend information.

Cycle

The cyclical component refers to the influence of business cycles on time-series data. *Business cycles* are upward and downward movements in general business activity. Sometimes we talk about "upswings" or "downturns" or "recessions." Business cycles are highly unpredictable in terms of amplitude (how far up or down the movement will be), duration (how long a movement will last), and shape (how fast the rise or drop will be). Generally, the downward movements are more rapid than the upswings; however, there are few other comments of a general nature which apply to business cycles. Figure 15-2 gives an example of a cycle, and Figure 15-3 shows the trend of Figure 15-1 after it is distorted by the cycle component depicted in Figure 15-2.

Business cycles indeed influence the variables which impact upon the performance of the firm. For instance, during a downward movement, firms marketing luxury items tend to slow down considerably while firms marketing necessity items slow down only slightly. Because of this impact, and the varying nature of it relative to the different industries, the cycle influence is an important component for time-series analysis.

Season

Season describes the upward and downward movements caused by the time or season of the year. Several industries offer prime examples of this influence.

Figure 15-2

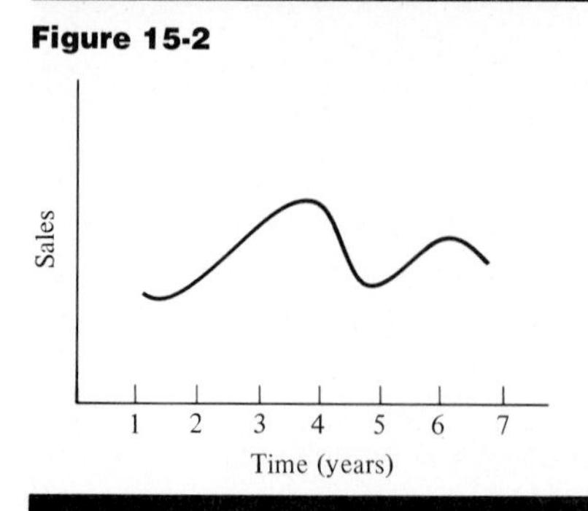

Figure 15-3

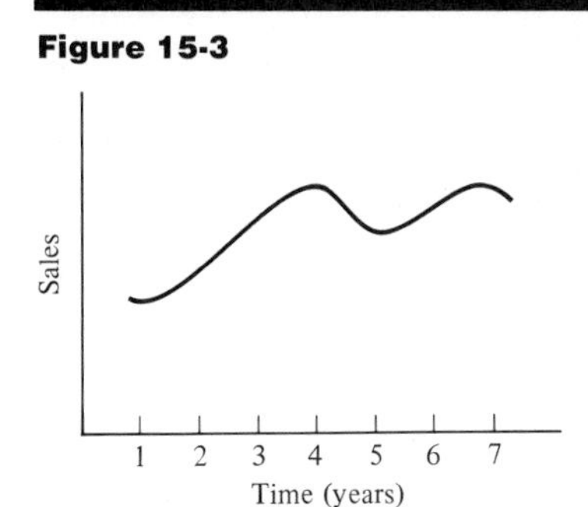

For instance, sales of ski equipment exhibit radical movements during the course of a year. Of course, other industries might experience little or no seasonal patterns. Although certain foods have their seasons, sales for grocery stores typically experience little or no seasonal pattern. Figure 15-4 illustrates a seasonal pattern.

The effects of the seasonal component average out over the course of the year. After one year, the process is back where it started—assuming, of course, no trend, cycle, or irregular influence. There are no net gains or losses in the time series which are attributable to this component.

Because the seasonal influence is linked to the time of the year, the time of occurrence is predictable. The influence exerted by the advent of spring on the women's fashions industry was experienced last spring and will be experienced again next spring. As customs change, however, the extent of the influence may change. Consequently, there may be slight variations in the measure of the seasonal effects over longer periods of time. The decision maker or analyst must select one of the following assumptions when treating the seasonal component:

1. The impact of the seasonal component is constant from year to year.
2. The seasonal effect is changing slightly from year to year.
3. The impact of the seasonal influence is changing dramatically.

The third assumption is seldom made. However, it is necessary at times. For instance, if the automobile industry were to change the timing of the introduction of the new models, we would expect a major shift in the effect of season.

Irregular

The irregular component incorporates two kinds of influences. The first is the occurrence of individual, unusual events which have impacted upon the time

Figure 15-4

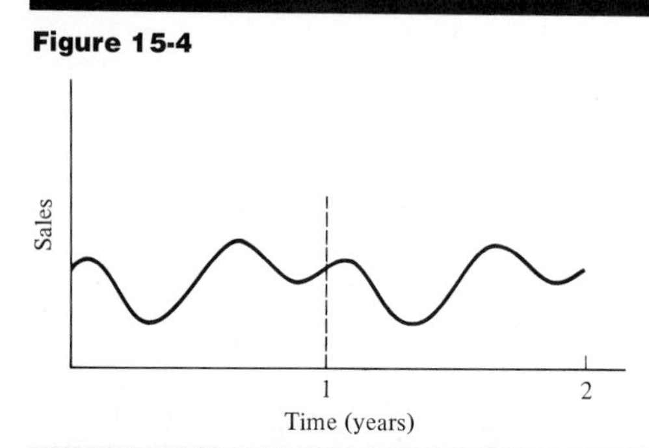

series. Possible events of this nature include strikes, high absenteeism due to a flu outbreak, or a major change in governmental economic policy. All these influences could cause major shifts in the random process which produced the time series. Typically, the process will exhibit new characteristics of behavior after the shift.

The second kind of influence incorporated in the irregular component is the random fluctuation. This fluctuation is usually small in nature, resulting in variations within, say, ±5 percent. As we assumed in regression analysis, random fluctuations are assumed to have a mean of zero, that is, their net effect is set equal to zero in the analysis.

The process of adjusting for the first kind of irregular influence is rather specialized statistical analysis. Also, Situation I gives no indication of individual, unusual economic events. Consequently, *for the remainder of the chapter we will treat the irregular component as including only random fluctuations*.

Figure 15-5 illustrates a series of random variations in a time series. Figure 15-6 is a composite of the trend of Figure 15-1, the cycle of Figure 15-2, the season of Figure 15-4, and the irregular variation of Figure 15-5. Thus, Figure 15-6 might illustrate a typical graph of time-series data.

15.3 FORECASTING: HOW?

The basic logic of forecasting is a two-step procedure. First, we build a model from the past observations of the process; then, we use that model to predict future occurrences. We must first estimate the influence of the various components and then combine the components to produce a forecasting model. As the method of estimating the component effects depends upon the form we choose for our model, we first discuss the form of our model.

The most common way of combining the components into a model is by the multiplication of the components. We use Y as the random variable which we are trying to forecast. The model we use is called the *multiplicative model* and suggests that

$$Y = T \cdot C \cdot S \cdot I$$

One important consequence of this model is that it dictates the units of measurement for the four components. If we are forecasting sales as measured by total revenue, the Y values are expressed in dollars. We let T be measured in dollars also. The other three components can be interpreted as percentages. When the seasonal component, for example, has *no* influence on a particular observation, the value of S is 1.0. Similar statements hold for the components C and I. If all three precentage components are equal to 1, namely,

$$C = S = I = 1.0$$

then the observed value might be called an "average" or "typical" occurrence. From Equation (15-1), it is also true that

$$Y = T$$

if $C = S = I = 1.0$. Thus, the multiplicative model gives us the trend component as the typical value, and the other components can be viewed as adjustments. A value of $C = 1.015$ indicates that the prevailing condition of the business cycle will produce an observed value 1.5 percent *higher* than the typical; a seasonal component value of $S = .970$ indicates that the time of the year will produce an observed value of 3 percent *lower* than the typical. All observations, then, can be viewed as adjustments from the typical because of business cycles, because of the time of the year, and because of random fluctuations.

One plan of attack for the estimation problem can be the following:

1. Estimate the trend component T using annual data. (Monthly or quarterly observations can be used if they are seasonally adjusted, that is, if the seasonal component has been removed.)

Figure 15-5

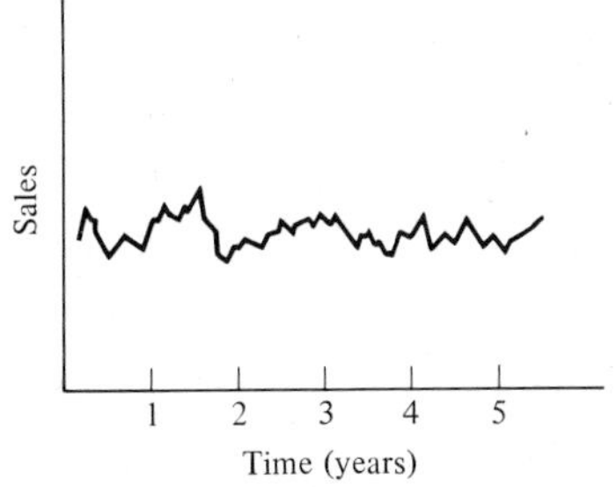

Figure 15-6

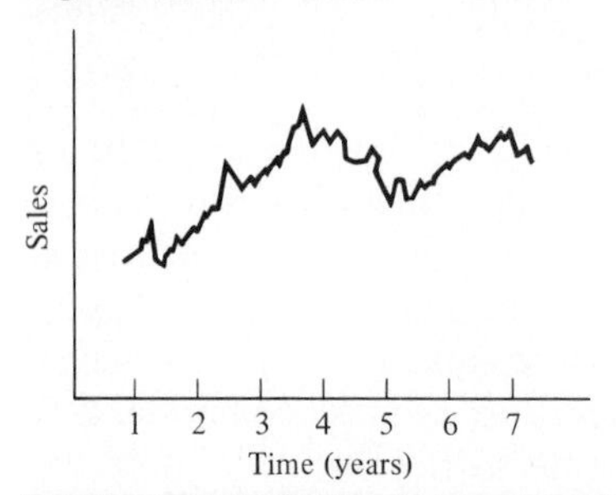

2. Estimate the seasonal component S using the "raw," that is, unaltered, monthly data. (Quarterly data can be used, if appropriate to the decision problem.)
3. Estimate the irregular component I by "smoothing" the seasonally adjusted data.
4. Estimate the cyclical component C using the original raw data and the estimates of T, S, and I.

At this stage we will be able to examine the goodness of the model by analyzing the irregular component values estimated in 3 above. Our examination would focus on whether the values for I are indicative of random fluctuations. This would be determined by deciding whether the values are independent of each other. If they are, we can judge the model to be acceptable.

These four steps would be followed for the complete analysis. That is, the analyst or decision maker can accomplish (*a*) the forecasting of future observations, and (*b*) the analysis of past observations to determine economic causation, that is, *why* we observed what we have observed. However, our purpose is only the former at this time. To accomplish the forecasting of future monthly revenues, it is necessary only to estimate T and S, for in the interest of brevity, we will make an assumption about C and simply assert that the values of I constitute random fluctuations.

To illustrate the procedures involved in the above estimation problems, let's return to the setting of Situation I. Our first step is to estimate the trend component using annual data. Suppose Table 15-1 gives the annual revenues for the past eight years for Southern Airways.

We are looking for some relationship between time in years and revenues. Following the curve-fitting procedures of Chapters 10 and 11, we plot the data from Table 15-1 on a scatter diagram. (See Figure 15-7.) It appears that a straight line might describe the trend component very well. That is, we might assume that, with t representing time in years from 19X0,

$$T = \beta_0 + \beta_1 t + \epsilon$$

and estimate β_0 and β_1 by the methods of Chapter 10.

The data are analyzed via a simple linear regression computer program. The output from that program gives

$$b_0 = 57.95$$
$$b_1 = 0.141$$
$$r = 0.9464$$
$$r^2 = 0.8957$$

The r^2 value indicates that 89.57 percent of the variation in the annual revenue figures is explained by the variation in t, the time in years from 19X0. Thus, we conclude that the model is an acceptable description of the relationship between annual revenue and time. Consequently, we choose as our trend component

$$T = 57.95 + .141t$$

where t is time in years after 19X0.

Now we have the trend component in terms of annual revenues. That is, this trend component is most useful for forecasting annual revenue. It can also be used, however, for monthly forecasts. Perhaps the easiest way to adapt these annual trend figures to monthly trend figures is to divide the annual forecast by 12. This assumes no trend influence *during* the year. That assumption may be somewhat unreasonable in some applications. It will suffice for our purposes, however.

The next step is to estimate the seasonal component using raw monthly data. Suppose Table 15-2 gives monthly revenues for the past four years for Southern.

Two issues need to be resolved at this stage. The first is whether we will develop monthly values for S or quarterly values for S. Examining Table 15-2 will help us resolve this issue. In each of the years, revenues for the month of

Table 15-1 Annual Revenue
(In Millions of Dollars)

Year	Revenue	Year	Revenue
19X1	58.03	19X5	58.84
19X2	58.27	19X6	58.72
19X3	58.29	19X7	58.74
19X4	58.58	19X8	59.16

Figure 15-7 (*Note:* t = 0 represents the year 19X0.)

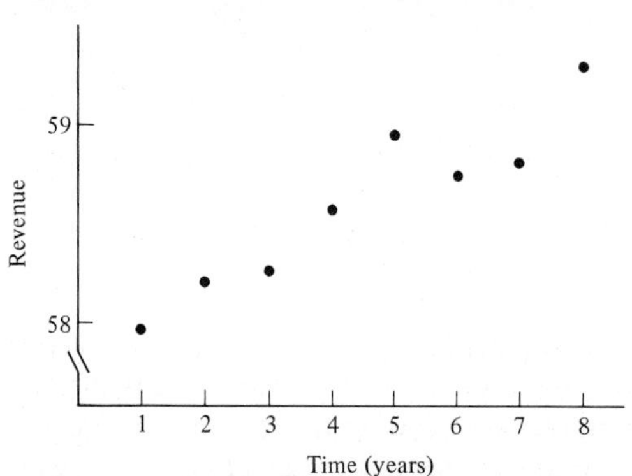

December are considerably higher than October revenues and November revenues. This is undoubtedly a result of increased travel over the Christmas holiday season. The point is, however, that using a quarterly seasonal component will tend to distort the modeling of this special influence, because one value of S will be associated with December and two other (October and November) very different months in terms of airline travel activity. Our need to forecast monthly revenues would not be satisfied appropriately because the October and the November projections would be too large.

The second issue we must resolve is whether to assume a constant seasonal component from year to year. Again, looking at Table 15-2 can help. We observe no major changes in the influence a particular month has on revenues from year to year. We would next look at economic conditions in an effort to decide if there might be a gradual change in S over time. For our purposes, we will assume a constant seasonal component.

The five steps we follow in estimating S are

1. Assign to each month the average of the five observations nearest that month. For instance, we would assign to April the average of the observations from February, March, April, May, and June. We would assign to May the average of the observations from March, April, May, June, and July. This is a smoothing process and calculates a *five-point moving average*.

2. For each month, calculate the ratio of the observed value divided by the five-point moving average. This ratio is called the *specific seasonal*, *SS*. Table 15-3 shows steps 1 and 2 performed on the data of Table 15-2.

3. Discard approximately the largest third and the smallest third of the specific seasonals. It is important to discard the same number of high values of specific seasonals as low values. This is done to remove the effects of outliers.

4. Estimate the seasonal component, or seasonal, S, to be the average of the remaining specific seasonals.
5. Adjust the seasonals so that they sum to 12. This is done so that there is no net influence of the seasonals over the course of the entire year. If quarterly seasonals were computed, their sum would be 4.

Table 15-4 shows the final calculations of the seasonals for the data of Table 15-2. The data in Table 15-2 can now be adjusted for the seasonal component. This would be accomplished by dividing each observation by the corresponding seasonal. These adjusted data can be used to estimate a trend component for monthly revenues in the same way that the annual data were used. We would likely get a different equation for the trend component than the adjustment of Equation (15-2) which we will use.

We now have the values for T and S, and all that remains is the estimation of C and I. We will assume at this stage that $C = 1.0$. This essentially represents the belief that the business cycle will have no influence on Southern's monthly revenues for the next year. For each of the 48 months in Table 15-2, the cyclical component is nearly 1.0.

The irregular component I can be estimated using the seasonally adjusted monthly data. We would smooth these data by using a variant of the five-point moving average used above. The purpose of estimating the irregular component is to determine the goodness of the model. If the I values are indicative of random variation, we accept our multiplicative model as an accurate description and therefore as a reasonable forecasting device. But testing the I values is an involved process using techniques not presented in this text. Let's just assert that our model is acceptable and go from there.

Table 15-2 Monthly Revenues
(In Millions of Dollars)

Month	19X5	19X6	19X7	19X8
January	4.69	4.85	4.86	4.76
February	4.78	4.56	4.43	4.67
March	4.36	4.41	4.65	4.39
April	4.50	4.70	4.52	4.25
May	5.09	5.00	5.21	5.02
June	5.22	5.66	5.29	5.36
July	5.44	5.13	5.41	5.47
August	5.24	5.09	5.00	5.48
September	4.89	5.05	5.11	4.87
October	4.68	4.64	4.38	4.71
November	4.22	4.23	4.41	4.42
December	5.73	5.40	5.47	5.76

Table 15-3 Calculation of Specific Seasonals, SS

Month		19X5	19X6	19X7	19X8
January	Actual	4.690	4.850	4.860	4.760
	Average	. . .	4.760	4.720	4.740
	SS	. . .	1.019	1.030	1.004
February	Actual	4.780	4.560	4.430	4.670
	Average	. . .	4.850	4.780	4.710
	SS	. . .	0.940	0.927	0.992
March	Actual	4.360	4.410	4.650	4.390
	Average	4.680	4.710	4.740	4.620
	SS	0.932	0.912	0.981	0.950
April	Actual	4.500	4.700	4.520	4.250
	Average	4.790	4.870	4.820	4.740
	SS	0.939	0.965	0.938	0.897
May	Actual	5.090	5.000	5.210	5.020
	Average	4.920	4.980	5.020	4.900
	SS	1.035	1.004	1.038	1.024
June	Actual	5.220	5.660	5.290	5.360
	Average	5.100	5.120	5.090	5.120
	SS	1.024	1.105	1.039	1.024
July	Actual	5.440	5.130	5.410	5.470
	Average	5.180	5.190	5.210	5.240
	SS	1.050	0.988	1.038	1.044
August	Actual	5.240	5.090	5.000	5.480
	Average	5.090	5.120	5.040	5.180
	SS	1.029	0.994	0.992	1.058
September	Actual	4.890	5.050	5.110	4.870
	Average	4.890	4.830	4.870	4.990
	SS	1.000	1.046	1.049	0.976
October	Actual	4.680	4.640	4.380	4.710
	Average	4.950	4.890	4.880	5.050
	SS	0.945	0.949	0.898	0.933
November	Actual	4.220	4.230	4.410	4.420
	Average	4.870	4.840	4.830	
	SS	0.867	0.874	0.913	
December	Actual	5.730	5.400	5.470	5.760
	Average	4.810	4.720	4.740	
	SS	1.191	1.144	1.154	

Table 15-4 Calculation of S

Month	Specific Seasonals				Average of Middle Third	Adjusted S
January	...	1.019	1.030	1.004	1.019	1.02
February	...	0.940	0.927	0.992	0.940	0.94
March	0.932	0.912	0.981	0.950	0.941	0.94
April	0.939	0.965	0.938	0.897	0.938	0.94
May	1.035	1.004	1.038	1.024	1.030	1.03
June	1.024	1.105	1.039	1.047	1.043	1.05
July	1.050	0.988	1.038	1.044	1.041	1.05
August	1.029	0.994	0.992	1.058	1.012	1.02
September	1.000	1.046	1.049	0.976	1.023	1.03
October	0.945	0.949	0.898	0.933	0.939	0.94
November	0.869	0.874	0.913	...	0.874	0.88
December	1.191	1.144	1.154	...	1.154	1.16
Total					11.948	12.00

Once we have T, C, and S, we are ready to forecast. Our forecast for any month is simply the product of the trend, cyclical, and seasonal components for that month. That is, our forecast is

$$Y = T \cdot C \cdot S$$

This forecast is comparable to the estimation done in Chapters 10 and 11. The analogy lies in the assumption of no random error. In Chapter 10, for instance,

$$\hat{y} = b_0 + b_1 x$$

was the estimate for a particular value of the random variable y. The error term was taken to be zero. In our case, we take $I = 1.0$, indicating, again, no random error.

In any particular application, however, we may want our estimate to be conservative. We saw this in Chapter 7, when we were estimating mean tread life for the truck tires. Recall that, in that example, the estimate used to calculate a price per tire was the lower confidence limit for the mean. We could incorporate this same attitude in estimation via regression by putting some value for ϵ in our regression equation so that

$$\hat{y} = b_0 + b_1 x + \epsilon$$

would be a conservative estimate. In the forecasting model, conservative estimates can be achieved by inserting a value for the irregular component I.

15.4 SITUATION I REVISITED

We have completed the estimation phase and are now able to forecast monthly costs and revenues for the next 18 months. Monthly costs are projected by increasing the current total cost of $4,320,000 by 1/2 percent per month. These projections are given below in millions.

Jan.	$4.34	June	$4.45	Nov.	$4.56	Mar.	$4.65
Feb.	$4.36	July	$4.47	Dec.	$4.58	Apr.	$4.68
Mar.	$4.39	Aug.	$4.49	Jan.	$4.61	May	$4.70
Apr.	$4.41	Sept.	$4.52	Feb.	$4.63	June	$4.72
May	$4.43	Oct.	$4.54				

These are projected monthly costs if the two airplanes are *not* purchased. If the two planes *are* purchased, Southern incurs an additional monthly cost of $210,000 beginning in July. This addition in cost, of course, results from the loan required to purchase the airplanes. This changes the monthly cost projections, in millions, to

Jan.	$4.34	June	$4.45	Nov.	$4.77	Mar.	$4.86
Feb.	$4.36	July	$4.68	Dec.	$4.79	Apr.	$4.89
Mar.	$4.39	Aug.	$4.70	Jan.	$4.82	May	$4.91
Apr.	$4.41	Sept.	$4.73	Feb.	$4.84	June	$4.93
May	$4.43	Oct.	$4.75				

The revenue projections make use of the multiplicative model

$$Y = T \cdot C \cdot S$$

where $T = 57.95 + .141t$

t = time in years after 19X0

$C = 1.0$

S as given in Table 15-4

We want to forecast revenues for 18 months. The first 12 of these will make up year 19X9, so $t = 9$. For the subsequent 6 months, we use $t = 10$. With $t = 9$, we forecast the trend component of annual revenue as

$$T = 57.95 + .141(9) = \$59.22$$

and of monthly revenue as

$$T/12 = \$4.94$$

Using the seasonals in Table 15-4, we obtain the following forecasts expressed in millions.

Jan.	$5.04	Apr.	$4.64	July	$5.19	Oct.	$4.64
Feb.	$4.64	May	$5.09	Aug.	$5.04	Nov.	$4.25
Mar.	$4.64	June	$5.19	Sept.	$5.09	Dec.	$5.73

For the subsequent 6 months, we use $t = 10$, and the seasonal factors of Table 15-4. This gives

$T = 57.95 + .141(10) = \$59.36$

and

$T/12 = \$4.95$

Finally, forecasted revenues, in millions, are

Jan.	$5.05	Apr.	$4.65
Feb.	$4.65	May	$5.10
Mar.	$4.65	June	$5.20

These projections have used

$Y = T \cdot C \cdot S$

but the actual revenues will be

$Y = T \cdot C \cdot S \cdot I$

that is, subject to random variation. This variation is typically between ±5 percent. However, it should be clear that the random fluctuations can have a tremendous impact. In fact, they may mean the difference between defaulting and not defaulting on the loan. Because the consequences of default are so serious, we might want to be conservative in our forecasts of future revenues. This is accomplished by incorporating into our forecasting model an I value less than 1.0. Since I can result in as much as a 5 percent decrease in revenue, we use $I = .95$. Incorporating this factor, we obtain the following forecasts, in millions, for the next 18 months.

Jan.	$4.79	June	$4.93	Nov.	$4.04	Mar.	$4.42
Feb.	$4.41	July	$4.93	Dec.	$5.44	Apr.	$4.42
Mar.	$4.41	Aug.	$4.79	Jan.	$4.80	May	$4.84
Apr.	$4.41	Sept.	$4.84	Feb.	$4.42	June	$4.94
May	$4.84	Oct.	$4.41				

The computational aspects of this problem are completed in Table 15-5. The important column for Southern's decision problem is the last column, "Cumulative Profit." All the values in that column must be positive before the plan to purchase two airplanes is feasible. In this case, we see that all these values *are* positive. Based upon our forecasts, it appears that Southern can service the loans required for the purchase of two airplanes.

Table 15-5 Forecasts of Profit
(In Millions of Dollars)

	Projected		Net	Cumulative
Month	Revenue	Cost	Profit	Profit
January	4.79	4.34	.45	.45
February	4.41	4.36	.05	.50
March	4.41	4.39	.02	.52
April	4.41	4.41	0	.52
May	4.84	4.43	.41	.93
June	4.93	4.45	.48	1.41
July	4.93	4.68	.25	1.66
August	4.79	4.70	.09	1.75
September	4.84	4.73	.11	1.86
October	4.41	4.75	−.34	1.52
November	4.04	4.77	−.73	.79
December	5.44	4.79	.65	1.44
January	4.80	4.82	−.02	1.42
February	4.42	4.84	−.42	1.00
March	4.42	4.86	−.44	.56
April	4.42	4.89	−.47	.09
May	4.84	4.91	−.07	.02
June	4.94	4.93	.01	.03

15.5 USING AUXILIARY VARIABLES

In Situation I, we constructed a model using only values of the variable we wanted to forecast. At times this approach is appropriate; the past observations when used to estimate the time-series components do provide sufficient explanation to allow accurate forecasts. The assumption here, of course, is that all economic factors impacting upon the process can be categorized into the time-series components and their effects on the process can be described as functions of time.

At other times, however, there may be better, more efficient ways to incorporate the influence of economic conditions on the process. In particular, a certain variable may have considerable impact but its effect is diluted by association with other variables in one component. For instance, the trend component incorporates, among other things, the size of the population. It may be that population size alone has considerable explanatory power in terms of describing movements of the process through time. The explanation of the process variation through the size of the population may even exceed the explanation provided by the trend component in its entirety.

We often hear government economists projecting movements in the economy. Typically, these are made on the basis of "leading indicators." The term "indicator" refers to a variable which "indicates" changes in the economic process because the indicator variable and the economic-process variable tend to move together. The word "leading" is used to mean that the indicator variable "leads" the process variable in the sense that changes in the indicator variable precede, in time, changes in the process variable.

In Chapters 10 and 11, we saw the value of capitalizing on relationships between variables in estimation problems. We were doing some forecasting, of course, in Chapters 10 and 11, when we developed estimates for a particular value of the dependent variable. In this section, we examine more closely the use of a regression model for forecasting.

Let's suppose Robert Smith is a building contractor specializing in one- and two-family dwellings. Smith owns some property on which he eventually intends to develop a housing subdivision. Before he invests the considerable sum necessary to begin construction, however, he wants to be confident that the houses he builds will be sold relatively soon after construction. In this way, he can acquire more capital to continue construction throughout the subdivision.

One factor that Smith feels is very important to the housing market and the activity in that market is the prevailing interest rate on first mortgage loans. In particular, when the interest rate is high, new-house sales are relatively slow, and when the interest rate is low, new-house sales are fairly active. His decision hinges on his forecast of the first mortgage interest rate.

Smith believes that interest-rate movements can be explained rather well by the prime interest rate and by the current interest rate on first mortgage loans. In other words, he hopes to be able to forecast the new interest rate on first mortgage loans by using a model based upon the current prime interest rate and the current first mortgage interest rate.

Smith gathers the following data to build a forecasting model for the first mortgage interest rate.

	Interest Rates, %	
Time	First Mortgage Loans	Prime Rate
1	9.00	8.00
2	9.25	8.25
3	9.25	8.00
4	9.25	7.75
5	9.00	7.75
6	9.00	7.50
7	9.00	7.25
8	8.75	6.75
9	8.75	6.75
10	9.00	6.50

We will use a multiple linear regression model with the variables defined as

Dependent variable = first mortgage interest rate this period
Independent variable = first mortgage interest rate last period
Independent variable = prime interest rate last period

This variable identification summarizes Smith's belief that the first mortgage interest rate for a given period can be described by a linear expression involving the first mortgage interest rate from the preceding period and the prime interest rate from the preceding period. Thus, for describing the behavior of the first mortgage interest rate at time t, y_t, we will use the model

$$y_t = \beta_0 + \beta_1 y_{t-1} + \beta_2 x_{t-1} + \epsilon$$

where y_{t-1} = the first mortgage interest rate at time $t - 1$
x_{t-1} = the prime rate at time $t - 1$

For this regression problem, we have the following data:

Time	y_t	y_{t-1}	x_{t-1}
2	9.25	9.00	8.00
3	9.25	9.25	8.25
4	9.25	9.25	8.00
5	9.00	9.25	7.75
6	9.00	9.00	7.75
7	9.00	9.00	7.50
8	8.75	9.00	7.25
9	8.75	8.75	6.75
10	9.00	8.75	6.75

When these data are analyzed by a computer program designed to do multiple linear regression analysis, the following values will be computed:

$$b_0 = 8.441$$
$$b_1 = -0.237$$
$$b_2 = 0.360$$
$$R^2 = 0.6634$$
$$R = 0.8145$$
$$F_{2,6} = 5.913$$

The F statistic value is significant at the $\alpha = .05$ level of significance, giving us confidence that the model provides a good description of the movements in first mortgage interest rates.

Using the description provided by the multiple regression model, we would forecast the first mortgage interest rates by the equation

$$\hat{y}_t = 8.441 - 0.237y_{t-1} + 0.360x_{t-1}$$

From the values that Smith has for period 10, we would forecast the first mortgage interest rate for the coming period, namely, period 11, as

$$\begin{aligned} y_{11} &= 8.441 - 0.237y_{10} + 0.360x_{10} \\ &= 8.441 - 0.237(9.00) + 0.360(6.50) \\ &= 8.648 \end{aligned}$$

Since this value is lower than any of the 10 rates observed in the recent past, Smith might want to begin construction.

We have called this forecasting procedure *forecasting with auxiliary variables*. That is, we have made use of information available in variables other than the one we are forecasting. Everything that was said in Chapters 10 and 11 about regression analysis and its use applies here. In particular, the comments made with regard to selecting the independent variables in a regression model apply here. We would select auxiliary variables that are believed to be related to the variable of interest.

One final note regarding forecasting using the regression model with auxiliary variables: The same general approach to forecasting is used here. That is, *past data are used to form a model which describes the behavior of the variable we are interested in forecasting. Then that descriptive model is used to forecast future occurrences or observations on the variable of interest.*

15.6 IDENTIFICATION OF APPLICATIONS: WHEN?

The need to forecast is pervasive in business decision making. The executive must always attempt to project the potential consequences of each of the possible courses of action. However, these projections are not always developed using the techniques of this chapter, for we have made assumptions about the interrelationships of economic conditions and the availability of data that may not hold in a particular application.

In time-series analysis, we identify two types of influences on the data—*systematic* influences and *random* influences. The random fluctuations are captured in the irregular component. The systematic fluctuations are classified into three groups—trend, cycle, and season—*but each group is assumed to be completely describable by the single variable time*. The trend, cyclical, and seasonal components are all functions of time, and the impact of any economic occurrence or condition is assumed to be modeled by one of these functions of time. This issue must be resolved if time-series analysis is to be applied.

A second requirement of time-series analysis is the amount of data required. We did not state this explicitly in the chapter, but our analysis of Situation I pointed out the need for a large amount of data. Estimating the trend component with annual data suggests that several years of data must be available. We used 48 monthly observations to estimate the seasonal components, but we would have profited considerably from an even larger sample. These data requirements raise another issue: In our rapidly changing economy and environment, how stable or stationary are these influences (that is, trend, cycle, season)? These issues must be resolved before time-series analysis can be applied.

In forecasting with auxiliary variables, we do not necessarily focus on time as the factor of prime importance. Rather, we try to capitalize upon known relationships between certain economic variables and the variable we wish to forecast. These variables may include past values of the variable we are forecasting. The typical approach is to use a regression model to describe the past observations of the variable of interest. That description is then used to forecast future observations. The assumptions necessary to apply regression must be satisfied here to allow this type of forecasting.

PROBLEMS

1. Identify an application of forecasting with the time-series model in your major field of study. Point out the characteristics which make this application appropriate and of value.

2. Identify a forecasting application in your major area of study in which the use of auxiliary variables would be more productive. Identify the auxiliary variables.

3. Suppose that you are a manager of a medium-sized local department store, and you have a set of monthly seasonals for the department store sales. If the monthly sales are divided by these seasonals, you obtain seasonally adjusted monthly sales. Of what value might these adjusted figures be?

4. As a member of the state's revenue department, you have been asked to forecast monthly construction completed in the state for the next two years. These estimates are needed to forecast the state's property tax revenues for the state's budget. You gather the monthly figures for the past five years as presented below.

Value of Completed Construction
(In Millions of Dollars)

Month	19X3	19X4	19X5	19X6	19X7
Jan.	131	148	111	146	166
Feb.	119	134	119	133	120
Mar.	124	158	126	136	139
Apr.	175	211	171	169	170
May	229	256	247	218	255
June	275	315	276	287	309
July	293	329	300	281	322
Aug.	313	321	322	326	350
Sept.	321	328	322	343	354
Oct.	318	295	279	326	400
Nov.	323	234	225	302	295
Dec.	199	205	235	245	248

It will be helpful if you proceed as follows:

a. Estimate the trend component T using the annual data for five years.

b. Estimate the seasonal component S using the unaltered data. (Use a five-point moving average.)

c. Estimate the irregular component I by smoothing the seasonally adjusted data.

d. Prepare the required forecast. (Use $C = 1.0$.)

5. Building Contractors, Inc., has recently received a government contract to build several small military observation stations in Alaska. The crew which will be sent to the site will stay approximately one year. Any additions to or subtractions from this crew will be costly and may mean the difference between financial success and failure of the contract. To determine the number of employees available for the project, executives must forecast labor force needs for construction in the continental United States for that year. The trend component for this problem has been estimated as

$$T = 400 + 3t$$

where T = number of employees required
t = number of months
t = 0 represents December 19X5

The seasonal component for each month follows.

Month	Index	Month	Index	Month	Index
Jan.	.871	May	1.017	Sept.	1.013
Feb.	.833	June	.895	Oct.	1.059
Mar.	1.004	July	.859	Nov.	1.118
Apr.	1.066	Aug.	.998	Dec.	1.267

The Alaskan construction will begin in April 19X7. Assume no cyclical influence and that any random variations in labor needs can be met by temporary help. Forecast the labor needs for the 12 months of Alaskan construction.

6. In May of this year, Bill Gossin, the newly appointed marketing director of Hudson Foods, utilized television advertising of the firm's frozen desserts for the first time and sales reached a record high. The frozen desserts line had been introduced 3½ years ago, but sales were never very high. Bill Gossin wrote to the director in his memo, "I would like to continue and expand the television campaign." The director, however, suspects that the record sales are attributable to trend and seasonal influences and random variation. The sales history for frozen desserts follows.

Sales
(In Thousands of Cases)

Month	First Year	Second Year	Third Year	Fourth Year
January	341	488	638	789
February	338	480	621	765
March	448	633	817	987

Month	First Year	Second Year	Third Year	Fourth Year
April	479	349	853	998
May	512	706	905	1107
June	577	791	994	
July	599	812	999	
August	576	776	978	
September	546	738	920	
October	511	679	846	
November	496	655	809	
December	536	701	867	

Assuming the cyclical component has no influence (that is, $C = 1.0$), make a recommendation to the director regarding Gossin's request for continued television advertising.

7. Big Mountain Beer Company

MEMO

TO: *Barbara Jarrew, Finance Manager*

FROM: *Gus Hadleychase, Director*

We are considering enlarging our plant. Scotty, the chief engineer, reports that the plant will take two years to complete. Our current production capacity is 12,000 barrels per month. Since beer has a relatively short storage life, production (including the aging process) and sales must occur within a few weeks of each other. Also, if monthly sales exceed production capacity for two months in a row, sales will be lost because of inventory shortages. On the other hand, if the plant operates at less than 50 percent of capacity in any month, we will suffer a loss. So, we need some accurate projections of sales.

We want the new plant to be adequate for the next five years. Sales data (in barrels) for the past six years follow.

Sales
(In Barrels)

Month	19X2	19X3	19X4	19X5	19X6	19X7
January	6,485	6,950	7,133	7,561	7,637	7,963
February	6,240	6,558	6,637	7,017	7,544	7,878
March	8,189	7,870	8,046	9,214	9,507	10,040
April	8,199	8,400	8,792	9,365	9,843	9,808
May	9,034	9,146	9,113	10,095	9,898	10,725
June	9,442	9,245	8,795	10,702	11,328	11,660
July	8,857	10,334	10,234	10,774	11,301	10.918
August	9,353	9,888	9,888	9,951	10,778	11,477
September	8,052	8,382	8,382	9,441	9,815	9,914
October	7,481	8,254	8,254	8,731	8,748	9,444
November	7,286	7,438	7,438	7,915	8,604	8,849
December	7,498	7,481	7,481	8,674	8,771	8,420

I will appreciate your opinion as to how large, in terms of the maximum monthly capacity, the enlarged plant should be.

(SIGNED) *Gus*

What should Jarrew respond?

8. To compete in the furniture business, it is necessary to maintain a relatively large inventory. Consequently inventory carrying costs are substantial. If a firm can reduce these costs somewhat, it may establish a competitive advantage. One way to do that is to better predict sales. As general manager of B and D Furniture, you are interested in developing a forecasting method for furniture sales in your area. It seems logical to you that your sales would be related to the number of new homes built in your area each month. You know that it takes about three months to complete a home, so you hope that your sales are closely related to housing started three months previously. A check of company records and county business statistics reveals the data shown below. The building starts are shown three months in advance. That is, the building starts shown next to June are for March, and so forth. Can these data be used to build a useful forecasting model? Explain.

Month	Sales*	Housing Starts†	Sales	Housing Starts	Sales	Housing Starts
Jan.	12	12	10	9	6	5
Feb.	17	15	18	16	10	8
Mar.	22	16	19	16	11	8
Apr.	26	20	24	21	21	16
May	29	22	28	23	22	17
June	34	28	33	29	25	19
July	41	30	43	32	28	22
Aug.	16	16	25	19	26	20
Sept.	15	14	18	16	19	15
Oct.	12	12	14	13	16	12
Nov.	8	6	9	8	8	6
Dec.	19	6	22	7	18	5

*Sales are in tens of thousands.
†Housing starts are in hundreds.

9. The bargaining strength of a labor union is directly related to its ability to sustain a long strike. During a strike, the affected union members collect payment from the union's strike fund. These payments are typically small but usually enough to enable the striking members to continue to insist on the union's major contract demands. Suppose you are the national president of a union which is facing the possibility of a major strike in two months. It is now July 19X9. As you make plans for these critical labor negotiations, it is important that you have a reliable estimate of the size of the union's strike fund in two months.

In the past, the size of this fund in any month has been observed to be dependent upon the number of members for the preceding month and the unemployment rate for the preceding month. The number of members for any month can be projected on the basis of the number of members in the preceding month. Using the data given below, (*a*) forecast the number of members in the union next month so that you can (*b*) forecast the size of the strike fund in two months.

Date	Size of Fund (In Millions of Dollars)	Number of Members (In Thousands)	Unemployment Rate, %
2/X8	8.5	215	5.6
3/X8	11.9	215	5.7
4/X8	12.0	190	5.9
5/X8	8.0	192	5.8
6/X8	8.4	190	5.6
7/X8	9.2	189	5.5
8/X8	9.6	191	5.5
9/X8	9.4	209	5.2
10/X8	10.5	200	5.1
11/X8	10.6	204	4.8
12/X8	11.5	217	4.6
1/X9	12.0	200	4.6
2/X9	11.9	196	4.6
3/X9	10.8	200	4.2
4/X9	12.0	191	4.8
5/X9	10.5	191	4.6
6/X9	9.2	186	4.7
7/X9	9.1	195	4.6

Next month's projected unemployment rate is 4.5 percent.

CASE 1: BURTON SHOE STORE

Burton Shoe Store is the largest shoe store in a city of 100,000. Burton carries a complete line of men's, women's, and children's shoes. Although the store carries a wide range of quality, its primary emphasis has been on the higher quality brands. Over the past two years, Burton's sales growth has slowed considerably. Ann Burton is concerned that the firm's emphasis on quality items has resulted in many people considering the store to be suitable only for "upper class" shoppers.

In order to determine what people think of Burton, 493 individuals have been interviewed. The questions and responses presented in the accompanying table are from these interviews. Ms. Burton wants to know if her customers have different impressions of the store than do noncustomers. In addition, she wants to know if Burton Shoe Store is seen as being equally appropriate for all types of shoppers. Ms. Burton feels that this information is critical for designing an appropriate merchandising strategy. Please advise Ms. Burton.

Question	Burton's Noncustomers (n = 305)	Burton's Customers (n = 188)
1. Where would a construction or mill worker buy:		
Dress shoes?		
Burton	10%	20%
Elsewhere	90	80
Casual shoes?		
Burton	5	7
Elsewhere	95	93

	Noncustomers	Customers
2. Where would a bank teller or draftsman buy:		
Dress shoes?		
Burton	39%	59%
Elsewhere	61	41
Casual shoes?		
Burton	39	50
Elsewhere	61	50
3. Where would a doctor or lawyer buy:		
Dress shoes?		
Burton	50	65
Elsewhere	50	35
Casual shoes?		
Burton	47	66
Elsewhere	53	34

CASE 2: COMMERCE CORPORATION

Commerce Corporation manufactures and markets a large number of small consumer appliances such as blenders, mixers, toasters, and so forth. It advertises heavily in magazines, in newspapers, and to a lesser extent on television. Its products are sold through department stores, discount stores, and appliance stores. It recently entered the market with a crock pot, an electric pot designed to cook soups, stews, and other items at a very slow speed.

In order to activate the one-year warranty associated with the product, purchasers were required to return a postcard on which several questions relating to the purchase were asked. After several thousand of these postcards were returned, Harold Watson, from the marketing research department, randomly selected 100 for detailed analysis. He was interested in two major questions. One, were the three media equally effective in informing purchasers of the item? This information would be used in making budget-allocation decisions for future new-product allocations. In addition, Watson wanted to know if the store at which the item was purchased was independent of the media in which the individual learned of the item. This information could be used to strengthen the firm's position in the various types of outlets.

Of the 100 cards examined, 20 reported learning of the product from television advertisements. Of these 20, 10 purchased at a department store, 5 purchased at a discount store, and 5 purchased at an appliance store. Fifty reported learning of the item from a newspaper advertisement. Of these 50, 15 purchased at a department store, 30 at a discount store, and 5 at an appliance store. Finally, 30 reported learning of the item from magazine advertisements. Five of these purchased at a department store, 5 at a discount store, and 20 at an appliance store.

How would you answer Watson's two questions?

CASE 3: LARSON'S TAX SERVICES

Ken Nalley, office manager of Larson's Tax Services, has been assigned responsibility for purchasing 50 new electronic desk calculators for the firm. Nalley has narrowed his choice down to two brands. The price of the Electro brand is 5 percent below the price of the Hutton brand. The sales representative for each brand claims that the brand has special features which are particularly useful for the requirements of Larson's office force. The result of these special features is that the computational work can be performed faster.

In order to compare the two brands, Nalley obtained a "loaner" from each firm. He had 20 of the office staff perform 8 common computational routines with the Electro brand and a separate group of 20 perform the same 8 routines with the Hutton brand. The 40 staff members involved were equally skilled in the operation of machines of this type. For the Electro machine, the mean time required per computational routine was 13.5 minutes with a standard deviation of 3.5. For the Hutton machine, the mean was 12.4 minutes with a standard deviation of 3.0.

What can Nalley conclude concerning the performance of the two brands?

CASE 4: WEST LYNN SCHOOL BOARD

The West Lynn School Board is convinced that the school district needs a new auditorium building. The existing auditorium is both outdated and too small to contain the school's growing student body. However, any capital improvement of this nature requires voter approval.

From past experience the board is convinced that at least 35 percent of the area voters must favor a proposal prior to any campaign or it will not pass. Past experience has also shown that once a proposal has been defeated, it is very difficult to pass in future elections.

An election will be held in three months and the board would like to place the auditorium issue on the ballot. To help determine whether or not the issue should go on the ballot, a random sample of 100 registered voters has been called and responses to the auditorium issue have been recorded. Thirty-two percent of the respondents stated that they would vote in favor of the auditorium.

Should the board place the auditorium proposal on the ballot? Justify your answer.

CASE 5: BEGGS DRUG STORES

Beggs Drug Stores is a chain of 15 "neighborhood" drugstores located in and around Seattle. Megan Tandy, merchandise manager for Beggs, decided to distribute flyers advertising a "summer's here at last" sale. The flyers also contained a coupon which allowed the bearer to purchase an ice chest which normally sold for $7.95 for $3.99. Since the ice chest cost Beggs $4.50, Tandy was hoping that the people buying the ice chest would buy enough additional items to offset the loss on the ice chest.

To measure the impact of the coupon, Tandy had all purchases which involved the redemption of the coupon recorded on the stores' cash register tapes by using a special key on the registers. At the end of the sale period she randomly selected 50 of the purchases marked with the special key. She wants to know the percentage of purchasers that bought only the ice chest. More importantly she wants to know the average purchase amount above the cost of the ice chest. She has calculated that this figure must be at least $3.75 for the promotion to be considered a success. The sample data are presented below. Can the promotion be considered a success? What percentage of the shoppers redeeming coupons purchased only the ice chest?

Purchase No.	Amount	Purchase No.	Amount
1	5.62	26	3.99
2	3.99	27	7.58
3	4.93	28	11.30
4	18.17	29	3.99
5	3.99	30	6.45
6	3.99	31	9.14
7	25.19	32	4.99
8	4.49	33	3.99
9	3.99	34	3.99
10	16.18	35	3.99
11	9.50	36	5.75
12	14.99	37	4.50
13	10.20	38	26.00
14	8.00	39	3.99
15	3.99	40	3.99
16	15.40	41	11.30
17	3.99	42	6.18
18	6.49	43	7.99
19	7.15	44	3.99
20	3.99	45	22.30
21	8.45	46	8.50
22	3.99	47	3.99
23	3.99	48	14.60
24	11.15	49	3.99
25	4.49	50	4.50

CASE 6: LIBERTY HILL INSURANCE CORPORATION

Joe Barton, sales manager of the life insurance division of Liberty Hill Insurance Corporation, was hoping to provide additional guidelines for isolating prospective customers for his sales force. One approach that he considered was to find the average amount of insurance that individuals with certain characteristics had. Then, any individual that had the same characteristics and less than the average amount of insurance could be considered a prime prospect.

Barton decided to implement this line of reasoning. He began by drawing a random sample of 20 policyholders from the firm's files. He started his analysis by focusing on income and family size. The relevant information taken from the files is presented below.

a. How would you analyze these data to isolate prospective consumers?

b. Which family or families appear to represent good prospects?

Family	Amount of Life Insurance (In Thousands of Dollars)	Income (In Thousands of Dollars)	Family Size
1	14	6	3
2	25	10	4
3	27	11	5
4	20	10	3
5	22	10	3
6	35	14	7
7	23	8	2
8	30	12	6
9	15	8	3
10	25	11	4
11	48	19	5
12	24	9	2
13	39	17	3
14	24	14	5
15	31	12	6
16	40	16	3
17	29	11	3
18	30	13	4
19	32	14	3
20	44	17	2

CASE 7: HIGHLAND COFFEE

Louise Halberry, advertising manager for Highland Coffee, was convinced that most consumers could not taste the difference between Highland Coffee and the best-selling national brands against which Highland competed. It was Halberry's belief that the image of the coffee generated by heavy national advertising created the "taste" preferences that consumers stated for those brands. Halberry hoped to convince the rest of Highland's management that she was correct. If she could do so, she felt certain that she would be allowed to increase the firm's advertising budget.

Both Halberry's idea of no taste differences and her desire for an increase in advertising expenditures were opposed by David Fuchs, the firm's production manager. Fuchs was convinced that consumers could taste the difference between brands and that the road to success lay in product improvements.

After 14 months of open hostility, the two managers decided to run a test of consumer taste. They took a random sample of 100 coffee drinkers from the area. Five cups of hot coffee were presented to each consumer. The cups were labeled L, M, N, O, and P and contained one of each of the four leading national brands and Highland. The cups were presented to each consumer in a random order. In the table below are shown the number of coffee drinkers who stated that they most preferred the indicated brand.

Brand Preferred		Number of Consumers
Folgers	(L)	25
Maxwell House	(M)	18
Nescafe	(N)	22
Highland	(O)	17
Sanka	(P)	18

a. If you were Halberry, how would you interpret the experiment?

b. If you were Fuchs, how would you criticize Halberry's interpretation?

CASE 8: EASTERN TRUST BANK

Eastern Trust Bank has experienced an unexpected high level of growth at its Third Street location. In fact, the branch is operating at a level the manager feels to be beyond optimal capacity. This has created a problem in finding a site on which to construct another branch. Previous branches have been located in a manner designed to avoid their drawing business from existing branches. However, now it is desired to create a branch that will draw some, but not too many, of the Third Street branch's customers. It is decided that the new branch should draw at least 15 percent of the Third Street branch's customers but no more than 25 percent. The following data have been collected.

	Would Transfer	Would Not Transfer
Site A	62	138
Site B	26	174
Site C	89	111

Which site(s), if any, meets Eastern's criteria?

CASE 9: UNITED IMPORT

Darrell Marrs, chief financial officer of United Import, is trying to decide on the best approach to securing a $7 million intermediate-term loan. Both the House and Senate passed a bill that Marrs was sure would lower the interest rate by at least .005. However, the President vetoed the bill. Now the Congress is attempting to override the veto. The Senate voted to override by the necessary two-thirds majority. The House vote is in a few days. If the House vote fails to override, it is agreed in financial circles that interest rates will rise by .0025. Thus, if Marrs were sure the veto would be overridden, he would wait to secure the loan. Otherwise, he would prefer to make the loan now.

A poll of 90 representatives taken by a major newspaper indicated that 58 of them planned to vote to override.

What should Marrs do?

CASE 10: INDUSTRIAL SAFETY PRODUCTS

Among the many products made by Industrial Safety is a line of insulated gloves designed for welders and others whose hands might be exposed to considerable heat. A month ago an industrial worker received severe burns on his hands and arms when a small engine on which he was welding burst into flames. The worker was wearing a pair of "heat resistant" insulated gloves manufactured by Industrial Safety.

An investigation of the accident by the state industrial safety board revealed that the gloves had performed as well as could be expected given the intense heat generated by the fire. However, the board purchased 10 pairs of the firm's gloves and found that 7 of them had less than the 0.75 centimeter of insulation that Industrial Safety advertised. The board turned this information over to the state attorney general's office.

Sheila Redding, a prosecutor in the attorney general's office, called Lloyd Dubbs, executive vice president of Industrial Safety, and offered him the chance to talk with her and "convince us why we shouldn't file a complaint against Industrial Safety Products for violation of the state's deceptive advertising laws."

Advise Mr. Dubbs.

CASE 11: HAUPT'S DEPARTMENT STORES

Haupt's Department Stores operates a major downtown store and eight shopping-center stores in a large Midwestern city. The firm issues its own credit card, which is honored at all nine outlets. Bad-debt losses resulting from credit-card sales have averaged almost 3 percent per year. These include both uncollectible accounts and purchases charged on stolen cards.

Recently the company installed a computerized credit-check system. For credit purchases over $10, the cashier was required to dial the credit-card number into the computer, which automatically searched the customer's payment record and gave an approval or refused credit.

Since the rental on the computer equipment plus the added cost of having the cashiers dial in the number amounted to 0.5 percent of the firm's credit sales, the equipment was initially installed on an experimental basis in one of the stores (all the stores had had similar bad-debt losses). After six months of operation, the outlet with the check system had experienced losses of 2.6 percent while the remaining stores had losses of 3.0 percent.

Should the equipment be placed in the remaining stores? Justify your answer.

CASE 12: MARCO, INC.

Marco, Inc., is a medium-sized producer of children's toys and games, plastic household items, and plastic component parts for the automotive and aircraft industries. The firm experienced relatively steady growth in both sales and profits until the early 1970s, when both sales and profits began to grow rapidly.

This growth in profits produced a visible strain among the small group of shareholders. Bob Tatham, majority stockholder and chief executive of the firm, insisted on paying minimum dividends and retaining the bulk of the increased profit for a reserve for equipment replacement and expansion. Most of the remaining stockholders wanted to increase the dividend rate substantially. They had no desire to weaken the position of the firm but felt that retaining most of the increased profits was not necessary.

As 1976 was drawing to a close, Tatham called you to his office and said: "Here are our sales, profits, and dividend figures for the past 20 years, as well as some information on inflation rates. Prepare me a report which I can use to convince those dummies that this recent jump in profits is more apparent than real. While you're at it, prepare a sales, profit, and inflation rate estimate for next year."

Prepare the report for Mr. Tatham. What cautions, if any, would you attach to your next year's estimates?

Year	Sales	Profits	Dividend per Share	Consumer Price Index
1956	$14,912,000	$511,920	$1.55	81.4
1957	16,329,600	485,570	1.55	84.3
1958	16,982,400	615,960	2.00	86.6
1959	16,008,000	387,610	1.80	87.3
1960	17,586,400	384,610	1.85	88.7
1961	16,920,000	496,800	1.95	89.6
1962	17,581,200	673,200	2.30	90.6
1963	19,154,400	533,600	1.75	91.7
1964	16,740,000	466,860	1.75	92.9
1965	20,185,200	609,120	2.00	94.5
1966	19,400,000	686,760	2.25	97.2
1967	21,401,000	671,690	2.10	100.0
1968	22,630,400	686,400	2.20	104.2

Year	Sales	Profits	Dividend per Share	Consumer Price Index
1969	19,851,800	529,100	2.10	109.8
1970	21,158,400	612,480	2.20	116.3
1971	24,974,400	696,960	2.40	121.3
1972	25,862,000	830,000	2.45	125.3
1973	26,233,500	633,080	2.20	133.1
1974	30,488,000	893,920	2.35	147.7
1975	32,390,000	985,320	2.35	161.2

CASE 13: TRIO BUICK

Trio Buick is a medium-sized Buick dealer. Its service department makes approximately 10,000 repairs on customers' cars per year. These repairs are made on all types of automobiles and involve a wide range of maintenance and repair services. The replacement parts used in the service department come from the Buick division of General Motors, regional wholesale distributors, and local auto supply houses. The gross margin realized by Trio varies depending on the source of the replacement part and the nature of the specific item involved.

Rather than calculate the exact gross margin on each of the 10,000 invoices, Trio uses a rule of thumb that the average gross margin is 30 percent. Using this rule, total sales of replacement parts are summed and multiplied by 30 percent to provide the gross margin figure for accounting purposes. The total dollar value of the parts inventory at the end of a year is calculated as: Beginning Inventory + [Cost of Parts Purchased − 0.7 (Dollar Sales of Retail Parts)].

A recent inventory audit found a substantial difference between the firm's estimated inventory value and the value obtained by the audit. A number of possible explanations for the difference exist. Mike Hayden, the parts department manager, feels that the 30 percent rule of thumb is inaccurate. However, although no direct accusations have been made as yet, several members of the firm have implied that the difference is due to mismanagement by Hayden.

What should Hayden do to support his position?

CASE 14: STANDARD PHARMACEUTICALS

Standard Pharmaceuticals is a large corporation that produces and markets both prescription and over-the-counter health care items. One of its major product lines is diet supplements, primarily vitamin and mineral capsules. As a result of its own research activities as well as a recent nationwide study of dietary habits among American teenagers, the firm was considering the introduction of a new product. The product would be low in calories and high in protein and a variety of vitamins and minerals that research had shown teenagers often to be deficient in. The product development department had man-

City A

Respondent No.	Consumption during Test	Respondent No.	Consumption during Test
1	0	51	0
2	0	52	0
3	1	53	0
4	0	54	8
5	15	55	0
6	4	56	1
7	0	57	3
8	0	58	7
9	8	59	0
10	0	60	9
11	5	61	2
12	3	62	0
13	7	63	10
14	0	64	2
15	9	65	0
16	0	66	12
17	0	67	10
18	22	68	5
19	0	69	0
20	2	70	0
21	6	71	4
22	0	72	0
23	0	73	14
24	11	74	6
25	4	75	0
26	0	76	7
27	0	77	0
28	13	78	0
29	0	79	0
30	3	80	3
31	0	81	9
32	0	82	4
33	12	83	0
34	1	84	8
35	0	85	0
36	2	86	0
37	0	87	10
38	20	88	2
39	0	89	0
40	0	90	11
41	0	91	0
42	1	92	0
43	9	93	1
44	0	94	0
45	16	95	1
46	4	96	0
47	0	97	10
48	0	98	2
49	5	99	0
50	8	100	1

City B

Respondent No.	Consumption during Test	Respondent No.	Consumption during Test
1	8	51	8
2	0	52	0
3	10	53	0
4	16	54	8
5	0	55	4
6	0	56	14
7	1	57	0
8	10	58	7
9	0	59	9
10	3	60	1
11	20	61	0
12	2	62	0
13	0	63	2
14	0	64	9
15	0	65	0
16	7	66	0
17	11	67	11
18	6	68	13
19	0	69	3
20	5	70	0
21	10	71	26
22	0	72	1
23	0	73	0
24	9	74	0
25	3	75	0
26	1	76	6
27	12	77	0
28	5	78	10
29	0	79	0
30	2	80	0
31	10	81	7
32	1	82	12
33	0	83	0
34	0	84	1
35	8	85	16
36	9	86	0
37	3	87	4
38	0	88	0
39	0	89	20
40	0	90	0
41	15	91	0
42	0	92	11
43	13	93	0
44	2	94	10
45	0	95	0
46	1	96	3
47	11	97	0
48	0	98	0
49	1	99	12
50	9	100	9

aged to incorporate these features in a slightly carbonated drink which could be flavored to match the taste characteristics of a variety of soft drinks.

The firm decided to position the product as a special soft drink rather than as a medicinal product. Primary discussions indicated that they would have no trouble gaining the distribution support of a major soft drink firm. However, the national advertising of the new product would be Standard's responsibility. Furthermore, the soft drink firm, while impressed with the concept and the product, wanted additional evidence that the product would sell before committing itself to distributing the product.

In order to provide evidence of the product's potential as well as to gain some insights into the appropriate advertising strategy, Standard decided to test-market the product. Two medium-sized cities with similar population characteristics were selected and the product was distributed in each city. Prior to distribution, 100 teenagers in each city were randomly selected and asked to form a panel. Each teenager was to keep a daily diary in which he or she listed all the beverages consumed in the past 24 hours. In city A, the product was heavily advertised as a healthy, vitamin- and protein-rich drink. Flavor and the fun of consuming the drink received secondary emphasis. In city B, this procedure was reversed.

The product manager of "Zing," the name given the new drink, and the marketing vice president of the soft drink firm both agreed that at least 60 percent of the total teenage market must try the product during its introductory period of two months (the test market is designed to duplicate the introductory period) and at least 30 percent must have consumed the product eight times or more during this period in order for the product to justify national introduction. The results of the test are presented on pages 394 and 395.

a. Should the product be introduced nationally?

b. Which advertising theme should be used?

CASE 15: FASHION STREET BOUTIQUE

The Fashion Street Boutique was opened in August of 1972 by Jane Dryer and Christie Lopez. Dryer and Lopez had been roommates during their last two years in college. Lopez had received a degree in business administration and Dryer a degree in clothing and textiles. Both women took jobs with large retail department chains in 1968. Although they lived in separate cities, they maintained close contact and by early 1970 were discussing the possibility of opening their own retail clothing store. In the spring of 1971, Lopez's parents sold the small farm which they had owned for almost 30 years to a real estate development firm for a considerable profit. Christie was presented with $20,000 of this as her share since her parents had been unable to provide her any support while she was in college. Within a year, Dryer raised $10,000, primarily by a no-interest loan from her father. With $30,000 in capital, and four years of retailing experience, they set out to develop a retail clothing store.

It had always been their intention to develop a full-line, fashion-oriented women's clothing store. Based on their experiences, they felt that a store catering to the increasing numbers of working women with fashion-oriented clothes of moderate quality and price could be quite successful. Therefore, they used census data to isolate moderate to large cities with a relatively high percentage of working women. They then consulted a variety of sources to develop estimates of the level of competition they would face in each city. It soon became apparent that as the climate improved, the number of independent women's clothing stores increased. They selected three medium-sized cities with relatively few competitive outlets. Each of these was visited and the most promising was selected.

After selecting the city, they acquired space in a small, covered mall shopping center located near the edge of the downtown area. By August 1972 they were able to open for business. Despite working 70 hours a week, they were almost bankrupt by December. Only loans from both parents kept them in operation until Easter. However, their Easter sale was a success and the operation showed a small profit for 1973. Sales and profits have grown since then. Monthly sales figures for the past 3½ years are shown below.

July had always been a slow month for Fashion Street Boutique. In July there was only limited demand for summer items and yet the interest in fall clothing hadn't picked up yet. This year, Dryer and Lopez decided to engage in a major advertising push during July. The normal July advertising budget was increased tenfold.

Local television was used for the first time except for previous Christmas and Easter sales. In addition the local newspaper was used heavily, as were bus signs. Sales for July were $15,692. On seeing the sales figure, both Lopez and Dryer felt that their advertising had increased sales. However, neither was completely confident but that the sales increase might have occurred anyway.

Did the advertising increase sales?

	Sales			
Month	1974	1975	1976	1977
January	13,225	14,910	16,804	18,935
February	10,841	12,223	13,774	15,524
March	17,867	20,148	22,697	25,580
April	15,703	17,706	19,948	22,485
May	14,251	16,067	18,105	20,404
June	13,449	15,163	17,085	19,258
July	10,757	12,127	13,664	
August	18,447	20,799	23,436	
September	19,833	22,361	25,195	
October	17,854	20,122	25,677	
November	16,983	19,145	21,578	
December	21,713	24,471	27,576	

CASE 16: ACADEMIC SERVICES, INCORPORATED

Academic Services, Incorporated, was formed in 1954 by William Scheer. From 1946 to 1954, Scheer had operated a small chain of campus bookstores on the East Coast. Scheer, who had a master's degree in English literature, had always been fascinated by books and had developed a significant collection of rare books and manuscripts. After his release from military service in 1945, operating a bookstore seemed like the best possible occupation. His eight bookstores enjoyed a moderate amount of success. However, Scheer soon found that most of his time was devoted to helping the local academic libraries obtain copies of scarce books and manuscripts. In 1954, he decided to separate this activity from his retail bookstores and attempt to put it on a paying basis. Thus, Academic Services was formed. Over the next 20 years, Academic Services evolved into the largest book wholesaler serving academic libraries in the United States.

In 1965, Scheer removed himself from active management of the operation. He remained the principal stockholder and ran the small section devoted to procuring rare books for libraries and a few individual collectors. This operation consistently operated at a loss.

Under the direction of Thomas Cole, who became president of the firm in 1965, both sales and profits increased except for the years 1970, 1971, and 1972 when many college budgets were reduced. In 1975, the continued growth of the firm appeared threatened by increasing price competition from two smaller firms. Each of these smaller firms had been acquired by expansion-oriented conglomerates the previous year. Cole, who hoped to avoid a costly price war, asked Gerald Matsui, his sales manager, for suggestions. Matsui made the following proposal:

We have been the major force in this field for 15 years and we have become complacent. Our sales force is the same size now as it was 10 years ago despite vastly increased accounts and sales. I think our superior services more than offset the price reductions being offered by our competitors. The problem is to convince our customers of this.

I see four major alternatives: (1) make no change, (2) have a moderate increase in the number of salespersons, (3) have a large increase in the number of salespersons, or (4) have a small increase in the number of salespersons but have the increased effort devoted entirely to our major accounts. I would suggest that we try each of these approaches for one year in a few of our sales territories and see which one works best.

Cole accepted Matsui's recommendations. The firm's 20 sales territories were randomly divided into 4 groups and one of the four alternatives was implemented in each of the groups. The results of the year's trial are shown below. The figures represent the change in gross profit minus direct selling expense from the previous year in each territory.

Which technique, if any, should be adopted? Why?

Alternative 1	Alternative 2	Alternative 3	Alternative 4
24,000	28,000	36,000	42,500
29,500	42,000	42,000	41,500
28,500	37,000	38,000	33,000
30,000	38,000	37,500	41,000
28,000	35,000	36,500	42,000

CASE 17: WELLS AND ASSOCIATES

Wells and Associates is one of the largest financial counsulting firms in the United States. It provides financial advice and services to both private firms and state and local governments. It has 86 offices located in 42 states. There are six regional managers who, with their staffs, supervise and support the local officers. Until three years ago, each region was operated as an almost completely autonomous unit. The national office set revenue and service goals in conjunction with the regional managers and provided technical services, computer facilities, economic forecasts, and so forth.

Three and a half years ago, the firm hired a management consulting firm in response to stable profits despite a continuing increase in total revenue. After six months of investigation, the consultants recommended a number of major changes in the firm's operations. A consistent theme in these recommendations was the need for more centralization of routine tasks to increase both the effectiveness and efficiency with which they are performed.

In response to the consultants' recommendations, a personnel division was established at the national headquarters to supervise all personnel matters, including recruitment. Bill Koehler, who had been in charge of personnel in the largest regional office, was placed in charge of this department. Lori Muller, who had just been placed in charge of the personnel department in the next-to-smallest division, was named as his assistant. The department was also assigned four secretaries. In the past three years, another assistant and two more secretaries have been added. Six weeks ago, the department added a newly graduated statistician.

Lori Muller recently began a review of the department's hiring practices. She started the review by examining the most critical area—financial trainees. The firm hires between 60 and 130 financial trainees per year depending on the growth of the firm, turnover, and the number of "outstanding" prospects that it encounters. Virtually all the financial trainees are recruited from graduating college seniors with a financial specialization.

Ms. Muller randomly selected 64 files from the 117 candidates who had been hired two years ago and were still on the job. Each file contained the following information (the data are in the accompanying appendix):

1. Sex.

2. Age when hired.
3. College grade point average.
4. "Quality" of the college attended—ranked from 1 (excellent).
5. Company index—score on a test devised for the firm. The consultants that designed the test state that it "will rank-order individuals but it probably is not accurate in assessing the magnitude of the differences between individuals." The test produces a score ranging from 0 (very unlikely to succeed at the job) to 100 (very likely to succeed at the job).
6. Second-year performance evaluation. This evaluation produces a numerical score from 0 (very poor) to 100 (excellent). Both Muller and Koehler are certain that the scale is interval in nature. They have also, based on three years' experience with the scale, decided that a score of less than 50 is unsatisfactory, 50–69 is satisfactory, 70–89 is above average, and above 89 is excellent.

Appendix

	Variable Number					
Trainee Number	1	2	3	4	5	6
2	F	35	3.7	1	94	95
3	F	27	2.9	3	67	54
5	M	22	2.4	2	73	67
6	F	22	2.9	2	74	63
8	F	22	2.8	1	97	90
9	M	22	2.6	2	79	73
12	M	21	2.9	3	91	93
13	F	23	3.5	2	78	45
14	M	23	2.5	3	69	64
16	M	23	2.7	3	65	56
19	F	22	2.6	1	61	45
21	M	22	2.4	3	50	40
22	F	22	2.8	3	85	77
23	M	21	2.2	2	60	52
27	F	22	2.4	1	61	55
29	F	22	2.1	2	90	80
30	F	25	3.2	3	83	74
32	M	21	2.1	3	49	37
33	M	22	2.6	1	89	91
34	M	23	2.2	3	40	30
36	M	21	2.4	3	85	77
39	M	24	2.3	2	77	66
40	M	22	3.1	2	88	92
42	M	23	3.6	2	72	79
45	F	23	2.7	2	82	79
46	M	23	2.9	2	82	72
47	F	22	3.1	1	54	54
50	F	21	3.6	2	46	60
51	M	22	2.8	3	67	62

Appendix

Trainee Number	Variable Number 1	2	3	4	5	6
53	M	22	2.9	1	76	69
56	F	22	2.1	2	52	46
57	F	22	3.5	1	95	93
59	F	22	2.8	2	93	99
60	M	30	3.0	2	85	82
62	F	22	2.8	3	76	67
65	F	21	3.0	3	88	90
66	M	21	2.8	2	62	59
68	F	22	2.6	1	73	67
73	M	24	3.1	1	55	49
74	F	23	3.1	1	81	87
75	F	22	3.0	2	64	59
76	M	21	2.9	2	76	66
77	F	22	3.1	2	96	97
79	M	23	2.8	1	85	71
82	M	22	2.3	2	41	43
83	F	22	2.7	2	84	85
85	M	22	2.9	1	87	88
86	F	23	3.2	1	80	65
87	F	22	2.5	2	58	50
90	F	23	2.8	3	56	49
93	F	22	2.5	2	44	30
94	F	23	2.7	1	92	81
96	F	29	2.3	1	46	42
97	M	22	3.0	1	86	75
99	M	21	3.4	1	81	84
103	M	22	3.0	1	75	65
104	F	23	3.0	2	90	91
105	F	22	2.2	3	58	47
107	M	23	3.1	3	67	57
108	M	25	2.4	1	68	61
109	F	27	2.1	1	79	71
111	M	22	2.4	2	91	94
113	F	24	2.5	2	79	68
116	M	21	3.0	3	71	70

Ms. Muller has called you into her office and made the following statement: "I'm glad we have a statistician to help us. We're not yet ready to develop a full-blown statistical model of what makes a good recruit. However, it is time to start evaluating some of the variables we have information on. The large number of people we recruit, the high cost of training them, and the fact that we can't really assess performance until the end of the second year mean that any improvement in our recruiting effectiveness will result in substantial savings for Wells and Associates. Would you prepare answers for the following questions for me?"

1. Would you summarize the data so that I can get a feel for it?
2. Do a higher percentage of males obtain a performance score above 89 than do females?
3. Do females score higher on the performance evaluation than do males?
4. Is there a difference in the performance scores obtained by graduates from the various categories of colleges?
5. What is the relationship between grade point average and performance score?
6. What is the relationship between age when hired and performance score?
7. What is the relationship between the company index score and the performance score?
8. Is the performance evaluation rating (excellent, above average, etc.) independent of the category of school attended?
9. Is the performance evaluation rating (excellent, above average, etc.) independent of the sex of the trainee?
10. How accurately can we predict performance scores using both age and college grade point average?

a. Describe the procedure you would use to develop an answer to each of these questions. Justify your approach.

b. Utilizing the material in the appendix, answer each question.

CASE 18: BIG MOUNTAIN BREWERY

Big Mountain Brewery is the producer and marketer of Big Mountain beer. Big Mountain is distributed throughout a five-state region. It is one of a limited number of regional breweries in the United States that have been able to hold their own against the national brands. Big Mountain advertises heavily throughout the five-state region.

Some months ago one of the production supervisors sent Fran Luchini, the general manager of Big Mountain, a copy of a letter his neighbor had received and shown him. The letter contained a request that the neighbor join a national "consumer panel." The neighbor would have to keep a daily "diary" of all the beverages he consumed during the next six months. In addition, he would be asked to complete several questionnaires concerning his attitudes toward various beverages and his consumption preferences.

A follow-up investigation by Ms. Luchini's assistant found that the study was sponsored by a pharmaceutical firm. The sponsoring firm was considering developing a new liquid food supplement and needed consumption and attitude data to aid in the product design. Ms. Luchini felt that some of the data being generated in this study could prove extremely useful in designing a marketing strategy for Big Mountain.

Negotiations with the sponsoring firm resulted in Big Mountain acquiring part of the data for $10,000. The data were supplied on 400 computer cards, one for each of the 400 respondents in the five states served by Big Mountain. The general nature of the information contained on the computer cards is listed below. Ms. Luchini has forwarded you the following list of questions concerning the newly acquired data.

1. Can you summarize the data so that I can get a feel for it?
2. What percent of the individuals over 18 in our trade area consume beer at least once a month?
3. What percent of the males over 18 in our trade area consume beer at least once a month?
4. How much beer is consumed per week in our trade area?
5. How much beer is consumed per week by females in our trade area?
6. Do males consume more beer than females?
7. Does a higher percentage of males consume beer than females?
8. Are there significant differences in the average amount of beer consumed among the four occupational categories?
9. Is the amount of beer consumed related to the age of the respondent?
10. Do married individuals consume more or less than single individuals?
11. Is the amount of beer consumed related to the educational level of the respondent?
12. Consider four categories of beer drinkers: (1) nondrinkers, (2) light drinkers, (3) moderate drinkers, and (4) heavy drinkers. Is the category of drinker independent of:
    a. sex category?
    b. occupational category?
    c. marital status?
    d. income category?
13. How do the responses to each of the attitude statements relate to the amount of beer consumed?
14. Do the responses to each of the attitude statements differ between drinkers and nondrinkers?
15. Do males and females differ in the rank they assign to Big Mountain?
16. Are the rankings of product attributes by heavy and light consumers similar?
17. Can we predict whether or not an individual will consume beer at least once a month based on his or her responses to the attitude questions?

Describe the procedure you would use to develop an answer to each of these questions. Justify your approach.

Data Available from Each Respondent

1. Sex.
2. Age (in years).
3. Occupational category (student, blue collar, white collar, managerial-professional).

4. Educational level (years of schooling completed).
5. Marital status (married, not married).
6. Income category (eight categories).
7. Whether the respondent drinks at least one beer per month (yes, no).
8. Average consumption of beer per week during the panel (in ounces).

9–18. Ten Likert-type attitude statements concerning beverages and beverage consumption. The respondent expresses his or her degree of agreement or disagreement with each statement on a scale from 1 to 7. The answers can be considered to represent interval data.

19. Brand rankings. The respondents rank-ordered (with no ties allowed) the 12 brands available in the area in terms of their preferences.
20. Attribute rankings. The respondents rank-ordered (with no ties allowed) in terms of their preferences 20 attributes that could be associated with a brand of beer, such as taste, price, etc.

CASE 19: FALLS CITY BANK

Bill LaBaume graduated from Falls City College in June and immediately went to work full time for the local bank at which he had been employed part time during his last two years in school. His first assignment was as assistant to the chief loan officer, Linda Finstad. During his second week on the job, Ms. Finstad made the following request:

"The default rate on our automobile loans is too high and it's costing the bank a substantial amount of money. However, I think I've figured out an answer. A lot of people move into this community, stay a year or two, and are transferred out. I think these are the ones causing the problem. Basically, they move around so often that it's easier for them to default on the loan than to take the car with them. I've gathered data on the number of communities a random group of our customers lived in prior to applying for a loan and whether or not they defaulted on the loan. Take a look at the data and tell me if my hunch is correct. The defaults are marked with a 0 and the others with a 1."

Prepare a report for the loan officer.

Individual	Number of Moves in Past 3 Years	Default	Individual	Number of Moves in Past 3 Years	Default
1	0	1	29	0	1
2	1	1	30	0	1
3	1	0	31	1	1
4	0	1	32	0	1
5	3	1	33	2	0
6	0	1	34	3	0
7	5	0	35	0	1

Individual	Number of Moves in Past 3 Years	Default	Individual	Number of Moves in Past 3 Years	Default
8	2	1	36	5	1
9	0	1	37	0	0
10	0	0	38	0	1
11	6	1	39	0	1
12	1	1	40	2	1
13	0	1	41	1	0
14	2	1	42	0	1
15	1	1	43	3	1
16	1	0	44	0	1
17	0	0	45	0	1
18	3	1	46	1	1
19	0	1	47	0	1
20	0	1	48	1	0
21	0	1	49	5	1
22	0	1	50	0	1
23	2	1	51	1	1
24	1	1	52	0	1
25	4	1	53	3	1
26	0	1	54	4	0
27	2	1	55	1	1
28	1	1			

CASE 20: GENERAL ACCOUNTING OFFICE

The General Accounting Office (GAO) is responsible for evaluating the performance of various components of the federal government. Assume that as an employee of GAO you have been asked to evaluate the performance of four major defense contractors with respect to cost overruns. The government selects the lowest competitive bid that meets the specification required for the project assuming that the firm also meets certain other standards.

Since many of the projects commissioned by the Department of Defense involve unique, new, and unproved technologies, accurate cost projections are difficult. Therefore, the government often pays for cost overruns by the contractors. Your agency is concerned that some of the firms are underestimating costs to ensure getting the contracts. They hope that the government will cover their overrun.

To test this, you randomly select 10 percent of the contracts awarded to each of the four major firms in the past two years and record the number of contracts involving overruns and the number completed within budget. Company A completed 25 of 36 within budget. Company B completed 32 of 44 within budget. Company C had 11 of 15 within budget, while D had 20 of 29 within budget.

What conclusion, if any, can you draw from these data?

CASE 21: WESTERN WELL SERVICE

Western Well Service operates in the oil-field regions of West Texas, New Mexico, and Oklahoma. Although it does not engage in the drilling of oil wells, it provides a complete program of oil-well service on a contract basis. It operates a fleet of over 200 trucks and maintains a substantial inventory of parts required for the maintenance of this fleet. In addition, it maintains a large inventory of supplies required for servicing the wells, such as tubing, valves, gaskets, and so forth.

Alton Dancy, controller for Western Well, has experienced considerable frustration over the failure of the firm's inventory system. Due to the makeshift nature of much of the firm's equipment and the emergency nature of many of the repairs, it has been common practice to send, for example, several pressure valves out to a well to ensure that the proper one is available. Often, the extra valves remain in the service truck and are used in a subsequent repair without being removed from the inventory records. Furthermore, it has been impossible to prevent mechanics and service hands from entering the parts-inventory area. This also results in parts being removed from inventory without being charged.

In order to maintain accurate financial records, Dancy has decided to implement a supplementary bimonthly inventory in addition to the firm's annual inventory. However, the firm carries over 7,000 different items in inventory and the last annual inventory cost nearly $5,000. Dancy cannot justify spending that much money on a bimonthly basis. He recently summarized his problem as follows:

"We carry over 7,000 different items (42,000 individual pieces) in inventory. Our last complete inventory value was nearly $1,200,000. However, 500 of the different items accounted for almost half the value. We can easily inventory all these items, since they are not only very expensive but also generally quite large. The problem is the remaining 6,500 items (about 41,000 individual pieces). The average value of an individual piece of this group was approximately $17 with a standard deviation of $10 according to our last inventory. Since the total value of these items is approximately $700,000, we need a good estimate of the total value at a regular interval. However, we can't afford a complete count more than once a year."

Provide Mr. Dancy with a detailed solution to his problem.

CASE 22: METROPOLITAN TAXI CORPORATION: A

Metropolitan Taxi Corporation maintains a fleet of over 500 taxis in three Midwestern cities. The size of the fleet and the large number of miles that each taxi is driven per year make maintenance a particularly critical area. Recently, E.A. Walton Manufacturing, a small manufacturer of brake linings, approached the firm with a newly developed brake lining that would last as long as traditional linings but cost 10 percent less.

Although confident that the lining would perform as indicated, Walton Manufacturing wanted to subject it to a severe test prior to full-scale market introduction. For this reason the company offered to place the lining on 75 of Metropolitan's taxis and to renew the existing brand of lining on a second 75 taxis. Walton offered to pay all costs of replacing the lining and asked permission to publish the results of the test.

Ron Alpert, president of Metropolitan, agreed to the test subject to one condition: The new lining must produce at least the same braking ability as the current linings. A previous test of the braking ability of the current linings involving 50 taxis produced an average stopping distance of 116 feet with a standard deviation of 10 when the brakes were applied at 40 miles per hour.

Fifty of the Metropolitan taxis were equipped with the new linings and their braking ability tested. The results are presented below. Should Alpert allow the usage test?

Braking Distances (in Feet)				
122	115	114	116	119
114	117	117	115	117
117	113	115	117	118
116	116	116	118	117
115	114	118	114	115
119	118	117	116	120
118	112	119	117	118
120	116	120	115	117
111	115	116	119	119
116	118	114	113	116

CASE 23: METROPOLITAN TAXI CORPORATION: B

Ron Alpert decided to let E.A. Walton Manufacturing utilize his taxis for the brake-lining performance test. (See Case 22.) The miles driven prior to replacement follow.

Mileage to Replacement—Old Lining (In Thousands)				
37	43	34	51	33
42	45	39	36	39
36	43	39	38	38
39	36	38	39	39
49	47	39	31	42

Mileage to Replacement—New Linings (In Thousands)				
41	41	43	40	35
45	34	35	33	34
40	36	40	38	32
35	30	37	45	48
40	37	36	34	38

Mileage to Replacement—Old Lining (In Thousands)				
33	31	42	38	40
36	35	39	43	43
37	42	45	47	36
38	42	47	42	38
42	43	34	36	39
41	35	39	41	46
39	41	33	28	39
37	37	42	37	38
31	44	43	41	40
46	42	41	44	41

Mileage to Replacement—New Linings (In Thousands)				
27	36	37	41	35
34	34	36	43	41
39	49	32	41	35
42	39	40	45	40
31	41	42	30	38
37	41	36	36	37
36	32	39	38	38
38	37	34	40	44
41	32	43	41	38
39	45	41	37	37

If you were the president of E.A. Walton, what would you conclude from this information? Should Alpert continue to use the new linings?

CASE 24: EASTMAN CORPORATION

Suppose Eastman Corporation has started a training program to train "hard-core" unemployed persons. Those who complete the program are then hired by Eastman Corporation. The program is extremely expensive, since the student-teacher ratio is almost one to one. Given the expense, only 50 applicants are admitted to each six-month session. There are numerous applicants, and standard screening methods are inappropriate; therefore the 50 positions have to be filled by what amounts to random procedures. Approximately 40 percent of those accepted fail to complete the course. This costs the firm funds and, more importantly, denies others of a chance to obtain the training.

A psychologist in Eastman's personnel department has suggested a relatively simple hypothesis to explain the high dropout rate and a screening device to improve the percentage completing the training. Her idea is that all the recruits find the structure imposed by the training sessions to be intensely frustrating. Those who drop out are unable to cope with this frustration. To screen out those least able to deal with frustration, the psychologist proposes using a manual puzzle which has no solution. The individuals will be observed as they attempt the puzzle and rated according to their tolerance for frustration. This can then serve as a screening device.

To test her ideas, you administer the test to the next group of 50 selected applicants. Each applicant is rated on a scale from 1 (high tolerance) to 8 (low tolerance) according to his or her apparent tolerance of frustration. By the end of the session, 21 have dropped out. You now want to see if the proposed screening device appears related to the decision to drop out. The ratings of those that completed the program and those that dropped out are presented below. What would you conclude?

Frustration-Tolerance Ratings

Completed Training			Dropped Out		
3	2	3	3	7	1
2	1	2	4	6	8
4	4	1	6	8	7
1	3	5	5	5	6
5	6	4	1	2	5
3	1	3	7	3	5
2	8	2	8	6	4
2	2	1			
4	1	2			
7	3				

CASE 25: BRONSON'S SHOE STORE

Bronson's Northwest Mall store has, after only two years' operation, approached the maximum capacity of its current location. Management believes that the maximum capacity of the current outlet is $180,000 in sales in any month. When sales exceed that level, potential sales and customer goodwill are lost because of overcrowding and slow service.

Bronson's was located in the Northwest Mall when it opened two years ago. The entire mall has experienced rapid growth but not quite as rapid as Bronson's. The developers of the Northwest Mall have decided to add a major new wing to the center, and Bronson's has been given the opportunity to lease as much space in the new wing as its management desires.

As can be seen in the figures below, Bronson's sales have grown rapidly during the past two years. However, management is concerned about several things. One is the fact that the new wing will result in a 20 percent reduction in parking spaces. Second, there is a rumor that construction will soon begin on the area's third mall-type shopping center. If true, the new center will have some impact on sales at the Northwest Mall. However, it will not be possible to estimate the total impact of the new center until final details concerning its size, design, and so forth are known. In the meantime, Bronson's management must decide whether to acquire a larger space in the Northwest Mall and, if so, how much larger than the present space. Advise Bronson's management.

Monthly Sales Figures for Bronson's Shoe Store

Date	Sales	Date	Sales
April	$ 80,553	April	$107,423
May	101,543	May	111,156
June	98,100	June	124,724
July	81,296	July	96,903
August	165,044	August	201,156

Date	Sales	Date	Sales
September	152,985	September	180,164
October	124,306	October	160,557
November	122,934	November	147,916
December	171,915	December	194,784
January	109,958	January	121,350
February	78,266	February	96,162
March	114,829	March	138,762

Bibliography

Probability

Frank, Harry: *Introduction to Probability and Statistics: Concepts and Principles,* John Wiley & Sons, Inc., New York, 1974.

Goldberg, Samuel: *Probability: An Introduction,* Prentice-Hall, Inc., Englewood Cliffs, NJ, 1960.

Shook, Robert, and Harold Highland: *Probability Models with Business Applications,* Richard D. Irwin, Inc., Homewood, IL, 1969.

Decision Theory

Jedamus, Paul, and Robert Frame: *Business Decision Theory,* McGraw-Hill Book Company, New York, 1969.

Lindgren, B. W.: *Elements of Decision Theory,* The Macmillan Company, New York, 1971.

Lindley, D. V.: *Making Decisions,* Wiley-Interscience, London, 1971.

Raiffa, Howard: *Decision Analysis,* Addison-Wesley Publishing Company, Inc., Reading, MA, 1968.

Regression Analysis

Draper, N. R., and H. Smith: *Applied Regression Analysis,* John Wiley & Sons, Inc., New York, 1966.

Neter, J., and W. Wasserman: *Applied Linear Statistical Models,* Richard D. Irwin, Inc., Homewood, IL, 1974.

Analysis of Variance

Mendenhall, W.: *Introduction to Linear Models and the Design and Analysis of Experiments,* Wadsworth Publishing Company, Inc., Belmont, CA, 1968.

Neter, J., and W. Wasserman: *Applied Linear Statistical Models,* Richard D. Irwin, Homewood, IL, 1974.

Nonparametric Methods

Conover, W. J.: *Practical Nonparametric Statistics,* John Wiley & Sons, Inc., New York, 1971.

Gibbons, Jean D.: *Nonparametric Methods for Quantitative Analysis,* Holt, Rinehart and Winston, Inc., New York, 1976.

Siegel, S.: *Nonparametric Statistics for the Behavioral Sciences,* McGraw-Hill Book Company, New York, 1956.

Forecasting

Nelson, Charles R.: *Applied Time Series Analysis for Managerial Forecasting,* Holden-Day, Inc., Publisher, San Francisco, 1973.

Wheelwright, S. C., and S. Makridakis: *Forecasting Methods for Management,* 2 Ed., Wiley-Interscience, New York, 1977.

Appendix

Table A-6 Critical Values of χ^2

Table A-7 Critical Values of F_{ν_1, ν_2} for α = .05

Table A-8 Critical Values of F_{ν_1, ν_2} for α = .01

Table A-1 Square Roots

n	$\sqrt{n}$	$\sqrt{10n}$	n	$\sqrt{n}$	$\sqrt{10n}$	n	$\sqrt{n}$	$\sqrt{10n}$
1.00	1.0000	3.1623	1.50	1.2247	3.8730	2.00	1.4142	4.4721
1.01	1.0050	3.1780	1.51	1.2288	3.8859	2.01	1.4177	4.4833
1.02	1.0100	3.1937	1.52	1.2329	3.8987	2.02	1.4213	4.4944
1.03	1.0149	3.2094	1.53	1.2369	3.9115	2.03	1.4248	4.5056
1.04	1.0198	3.2249	1.54	1.2410	3.9243	2.04	1.4283	4.5166
1.05	1.0247	3.2404	1.55	1.2450	3.9370	2.05	1.4318	4.5277
1.06	1.0296	3.2558	1.56	1.2490	3.9497	2.06	1.4353	4.5387
1.07	1.0344	3.2711	1.57	1.2530	3.9623	2.07	1.4387	4.5497
1.08	1.0392	3.2863	1.58	1.2570	3.9749	2.08	1.4422	4.5607
1.09	1.0440	3.3015	1.59	1.2610	3.9875	2.09	1.4457	4.5717
1.10	1.0488	3.3166	1.60	1.2649	4.0000	2.10	1.4491	4.5826
1.11	1.0536	3.3317	1.61	1.2689	4.0125	2.11	1.4526	4.5935
1.12	1.0583	3.3466	1.62	1.2728	4.0249	2.12	1.4560	4.6043
1.13	1.0630	3.3615	1.63	1.2767	4.0373	2.13	1.4595	4.6152
1.14	1.0677	3.3764	1.64	1.2806	4.0497	2.14	1.4629	4.6260
1.15	1.0724	3.3912	1.65	1.2845	4.0620	2.15	1.4663	4.6368
1.16	1.0770	3.4059	1.66	1.2884	4.0743	2.16	1.4697	4.6476
1.17	1.0817	3.4205	1.67	1.2923	4.0866	2.17	1.4731	4.6583
1.18	1.0863	3.4351	1.68	1.2961	4.0988	2.18	1.4765	4.6690
1.19	1.0909	3.4496	1.69	1.3000	4.1110	2.19	1.4799	4.6797
1.20	1.0954	3.4641	1.70	1.3038	4.1231	2.20	1.4832	4.6904
1.21	1.1000	3.4785	1.71	1.3077	4.1352	2.21	1.4866	4.7011
1.22	1.1045	3.4928	1.72	1.3115	4.1473	2.22	1.4900	4.7117
1.23	1.1091	3.5071	1.73	1.3153	4.1593	2.23	1.4933	4.7223
1.24	1.1136	3.5214	1.74	1.3191	4.1713	2.24	1.4967	4.7329
1.25	1.1180	3.5355	1.75	1.3229	4.1833	2.25	1.5000	4.7434
1.26	1.1225	3.5496	1.76	1.3266	4.1952	2.26	1.5033	4.7539
1.27	1.1269	3.5637	1.77	1.3304	4.2071	2.27	1.5067	4.7645
1.28	1.1314	3.5777	1.78	1.3342	4.2190	2.28	1.5100	4.7749
1.29	1.1358	3.5917	1.79	1.3379	4.2308	2.29	1.5133	4.7854
1.30	1.1402	3.6056	1.80	1.3416	4.2426	2.30	1.5166	4.7958
1.31	1.1446	3.6194	1.81	1.3454	4.2544	2.31	1.5199	4.8062
1.32	1.1489	3.6332	1.82	1.3491	4.2661	2.32	1.5232	4.8166
1.33	1.1533	3.6469	1.83	1.3528	4.2778	2.33	1.5264	4.8270
1.34	1.1576	3.6606	1.84	1.3565	4.2895	2.34	1.5297	4.8374
1.35	1.1619	3.6742	1.85	1.3601	4.3012	2.35	1.5330	4.8477
1.36	1.1662	3.6878	1.86	1.3638	4.3128	2.36	1.5362	4.8580
1.37	1.1705	3.7014	1.87	1.3675	4.3243	2.37	1.5395	4.8683
1.38	1.1747	3.7148	1.88	1.3711	4.3359	2.38	1.5427	4.8785
1.39	1.1790	3.7283	1.89	1.3748	4.3474	2.39	1.5460	4.8888
1.40	1.1832	3.7417	1.90	1.3784	4.3589	2.40	1.5492	4.8990
1.41	1.1874	3.7550	1.91	1.3820	4.3704	2.41	1.5524	4.9092
1.42	1.1916	3.7683	1.92	1.3856	4.3818	2.42	1.5556	4.9193
1.43	1.1958	3.7815	1.93	1.3892	4.3932	2.43	1.5588	4.9295
1.44	1.2000	3.7947	1.94	1.3928	4.4045	2.44	1.5620	4.9396
1.45	1.2042	3.8079	1.95	1.3964	4.4159	2.45	1.5652	4.9497
1.46	1.2083	3.8210	1.96	1.4000	4.4272	2.46	1.5684	4.9598
1.47	1.2124	3.8341	1.97	1.4036	4.4385	2.47	1.5716	4.9699
1.48	1.2166	3.8471	1.98	1.4071	4.4497	2.48	1.5748	4.9800
1.49	1.2207	3.8601	1.99	1.4107	4.4609	2.49	1.5780	4.9900

Table A-1 Square Roots (Continued)

n	$\sqrt{n}$	$\sqrt{10n}$	n	$\sqrt{n}$	$\sqrt{10n}$	n	$\sqrt{n}$	$\sqrt{10n}$
2.50	1.5811	5.0000	3.00	1.7321	5.4772	3.50	1.8708	5.9161
2.51	1.5843	5.0100	3.01	1.7349	5.4863	3.51	1.8735	5.9245
2.52	1.5875	5.0200	3.02	1.7378	5.4955	3.52	1.8762	5.9330
2.53	1.5906	5.0299	3.03	1.7407	5.5045	3.53	1.8788	5.9414
2.54	1.5937	5.0398	3.04	1.7436	5.5136	3.54	1.8815	5.9498
2.55	1.5969	5.0498	3.05	1.7464	5.5227	3.55	1.8841	5.9582
2.56	1.6000	5.0596	3.06	1.7493	5.5317	3.56	1.8868	5.9666
2.57	1.6031	5.0695	3.07	1.7521	5.5408	3.57	1.8894	5.9749
2.58	1.6062	5.0794	3.08	1.7550	5.5498	3.58	1.8921	5.9833
2.59	1.6093	5.0892	3.09	1.7578	5.5588	3.59	1.8947	5.9917
2.60	1.6125	5.0990	3.10	1.7607	5.5678	3.60	1.8974	6.0000
2.61	1.6155	5.1088	3.11	1.7635	5.5767	3.61	1.9000	6.0083
2.62	1.6186	5.1186	3.12	1.7664	5.5857	3.62	1.9026	6.0166
2.63	1.6217	5.1284	3.13	1.7692	5.5946	3.63	1.9053	6.0249
2.64	1.6248	5.1381	3.14	1.7720	5.6036	3.64	1.9079	6.0332
2.65	1.6279	5.1478	3.15	1.7748	5.6125	3.65	1.9105	6.0415
2.66	1.6310	5.1575	3.16	1.7776	5.6214	3.66	1.9131	6.0498
2.67	1.6340	5.1672	3.17	1.7804	5.6303	3.67	1.9157	6.0581
2.68	1.6371	5.1769	3.18	1.7833	5.6391	3.68	1.9183	6.0663
2.69	1.6401	5.1865	3.19	1.7861	5.6480	3.69	1.9209	6.0745
2.70	1.6432	5.1962	3.20	1.7889	5.6569	3.70	1.9235	6.0828
2.71	1.6462	5.2058	3.21	1.7916	5.6657	3.71	1.9261	6.0910
2.72	1.6492	5.2154	3.22	1.7944	5.6745	3.72	1.9287	6.0992
2.73	1.6523	5.2249	3.23	1.7972	5.6833	3.73	1.9313	6.1074
2.74	1.6553	5.2345	3.24	1.8000	5.6921	3.74	1.9339	6.1156
2.75	1.6583	5.2440	3.25	1.8028	5.7009	3.75	1.9365	6.1237
2.76	1.6613	5.2536	3.26	1.8055	5.7096	3.76	1.9391	6.1319
2.77	1.6643	5.2631	3.27	1.8083	5.7184	3.77	1.9416	6.1400
2.78	1.6673	5.2726	3.28	1.8111	5.7271	3.78	1.9442	6.1482
2.79	1.6703	5.2820	3.29	1.8138	5.7359	3.79	1.9468	6.1563
2.80	1.6733	5.2915	3.30	1.8166	5.7446	3.80	1.9494	6.1644
2.81	1.6763	5.3009	3.31	1.8193	5.7533	3.81	1.9519	6.1725
2.82	1.6793	5.3104	3.32	1.8221	5.7619	3.82	1.9545	6.1806
2.83	1.6823	5.3198	3.33	1.8248	5.7706	3.83	1.9570	6.1887
2.84	1.6852	5.3292	3.34	1.8276	5.7793	3.84	1.9596	6.1968
2.85	1.6882	5.3385	3.35	1.8303	5.7879	3.85	1.9621	6.2048
2.86	1.6912	5.3479	3.36	1.8330	5.7966	3.86	1.9647	6.2129
2.87	1.6941	5.3572	3.37	1.8358	5.8052	3.87	1.9672	6.2209
2.88	1.6971	5.3666	3.38	1.8385	5.8138	3.88	1.9698	6.2290
2.89	1.7000	5.3759	3.39	1.8412	5.8224	3.89	1.9723	6.2370
2.90	1.7029	5.3852	3.40	1.8439	5.8310	3.90	1.9748	6.2450
2.91	1.7059	5.3944	3.41	1.8466	5.8395	3.91	1.9774	6.2530
2.92	1.7088	5.4037	3.42	1.8493	5.8481	3.92	1.9799	6.2610
2.93	1.7117	5.4129	3.43	1.8520	5.8566	3.93	1.9824	6.2690
2.94	1.7146	5.4222	3.44	1.8547	5.8652	3.94	1.9849	6.2769
2.95	1.7176	5.4314	3.45	1.8574	5.8737	3.95	1.9875	6.2849
2.96	1.7205	5.4406	3.46	1.8601	5.8822	3.96	1.9900	6.2929
2.97	1.7234	5.4498	3.47	1.8628	5.8907	3.97	1.9925	6.3008
2.98	1.7263	5.4589	3.48	1.8655	5.8992	3.98	1.9950	6.3087
2.99	1.7292	5.4681	3.49	1.8682	5.9076	3.99	1.9975	6.3166

Table A-1 Square Roots (Continued)

n	$\sqrt{n}$	$\sqrt{10n}$	n	$\sqrt{n}$	$\sqrt{10n}$	n	$\sqrt{n}$	$\sqrt{10n}$
4.00	2.0000	6.3246	4.50	2.1213	6.7082	5.00	2.2361	7.0711
4.01	2.0025	6.3325	4.51	2.1237	6.7157	5.01	2.2383	7.0781
4.02	2.0050	6.3403	4.52	2.1260	6.7231	5.02	2.2405	7.0852
4.03	2.0075	6.3482	4.53	2.1284	6.7305	5.03	2.2428	7.0922
4.04	2.0100	6.3561	4.54	2.1307	6.7380	5.04	2.2450	7.0993
4.05	2.0125	6.3640	4.55	2.1331	6.7454	5.05	2.2472	7.1063
4.06	2.0149	6.3718	4.56	2.1354	6.7528	5.06	2.2494	7.1134
4.07	2.0174	6.3797	4.57	2.1378	6.7602	5.07	2.2517	7.1204
4.08	2.0199	6.3875	4.58	2.1401	6.7676	5.08	2.2539	7.1274
4.09	2.0224	6.3953	4.59	2.1424	6.7750	5.09	2.2561	7.1344
4.10	2.0248	6.4031	4.60	2.1448	6.7823	5.10	2.2583	7.1414
4.11	2.0273	6.4109	4.61	2.1471	6.7897	5.11	2.2605	7.1484
4.12	2.0298	6.4187	4.62	2.1494	6.7971	5.12	2.2627	7.1554
4.13	2.0322	6.4265	4.63	2.1517	6.8044	5.13	2.2650	7.1624
4.14	2.0347	6.4343	4.64	2.1541	6.8118	5.14	2.2672	7.1694
4.15	2.0372	6.4420	4.65	2.1564	6.8191	5.15	2.2694	7.1764
4.16	2.0396	6.4498	4.66	2.1587	6.8264	5.16	2.2716	7.1833
4.17	2.0421	6.4576	4.67	2.1610	6.8337	5.17	2.2738	7.1903
4.18	2.0445	6.4653	4.68	2.1633	6.8411	5.18	2.2760	7.1972
4.19	2.0469	6.4730	4.69	2.1656	6.8484	5.19	2.2782	7.2042
4.20	2.0494	6.4807	4.70	2.1679	6.8557	5.20	2.2804	7.2111
4.21	2.0518	6.4885	4.71	2.1703	6.8629	5.21	2.2825	7.2180
4.22	2.0543	6.4962	4.72	2.1726	6.8702	5.22	2.2847	7.2250
4.23	2.0567	6.5038	4.73	2.1749	6.8775	5.23	2.2869	7.2319
4.24	2.0591	6.5115	4.74	2.1772	6.8848	5.24	2.2891	7.2388
4.25	2.0616	6.5192	4.75	2.1794	6.8920	5.25	2.2913	7.2457
4.26	2.0640	6.5269	4.76	2.1817	6.8993	5.26	2.2935	7.2526
4.27	2.0664	6.5345	4.77	2.1840	6.9065	5.27	2.2956	7.2595
4.28	2.0688	6.5422	4.78	2.1863	6.9138	5.28	2.2978	7.2664
4.29	2.0712	6.5498	4.79	2.1886	6.9210	5.29	2.3000	7.2732
4.30	2.0736	6.5574	4.80	2.1909	6.9282	5.30	2.3022	7.2801
4.31	2.0761	6.5651	4.81	2.1932	6.9354	5.31	2.3043	7.2870
4.32	2.0785	6.5727	4.82	2.1954	6.9426	5.32	2.3065	7.2938
4.33	2.0809	6.5803	4.83	2.1977	6.9498	5.33	2.3087	7.3007
4.34	2.0833	6.5879	4.84	2.2000	6.9570	5.34	2.3108	7.3075
4.35	2.0857	6.5955	4.85	2.2023	6.9642	5.35	2.3130	7.3144
4.36	2.0881	6.6030	4.86	2.2045	6.9714	5.36	2.3152	7.3212
4.37	2.0905	6.6106	4.87	2.2068	6.9785	5.37	2.3173	7.3280
4.38	2.0928	6.6182	4.88	2.2091	6.9857	5.38	2.3195	7.3348
4.39	2.0952	6.6257	4.89	2.2113	6.9929	5.39	2.3216	7.3417
4.40	2.0976	6.6332	4.90	2.2136	7.0000	5.40	2.3238	7.3485
4.41	2.1000	6.6408	4.91	2.2159	7.0071	5.41	2.3259	7.3553
4.42	2.1024	6.6483	4.92	2.2181	7.0143	5.42	2.3281	7.3621
4.43	2.1048	6.6558	4.93	2.2204	7.0214	5.43	2.3302	7.3689
4.44	2.1071	6.6633	4.94	2.2226	7.0285	5.44	2.3324	7.3756
4.45	2.1095	6.6708	4.95	2.2249	7.0356	5.45	2.3345	7.3824
4.46	2.1119	6.6783	4.96	2.2271	7.0427	5.46	2.3367	7.3892
4.47	2.1142	6.6858	4.97	2.2293	7.0498	5.47	2.3388	7.3959
4.48	2.1166	6.6933	4.98	2.2316	7.0569	5.48	2.3409	7.4027
4.49	2.1190	6.7007	4.99	2.2338	7.0640	5.49	2.3431	7.4095

Table A-1 Square Roots *(Continued)*

n	$\sqrt{n}$	$\sqrt{10n}$	n	$\sqrt{n}$	$\sqrt{10n}$	n	$\sqrt{n}$	$\sqrt{10n}$
5.50	2.3452	7.4162	6.00	2.4495	7.7460	6.50	2.5495	8.0623
5.51	2.3473	7.4229	6.01	2.4515	7.7524	6.51	2.5515	8.0685
5.52	2.3495	7.4297	6.02	2.4536	7.7589	6.52	2.5534	8.0747
5.53	2.3516	7.4364	6.03	2.4556	7.7653	6.53	2.5554	8.0808
5.54	2.3537	7.4431	6.04	2.4576	7.7717	6.54	2.5573	8.0870
5.55	2.3558	7.4498	6.05	2.4597	7.7782	6.55	2.5593	8.0932
5.56	2.3580	7.4565	6.06	2.4617	7.7846	6.56	2.5612	8.0994
5.57	2.3601	7.4632	6.07	2.4637	7.7910	6.57	2.5632	8.1056
5.58	2.3622	7.4699	6.08	2.4658	7.7974	6.58	2.5652	8.1117
5.59	2.3643	7.4766	6.09	2.4678	7.8038	6.59	2.5671	8.1179
5.60	2.3664	7.4833	6.10	2.4698	7.8102	6.60	2.5690	8.1240
5.61	2.3685	7.4900	6.11	2.4718	7.8166	6.61	2.5710	8.1302
5.62	2.3707	7.4967	6.12	2.4739	7.8230	6.62	2.5729	8.1363
5.63	2.3728	7.5033	6.13	2.4759	7.8294	6.63	2.5749	8.1425
5.64	2.3749	7.5100	6.14	2.4779	7.8358	6.64	2.5768	8.1486
5.65	2.3770	7.5166	6.15	2.4799	7.8422	6.65	2.5788	8.1548
5.66	2.3791	7.5233	6.16	2.4819	7.8486	6.66	2.5807	8.1609
5.67	2.3812	7.5299	6.17	2.4839	7.8549	6.67	2.5826	8.1670
5.68	2.3833	7.5366	6.18	2.4860	7.8613	6.68	2.5846	8.1731
5.69	2.3854	7.5432	6.19	2.4880	7.8677	6.69	2.5865	8.1792
5.70	2.3875	7.5498	6.20	2.4900	7.8740	6.70	2.5884	8.1854
5.71	2.3896	7.5565	6.21	2.4920	7.8804	6.71	2.5904	8.1915
5.72	2.3917	7.5631	6.22	2.4940	7.8867	6.72	2.5923	8.1976
5.73	2.3937	7.5697	6.23	2.4960	7.8930	6.73	2.5942	8.2037
5.74	2.3958	7.5763	6.24	2.4980	7.8994	6.74	2.5962	8.2098
5.75	2.3979	7.5829	6.25	2.5000	7.9057	6.75	2.5981	8.2158
5.76	2.4000	7.5895	6.26	2.5020	7.9120	6.76	2.6000	8.2219
5.77	2.4021	7.5961	6.27	2.5040	7.9183	6.77	2.6019	8.2280
5.78	2.4042	7.6026	6.28	2.5060	7.9246	6.78	2.6038	8.2341
5.79	2.4062	7.6092	6.29	2.5080	7.9310	6.79	2.6058	8.2401
5.80	2.4083	7.6158	6.30	2.5100	7.9373	6.80	2.6077	8.2462
5.81	2.4104	7.6223	6.31	2.5120	7.9436	6.81	2.6096	8.2523
5.82	2.4125	7.6289	6.32	2.5140	7.9498	6.82	2.6115	8.2583
5.83	2.4145	7.6354	6.33	2.5159	7.9561	6.83	2.6134	8.2644
5.84	2.4166	7.6420	6.34	2.5179	7.9624	6.84	2.6153	8.2704
5.85	2.4187	7.6485	6.35	2.5199	7.9687	6.85	2.6173	8.2765
5.86	2.4207	7.6551	6.36	2.5219	7.9750	6.86	2.6192	8.2825
5.87	2.4228	7.6616	6.37	2.5239	7.9812	6.87	2.6211	8.2885
5.88	2.4249	7.6681	6.38	2.5259	7.9875	6.88	2.6230	8.2946
5.89	2.4269	7.6746	6.39	2.5278	7.9937	6.89	2.6249	8.3006
5.90	2.4290	7.6811	6.40	2.5298	8.0000	6.90	2.6268	8.3066
5.91	2.4310	7.6877	6.41	2.5318	8.0062	6.91	2.6287	8.3126
5.92	2.4331	7.6942	6.42	2.5338	8.0125	6.92	2.6306	8.3187
5.93	2.4352	7.7006	6.43	2.5357	8.0187	6.93	2.6325	8.3247
5.94	2.4372	7.7071	6.44	2.5377	8.0250	6.94	2.6344	8.3307
5.95	2.4393	7.7136	6.45	2.5397	8.0312	6.95	2.6363	8.3367
5.96	2.4413	7.7201	6.46	2.5417	8.0374	6.96	2.6382	8.3427
5.97	2.4434	7.7266	6.47	2.5436	8.0436	6.97	2.6401	8.3487
5.98	2.4454	7.7330	6.48	2.5456	8.0498	6.98	2.6420	8.3546
5.99	2.4474	7.7395	6.49	2.5475	8.0561	6.99	2.6439	8.3606

Table A-1 Square Roots *(Continued)*

n	$\sqrt{n}$	$\sqrt{10n}$	n	$\sqrt{n}$	$\sqrt{10n}$	n	$\sqrt{n}$	$\sqrt{10n}$
7.00	2.6458	8.3666	7.50	2.7386	8.6603	8.00	2.8284	8.9443
7.01	2.6476	8.3726	7.51	2.7404	8.6660	8.01	2.8302	8.9499
7.02	2.6495	8.3785	7.52	2.7423	8.6718	8.02	2.8320	8.9554
7.03	2.6514	8.3845	7.53	2.7441	8.6776	8.03	2.8337	8.9610
7.04	2.6533	8.3905	7.54	2.7459	8.6833	8.04	2.8355	8.9666
7.05	2.6552	8.3964	7.55	2.7477	8.6891	8.05	2.8373	8.9722
7.06	2.6571	8.4024	7.56	2.7495	8.6948	8.06	2.8390	8.9778
7.07	2.6589	8.4083	7.57	2.7514	8.7006	8.07	2.8408	8.9833
7.08	2.6608	8.4143	7.58	2.7532	8.7063	8.08	2.8425	8.9889
7.09	2.6627	8.4202	7.59	2.7550	8.7121	8.09	2.8443	8.9944
7.10	2.6646	8.4261	7.60	2.7568	8.7178	8.10	2.8460	9.0000
7.11	2.6665	8.4321	7.61	2.7586	8.7235	8.11	2.8478	9.0056
7.12	2.6683	8.4380	7.62	2.7604	8.7293	8.12	2.8496	9.0111
7.13	2.6702	8.4439	7.63	2.7622	8.7350	8.13	2.8513	9.0167
7.14	2.6721	8.4499	7.64	2.7641	8.7407	8.14	2.8531	9.0222
7.15	2.6739	8.4558	7.65	2.7659	8.7464	8.15	2.8548	9.0277
7.16	2.6758	8.4617	7.66	2.7677	8.7521	8.16	2.8566	9.0333
7.17	2.6777	8.4676	7.67	2.7695	8.7579	8.17	2.8583	9.0388
7.18	2.6796	8.4735	7.68	2.7713	8.7636	8.18	2.8601	9.0443
7.19	2.6814	8.4794	7.69	2.7731	8.7693	8.19	2.8618	9.0499
7.20	2.6833	8.4853	7.70	2.7749	8.7750	8.20	2.8636	9.0554
7.21	2.6851	8.4912	7.71	2.7767	8.7807	8.21	2.8653	9.0609
7.22	2.6870	8.4971	7.72	2.7785	8.7864	8.22	2.8671	9.0664
7.23	2.6889	8.5029	7.73	2.7803	8.7920	8.23	2.8688	9.0719
7.24	2.6907	8.5088	7.74	2.7821	8.7977	8.24	2.8705	9.0774
7.25	2.6926	8.5147	7.75	2.7839	8.8034	8.25	2.8723	9.0830
7.26	2.6944	8.5206	7.76	2.7857	8.8091	8.26	2.8740	9.0885
7.27	2.6963	8.5264	7.77	2.7875	8.8148	8.27	2.8758	9.0940
7.28	2.6981	8.5323	7.78	2.7893	8.8204	8.28	2.8775	9.0995
7.29	2.7000	8.5381	7.79	2.7911	8.8261	8.29	2.8792	9.1049
7.30	2.7019	8.5440	7.80	2.7928	8.8318	8.30	2.8810	9.1104
7.31	2.7037	8.5499	7.81	2.7946	8.8374	8.31	2.8827	9.1159
7.32	2.7055	8.5557	7.82	2.7964	8.8431	8.32	2.8844	9.1214
7.33	2.7074	8.5615	7.83	2.7982	8.8487	8.33	2.8862	9.1269
7.34	2.7092	8.5674	7.84	2.8000	8.8544	8.34	2.8879	9.1324
7.35	2.7111	8.5732	7.85	2.8018	8.8600	8.35	2.8896	9.1378
7.36	2.7129	8.5790	7.86	2.8036	8.8657	8.36	2.8914	9.1433
7.37	2.7148	8.5849	7.87	2.8054	8.8713	8.37	2.8931	9.1488
7.38	2.7166	8.5907	7.88	2.8071	8.8769	8.38	2.8948	9.1542
7.39	2.7185	8.5965	7.89	2.8089	8.8826	8.39	2.8965	9.1597
7.40	2.7203	8.6023	7.90	2.8107	8.8882	8.40	2.8983	9.1652
7.41	2.7221	8.6081	7.91	2.8125	8.8938	8.41	2.9000	9.1706
7.42	2.7240	8.6139	7.92	2.8142	8.8994	8.42	2.9017	9.1761
7.43	2.7258	8.6197	7.93	2.8160	8.9051	8.43	2.9034	9.1815
7.44	2.7276	8.6255	7.94	2.8178	8.9107	8.44	2.9052	9.1869
7.45	2.7295	8.6313	7.95	2.8196	8.9163	8.45	2.9069	9.1924
7.46	2.7313	8.6371	7.96	2.8213	8.9219	8.46	2.9086	9.1978
7.47	2.7331	8.6429	7.97	2.8231	8.9275	8.47	2.9103	9.2033
7.48	2.7350	8.6487	7.98	2.8249	8.9331	8.48	2.9120	9.2087
7.49	2.7368	8.6545	7.99	2.8267	8.9387	8.49	2.9138	9.2141

Table A-1 Square Roots *(Continued)*

n	$\sqrt{n}$	$\sqrt{10n}$	n	$\sqrt{n}$	$\sqrt{10n}$	n	$\sqrt{n}$	$\sqrt{10n}$
8.50	2.9155	9.2195	9.00	3.0000	9.4868	9.50	3.0822	9.7468
8.51	2.9172	9.2250	9.01	3.0017	9.4921	9.51	3.0838	9.7519
8.52	2.9189	9.2304	9.02	3.0033	9.4974	9.52	3.0854	9.7570
8.53	2.9206	9.2358	9.03	3.0050	9.5026	9.53	3.0871	9.7622
8.54	2.9223	9.2412	9.04	3.0067	9.5079	9.54	3.0887	9.7673
8.55	2.9240	9.2466	9.05	3.0083	9.5131	9.55	3.0903	9.7724
8.56	2.9257	9.2520	9.06	3.0100	9.5184	9.56	3.0919	9.7775
8.57	2.9275	9.2574	9.07	3.0116	9.5237	9.57	3.0935	9.7826
8.58	2.9292	9.2628	9.08	3.0133	9.5289	9.58	3.0952	9.7877
8.59	2.9309	9.2682	9.09	3.0150	9.5341	9.59	3.0968	9.7929
8.60	2.9326	9.2736	9.10	3.0166	9.5394	9.60	3.0984	9.7980
8.61	2.9343	9.2790	9.11	3.0183	9.5446	9.61	3.1000	9.8031
8.62	2.9360	9.2844	9.12	3.0199	9.5499	9.62	3.1016	9.8082
8.63	2.9377	9.2898	9.13	3.0216	9.5551	9.63	3.1032	9.8133
8.64	2.9394	9.2952	9.14	3.0232	9.5603	9.64	3.1048	9.8184
8.65	2.9411	9.3005	9.15	3.0249	9.5656	9.65	3.1064	9.8234
8.66	2.9428	9.3059	9.16	3.0265	9.5708	9.66	3.1081	9.8285
8.67	2.9445	9.3113	9.17	3.0282	9.5760	9.67	3.1097	9.8336
8.68	2.9462	9.3167	9.18	3.0299	9.5812	9.68	3.1113	9.8387
8.69	2.9479	9.3220	9.19	3.0315	9.5864	9.69	3.1129	9.8438
8.70	2.9496	9.3274	9.20	3.0332	9.5917	9.70	3.1145	9.8489
8.71	2.9513	9.3327	9.21	3.0348	9.5969	9.71	3.1161	9.8539
8.72	2.9530	9.3381	9.22	3.0364	9.6021	9.72	3.1177	9.8590
8.73	2.9547	9.3434	9.23	3.0381	9.6073	9.73	3.1193	9.8641
8.74	2.9563	9.3488	9.24	3.0397	9.6125	9.74	3.1209	9.8691
8.75	2.9580	9.3541	9.25	3.0414	9.6177	9.75	3.1225	9.8742
8.76	2.9597	9.3595	9.26	3.0430	9.6229	9.76	3.1241	9.8793
8.77	2.9614	9.3648	9.27	3.0447	9.6281	9.77	3.1257	9.8843
8.78	2.9631	9.3702	9.28	3.0463	9.6333	9.78	3.1273	9.8894
8.79	2.9648	9.3755	9.29	3.0480	9.6385	9.79	3.1289	9.8944
8.80	2.9665	9.3808	9.30	3.0496	9.6437	9.80	3.1305	9.8995
8.81	2.9682	9.3862	9.31	3.0512	9.6488	9.81	3.1321	9.9045
8.82	2.9698	9.3915	9.32	3.0529	9.6540	9.82	3.1337	9.9096
8.83	2.9715	9.3968	9.33	3.0545	9.6592	9.83	3.1353	9.9146
8.84	2.9732	9.4021	9.34	3.0561	9.6644	9.84	3.1369	9.9197
8.85	2.9749	9.4074	9.35	3.0578	9.6695	9.85	3.1385	9.9247
8.86	2.9766	9.4128	9.36	3.0594	9.6747	9.86	3.1401	9.9298
8.87	2.9783	9.4181	9.37	3.0610	9.6799	9.87	3.1417	9.9348
8.88	2.9799	9.4234	9.38	3.0627	9.6850	9.88	3.1432	9.9398
8.89	2.9816	9.4287	9.39	3.0643	9.6902	9.89	3.1448	9.9448
8.90	2.9833	9.4340	9.40	3.0659	9.6954	9.90	3.1464	9.9499
8.91	2.9850	9.4393	9.41	3.0676	9.7005	9.91	3.1480	9.9549
8.92	2.9866	9.4446	9.42	3.0692	9.7057	9.92	3.1496	9.9599
8.93	2.9883	9.4499	9.43	3.0708	9.7108	9.93	3.1512	9.9649
8.94	2.9900	9.4552	9.44	3.0725	9.7160	9.94	3.1528	9.9700
8.95	2.9917	9.4604	9.45	3.0741	9.7211	9.95	3.1544	9.9750
8.96	2.9933	9.4657	9.46	3.0757	9.7263	9.96	3.1559	9.9800
8.97	2.9950	9.4710	9.47	3.0773	9.7314	9.97	3.1575	9.9850
8.98	2.9967	9.4763	9.45	3.0790	9.7365	9.98	3.1591	9.9900
8.99	2.9983	9.4816	9.49	3.0806	9.7417	9.99	3.1607	9.9950

Table A-2 14,000 Random Units

Line/Col.	(1)	(2)	(3)	(4)	(5)	(6)	(7)	(8)	(9)	(10)	(11)	(12)	(13)	(14)
1	10480	15011	01536	02011	81647	91646	69179	14194	62590	36207	20969	99570	91291	90700
2	22368	46573	25595	85393	30995	89198	27982	53402	93965	34095	52666	19174	39615	99505
3	24130	48360	22527	97265	76393	64809	15179	24830	49340	32081	30680	19655	63348	58629
4	42167	93093	06243	61680	07856	16376	39440	53537	71341	57004	00849	74917	97758	16379
5	37570	39975	81837	16656	06121	91782	60468	81305	49684	60672	14110	06927	01263	54613
6	77921	06907	11008	42751	27756	53498	18602	70659	90655	15053	21916	81825	44394	42880
7	99562	72905	56420	69994	98872	31016	71194	18738	44013	48840	63213	21069	10634	12952
8	96301	91977	05463	07972	18876	20922	94595	56869	69014	60045	18425	84903	42508	32307
9	89579	14342	63661	10281	17453	18103	57740	84378	25331	12566	58678	44947	05585	56941
10	85475	36857	43342	53988	53060	59533	38867	62300	08158	17983	16439	11458	18593	64952
11	28918	69578	88231	33276	70997	79936	56865	05859	90106	31595	01547	85590	91610	78188
12	63553	40961	48235	03427	49626	69445	18663	72695	52180	20847	12234	90511	33703	90322
13	09429	93969	52636	92737	88974	33488	36320	17617	30015	08272	84115	27156	30613	74952
14	10365	61129	87529	85689	48237	52267	67689	93394	01511	26358	85104	20285	29975	89868
15	07119	97336	71048	08178	77233	13916	47564	81056	97735	85977	29372	74461	28551	90707
16	51085	12765	51821	51259	77452	16308	60756	92144	49442	53900	70960	63990	75601	40719
17	02368	21382	52404	60268	89368	19885	55322	44819	01188	65255	64835	44919	05944	55157
18	01011	54092	33362	94904	31273	04146	18594	29852	71585	85030	51132	01915	92747	64951
19	52162	53916	46369	58586	23216	14513	83149	98736	23495	64350	94738	17752	35156	35749
20	07056	97628	33787	09998	42698	06691	76988	13602	51851	46104	88916	19509	25625	58104
21	48663	91245	85828	14346	09172	30168	90229	04734	59193	22178	30421	61666	99904	32812
22	54164	58492	22421	74103	47070	25306	76468	26384	58151	06646	21524	15227	96909	44592
23	32639	32363	05597	24200	13363	38005	94342	28728	35806	06912	17012	64161	18296	22851
24	29334	27001	87637	87308	58731	00256	45834	15398	46557	41135	10367	07684	36188	18510
25	02488	33062	28834	07351	19731	92420	60952	61280	50001	67658	32586	86679	50720	94953
26	81525	72295	04839	96423	24878	82651	66566	14778	76797	14780	13300	87074	79666	95725
27	29676	20591	68086	26432	46901	20849	89768	81536	86645	12659	92259	57102	80428	25280
28	00742	57392	39064	66432	84673	40027	32832	61362	98947	96067	64760	64584	96096	98253
29	05366	04213	25669	26422	44407	44048	37937	63904	45766	66134	75470	66520	34693	90449
30	91921	26418	64117	94305	26766	25940	39972	22209	71500	64568	91402	42416	07844	69618
31	00582	04711	87917	77341	42206	35126	74087	99547	81817	42607	43808	76655	62028	76630
32	00725	69884	62797	56170	86324	88072	76222	36086	84637	93161	76038	65855	77919	88006
33	69011	65797	95876	55293	18988	27354	26575	08625	40801	59920	29841	80150	12777	48501
34	25976	57948	29888	88604	67917	48708	18912	82271	65424	69774	33611	54262	85963	03547
35	09763	83473	73577	12908	30883	18317	28290	35797	05998	41688	34952	37888	38917	88050
36	91567	42595	27958	30134	04024	86385	29880	99730	55536	84855	29080	09250	79656	73211
37	17955	56349	90999	49127	20044	59931	06115	20542	18059	02008	73708	83517	36103	42791
38	46503	18584	18845	49618	02304	51038	20655	58727	28168	15475	56942	53389	20562	87338
39	92157	89634	94824	78171	84610	82834	09922	25417	44137	48413	25555	21246	35509	20468
40	14577	62765	35605	81263	39667	47358	56873	56307	61607	49518	89656	20103	77490	18062
41	98427	07523	33362	64270	01638	92477	66969	98420	04880	45585	46565	04102	46880	45709
42	34914	63976	88720	82765	34476	17032	87589	40836	32427	70002	70663	88863	77775	69348
43	70060	28277	39475	46473	23219	53416	94970	25832	69975	94884	19661	72828	00102	66794
44	53976	54914	06990	67245	68350	82948	11398	42878	80287	88267	47363	46634	06541	97809
45	76072	29515	40980	07391	58745	25774	22987	80059	39911	96189	41151	14222	60697	59583
46	90725	52210	83974	29992	65831	38857	50490	83765	55657	14361	31720	57375	56228	41546
47	64364	67412	33339	31926	14883	24413	59744	92351	97473	89286	35931	04110	23726	51900
48	08962	00358	31662	25388	61642	34072	81249	35648	56891	69352	48373	45578	78547	81788
49	95012	68379	93526	70765	10593	04542	76463	54328	02349	17247	28865	14777	62730	92277
50	15664	10493	20492	38391	91132	21999	59516	81652	27195	48223	46751	22923	32261	85653

Table A-2 14,000 Random Units *(Continued)*

Line/Col.	(1)	(2)	(3)	(4)	(5)	(6)	(7)	(8)	(9)	(10)	(11)	(12)	(13)	(14)
51	16408	81899	04153	53381	79401	21438	83035	92350	36693	31238	59649	91754	72772	02338
52	18629	81953	05520	91962	04739	13092	97662	24822	94730	06496	35090	04822	86772	98289
53	73115	35101	47498	87637	99016	71060	88824	71013	18735	20286	23153	72924	35165	43040
54	57491	16703	23167	49323	45021	33132	12544	41035	80780	45393	44812	12515	98931	91202
55	30405	83946	23792	14422	15059	45799	22716	19792	09983	74353	68668	30429	70735	25499
56	16631	35006	85900	98275	32388	52390	16815	69298	82732	38480	73817	32523	41961	44437
57	96773	20206	42559	78985	05300	22164	24369	54224	35083	19687	11052	91491	60383	19746
58	38935	64202	14349	82674	66523	44133	00697	35552	35970	19124	63318	29686	03387	59846
59	31624	76384	17403	53363	44167	64486	64758	75366	76554	31601	12614	33072	60332	92325
60	78919	19474	23632	27889	47914	02584	37680	20801	72152	39339	34806	08930	85001	87820
61	03931	33309	57047	74211	63445	17361	62825	39908	05607	91284	68833	25570	38818	46920
62	74426	33278	43972	10119	89917	15665	52872	73823	73144	88662	88970	74492	51805	99378
63	09066	00903	20795	95452	92648	45454	09552	88815	16553	51125	79375	97596	16296	66092
64	42238	12426	87025	14267	20979	04508	64535	31355	86064	29472	47689	05974	52468	16834
65	16153	08002	26504	41744	81959	65642	74240	56302	00033	67107	77510	70625	28725	34191
66	21457	40742	29820	96783	29400	21840	15035	34537	33310	06116	95240	15957	16572	06004
67	21581	57802	02050	89728	17937	37621	47075	42080	97403	48626	68995	43805	33386	21597
68	55612	78095	83197	33732	05810	24813	86902	60397	16489	03264	88525	42786	05269	92532
69	44657	66999	99324	51281	84463	60563	79312	93454	68876	25471	93911	25650	12682	73572
70	91340	84979	46949	81973	37949	61023	43997	15263	80644	43942	89203	71795	99533	50501
71	91227	21199	31935	27022	84067	05462	35216	14486	29891	68607	41867	14951	91696	85065
72	50001	38140	66321	19924	72163	09538	12151	06878	91903	18749	34405	56087	82790	70925
73	65390	05224	72958	28609	81406	39147	25549	48542	42627	45233	57202	94617	23772	07896
74	27504	96131	83944	41575	10573	08619	64482	73923	36152	05184	94142	25299	84387	34925
75	37169	94851	39117	89632	00959	16487	65536	49071	39782	17095	02330	74301	00275	48280
76	11508	70225	51111	38351	19444	66499	71945	05422	13442	78675	84081	66938	93654	59894
77	37449	30362	06694	54690	04052	53115	62757	95348	78662	11163	81651	50245	34971	52924
78	46515	70331	85922	38329	57015	15765	97161	17869	45349	61796	66345	81073	49106	79860
79	30986	81223	42416	58353	21532	30502	32305	86482	05174	07901	54339	58861	74818	46942
80	63798	64995	46583	09765	44160	78128	83991	42865	92520	83531	80377	35909	81250	54238
81	82486	84846	99254	67632	43218	50076	21361	64816	51202	88124	41870	52689	51275	83556
82	21885	32906	92431	09060	64297	51674	64126	62570	26123	05155	59194	52799	28225	85762
83	60336	98782	07408	53458	13564	59089	26445	29789	85205	41001	12535	12133	14645	23541
84	43937	46891	24010	25560	86355	33941	25786	54990	71899	15475	95434	98227	21824	19585
85	97656	63175	89303	16275	07100	92063	21942	18611	47348	20203	18534	03862	78095	50136
86	03299	01221	05418	38982	55758	92237	26759	86367	21216	98442	08303	56613	91511	75928
87	79626	06486	03574	17668	07785	76020	79924	25651	83325	88428	85076	72811	22717	50585
88	85636	68335	47539	03129	65651	11977	02510	26113	99447	68645	34327	15152	55230	93448
89	18039	14367	61337	06177	12143	46609	32989	74014	64708	00533	35398	58408	13261	47908
90	08362	15656	60627	36478	65648	16764	53412	09013	07832	41574	17639	82163	60859	75567
91	79556	29068	04142	16268	15387	12856	66227	38358	22478	73373	88732	09443	82558	05250
92	92608	82674	27072	32534	17075	27698	98204	63863	11951	34648	88022	56148	34925	57031
93	23982	25835	40055	67006	12293	02753	14827	22235	35071	99704	37543	11601	35503	85171
94	09915	96306	05908	97901	28395	14186	00821	80703	70426	75647	76310	88717	37890	40129
95	50937	33300	26695	62247	69927	76123	50842	43834	86654	70959	79725	93872	28117	19233
96	42488	78077	69882	61657	34136	79180	97526	43092	04098	73571	80799	76536	71255	64239
97	46764	86273	63003	93017	31204	36692	40202	35275	57306	55543	53203	18098	47625	88684
98	03237	45430	55417	63282	90816	17349	88298	90183	36600	78406	06216	95787	42579	90730
99	86591	81482	52667	61583	14972	90053	89534	76036	49199	43716	97548	04379	46370	28672
100	38534	01715	94964	87288	65680	43772	39560	12918	86537	62738	19636	51132	25739	56947

Table A-2 14,000 Random Units *(Continued)*

Line/Col.	(1)	(2)	(3)	(4)	(5)	(6)	(7)	(8)	(9)	(10)	(11)	(12)	(13)	(14)
101	13284	16834	74151	92027	24670	36665	00770	22878	02179	51602	07270	76517	97275	45960
102	21224	00370	30420	03883	96648	89428	41583	17564	27395	63904	41548	49197	82277	24120
103	99052	47887	81085	64933	66279	80432	65793	83287	34142	13241	30590	97760	35848	91983
104	00199	50993	98603	38452	87890	94624	69721	57484	67501	77638	44331	11257	71131	11059
105	60578	06483	28733	37867	07936	98710	98539	27186	31237	80612	44488	97819	70401	95419
106	91240	18312	17441	01929	18163	69201	31211	54288	39296	37318	65724	90401	79017	62077
107	97458	14229	12063	59611	32249	90466	33216	19358	02591	54263	88449	01912	07436	50813
108	35249	38646	34475	72417	60514	69257	12489	51924	86871	92446	36607	11458	30440	52639
109	38980	46600	11759	11900	46743	27860	77940	39298	97838	95145	32378	68038	89351	37005
110	10750	52745	38749	87365	58959	53731	89295	59062	39404	13198	59960	70408	29812	83126
111	36247	27850	73958	20673	37800	63835	71051	84724	52492	22342	78071	17456	96104	18327
112	70994	66986	99744	72438	01174	42159	11392	20724	54322	36923	70009	23233	65438	59685
113	99638	94702	11463	18148	81386	80431	90628	52506	02016	85151	88598	47821	00265	82525
114	72055	15774	43857	99805	10419	76939	25993	03544	21560	83471	43989	90770	22965	44247
115	24038	65541	85788	55835	38835	59399	13790	35112	01324	39520	76210	22467	83275	32286
116	74976	14631	35908	28221	39470	91548	12854	30166	09073	75887	36782	00268	97121	57676
117	35553	71628	70189	26436	63407	91178	90348	55359	80392	41012	36270	77786	89578	21059
118	35676	12797	51434	82976	42010	26344	92920	92155	58807	54644	58581	95331	78629	73344
119	74815	67523	72985	23183	02446	63594	98924	20633	58842	85961	07648	70164	34994	67662
120	45246	88048	65173	50989	91060	89894	36063	32819	68559	99221	49475	50558	34698	71800
121	76509	47069	86378	41797	11910	49672	88575	97966	32466	10083	54728	81972	58975	30761
122	19689	90332	04315	21358	97248	11188	39062	63312	52496	07349	79178	33692	57352	72862
123	42751	35318	97513	61537	54955	08159	00337	80778	27507	95478	21252	12746	37554	97775
124	11946	22681	45045	13964	57517	59419	58045	44067	58716	58840	45557	96345	33271	53464
125	96518	48688	20996	11090	48396	57177	83867	86464	14342	21545	46717	72364	86954	55580
126	35726	58643	76869	84622	39098	36083	72505	92265	23107	60278	05822	46760	44294	07672
127	39737	42750	48968	70536	84864	64952	38404	94317	65402	13589	01055	79044	19308	83623
128	97025	66492	56177	04049	80312	48028	26408	43591	75528	65341	49044	95495	81256	53214
129	62814	08075	09788	56350	76787	51591	54509	49295	85830	59860	30883	89660	96142	18354
130	25578	22950	15227	83291	41737	79599	96191	71845	86899	70694	24290	01551	80092	82118
131	68763	69576	88991	49662	46704	63362	56625	00481	73323	91427	15264	06969	57048	54149
132	17900	00813	64361	60725	88974	61005	99709	30666	26451	11528	44323	34778	60342	60388
133	71944	60227	63551	71109	05624	43836	58254	26160	32116	63403	35404	57146	10909	07346
134	54684	93691	85132	64399	29182	44324	14491	55226	78793	34107	30374	48429	51376	09559
135	25946	27623	11258	65204	52832	50880	22273	05554	99521	73791	85744	29276	70326	60251
136	01353	39318	44961	44972	91766	90262	56073	06606	51826	18893	83448	31915	97764	75091
137	99083	88191	27662	99113	57174	35571	99884	13951	71057	53961	61448	74909	07322	80960
138	52021	45406	37945	75234	24327	86978	22644	87779	23753	99926	63898	54886	18051	96314
139	78755	47744	43776	83098	03225	14281	83637	55984	13300	52212	58781	14905	46502	04472
140	25282	69106	59180	16257	22810	43609	12224	25643	89884	31149	85423	32581	34374	70873
141	11959	94202	02743	86847	79725	51811	12998	76844	05320	54236	53891	70226	38632	84776
142	11644	13792	98190	01424	30078	28197	55583	05197	47714	68440	22016	79204	06862	94451
143	06307	97912	68110	59812	95448	43244	31262	88880	13040	16458	43813	89416	42482	33939
144	76285	75714	89585	99296	52640	46518	55486	90754	88932	19937	57119	23251	55619	23679
145	55322	07589	39600	60866	63007	20007	66819	84164	61131	81429	60676	42807	78286	29015
146	78017	90928	90220	92503	83375	26986	74399	30885	88567	29169	72816	53357	15428	86932
147	44768	43342	20696	26331	43140	69744	82928	24988	94237	46138	77426	39039	55596	12655
148	25100	19336	14605	86603	51680	97678	24261	02464	86563	74812	60069	71674	15478	47642
149	83612	46623	62876	85197	07824	91392	58317	37726	84628	42221	10268	20692	15699	29167
150	41347	81666	82961	60413	71020	83658	02415	33322	66036	98712	46795	16308	28413	05417

Table A-2 14,000 Random Units *(Continued)*

Line/Col.	(1)	(2)	(3)	(4)	(5)	(6)	(7)	(8)	(9)	(10)	(11)	(12)	(13)	(14)
151	38128	51178	75096	13609	16110	73533	42564	59870	29399	67834	91055	89917	51096	89011
152	60950	00455	73254	96067	50717	13878	03216	78274	65863	37011	91283	33914	91303	49326
153	90524	17320	29832	96118	75792	25326	22940	24904	80523	38928	91374	55597	97567	38914
154	49897	18278	67160	39408	97056	43517	84426	59650	20247	19293	02019	14790	02852	05819
155	18494	99209	81060	19488	65596	59787	47939	91225	98768	43688	00438	05548	09443	82897
156	65373	72984	30171	37741	70203	94094	87261	30056	58124	70133	18936	02138	59372	09075
157	40653	12843	04213	70925	95360	55774	76439	61768	52817	81151	52188	31940	54273	49032
158	51638	22238	56344	44587	83231	50317	74541	07719	25472	41602	77318	15145	57515	07633
159	69742	99303	62578	83575	30337	07488	51941	84316	42067	49692	28616	29101	03013	73449
160	58012	74072	67488	74580	47992	69482	58624	17106	47538	13452	22620	24260	40155	74716
161	18348	19855	42887	08279	43206	47077	42637	45606	00011	20662	14642	49984	94509	56380
162	59614	09193	58064	29086	44385	45740	70752	05663	49081	26960	57454	99264	24142	74648
163	75688	28630	39210	52897	62748	72658	98059	67202	72789	01869	13496	14663	87645	89713
164	13941	77802	69101	70061	35460	34576	15412	81304	58757	35498	94830	75521	00603	97701
165	96656	86420	96475	86458	54463	96419	55417	41375	76886	19008	66877	35934	59801	00497
166	03363	82042	15942	14549	38324	87094	19069	67590	11087	68570	22591	65232	85915	91499
167	70366	08390	69155	25496	13240	57407	91407	49160	07379	34444	94567	66035	38918	65708
168	47870	36605	12927	16043	53257	93796	52721	73120	48025	76074	95605	67422	41646	14557
169	79504	77606	22761	30518	28373	73898	30550	76684	77366	32276	04690	61667	64798	66276
170	46967	74841	50923	15339	37755	98995	40162	89561	69199	42257	11647	47603	48779	97907
171	14558	50769	35444	59030	87516	48193	02945	00922	48189	04724	21263	20892	92955	90251
172	12440	25057	01132	38611	28135	68089	10954	10097	54243	06460	50856	65435	79377	53890
173	32293	29938	68653	10497	98919	46587	77701	99119	93165	67788	17638	23097	21468	36992
174	10640	21875	72462	77981	56550	55999	87310	69643	45124	00349	25748	00844	96831	30651
175	47615	23169	39571	56972	20628	21788	51736	33133	72696	32605	41569	76148	91544	21121
176	16948	11128	71624	72754	49084	96303	27830	45817	67867	18062	87453	17226	72904	71474
177	21258	61092	66634	70335	92448	17354	83432	49608	66520	06442	59664	20420	39201	69549
178	15072	48853	15178	30730	47481	48490	41436	25015	49932	20474	53821	51015	79841	32405
179	99154	57412	09858	65671	70655	71479	63520	31357	56968	06729	34465	70685	04184	25250
180	08759	61089	23706	32994	35426	36666	63988	98844	37533	08269	27021	45886	22835	78451
181	67323	57839	61114	62192	47547	58023	64630	34886	98777	75442	95592	06141	45096	73117
182	09255	13986	84834	20764	72206	89393	34548	93438	88730	61805	78955	18952	46436	58740
183	36304	74712	00374	10107	85061	69228	81969	92216	03568	39630	81869	52824	50937	27954
184	15884	67429	86612	47367	10242	44880	12060	44309	46629	55105	66793	93173	00480	13311
185	18745	32031	35303	08134	33925	03044	59929	95418	04917	57596	24878	61733	92834	64454
186	72934	40086	88292	65728	38300	42323	64068	98373	48971	09049	59943	36538	05976	82118
187	17626	02944	20910	57662	80181	38579	24580	90529	52303	50436	29401	57824	86039	81062
188	27117	61399	50967	41399	81636	16663	15634	79717	94696	59240	25543	97989	63306	90946
189	93995	18678	90012	63645	85701	85269	62263	68331	00389	72571	15210	20769	44686	96176
190	67392	89421	09623	80725	62620	84162	87368	29560	00519	84545	08004	24526	41252	14521
191	04910	12261	37566	80016	21245	69377	50420	85658	55263	68667	78770	04533	14513	18099
192	81453	20283	79929	59839	23875	13245	46808	74124	74703	35769	95588	21014	37078	39170
193	19480	75790	48539	23703	15537	48885	02861	86587	74539	65227	90799	58789	96257	02708
194	21456	13162	74608	81011	55512	07481	93551	72189	76261	91206	89941	15132	37738	59284
195	89406	20912	46189	76376	25538	87212	20748	12831	57166	35026	16817	79121	18929	40628
196	09866	07414	55977	16419	01101	69343	13305	94302	80703	57910	36933	57771	42546	03003
197	86541	24681	23421	13521	28000	94917	07423	57523	97234	63951	42876	46829	09781	58160
198	10414	96941	06205	72222	57167	83902	07460	69507	10600	08858	07685	44472	64220	27040
199	49942	06683	41479	58982	56288	42853	92196	20632	62045	78812	35895	51851	83534	10689
200	23995	68882	42291	23374	24299	27024	67460	94783	40937	16961	26053	78749	46704	21983

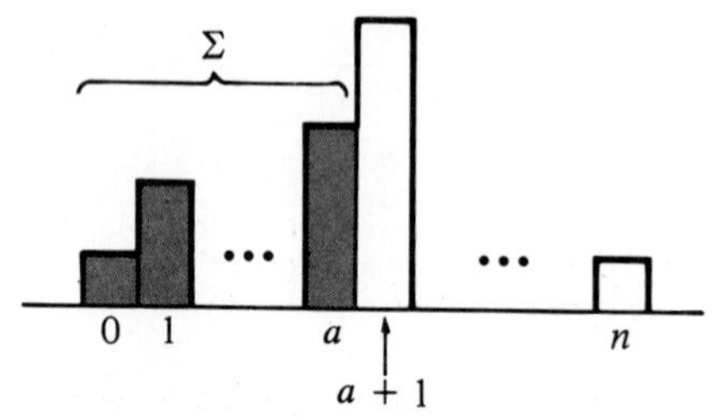

Table A-3 Binomial Distribution

n	a	$\pi = .05$	.10	.15	.20	.25	.30	.35	.40	.45
1	0	.9500	.9000	.8500	.8000	.7500	.7000	.6500	.6000	.5500
	1	1.0000	1.0000	1.0000	1.0000	1.0000	1.0000	1.0000	1.0000	1.0000
2	0	.9025	.8100	.7225	.6400	.5625	.4900	.4225	.3600	.3025
	1	.9975	.9900	.9775	.9600	.9375	.9100	.8775	.8400	.7975
	2	1.0000	1.0000	1.0000	1.0000	1.0000	1.0000	1.0000	1.0000	1.0000
3	0	.8574	.7290	.6141	.5120	.4219	.3430	.2746	.2160	.1664
	1	.9928	.9720	.9392	.8960	.8438	.7840	.7182	.6480	.5748
	2	.9999	.9990	.9966	.9920	.9844	.9730	.9571	.9360	.9089
	3	1.0000	1.0000	1.0000	1.0000	1.0000	1.0000	1.0000	1.0000	1.0000
4	0	.8145	.6561	.5220	.4096	.3164	.2401	.1785	.1296	.0915
	1	.9860	.9477	.8905	.8192	.7383	.6517	.5630	.4752	.3910
	2	.9995	.9963	.9880	.9728	.9492	.9163	.8735	.8208	.7585
	3	1.0000	.9999	.9995	.9984	.9961	.9919	.9850	.9743	.9590
	4	1.0000	1.0000	1.0000	1.0000	1.0000	1.0000	1.0000	1.0000	1.0000
5	0	.7738	.5905	.4437	.3277	.2373	.1681	.1160	.0778	.0503
	1	.9774	.9185	.8352	.7373	.6328	.5282	.4284	.3370	.2562
	2	.9988	.9914	.9734	.9421	.8965	.8369	.7648	.6826	.5931
	3	1.0000	.9995	.9978	.9933	.9844	.9692	.9460	.9130	.8688
	4	1.0000	1.0000	.9999	.9997	.9990	.9976	.9947	.9898	.9815
	5	1.0000	1.0000	1.0000	1.0000	1.0000	1.0000	1.0000	1.0000	1.0000
6	0	.7351	.5314	.3771	.2621	.1780	.1176	.0754	.0467	.0277
	1	.9672	.8857	.7765	.6554	.5339	.4202	.3191	.2333	.1636
	2	.9978	.9842	.9527	.9011	.8306	.7443	.6471	.5443	.4415
	3	.9999	.9987	.9941	.9830	.9624	.9295	.8826	.9208	.7447
	4	1.0000	.9999	.9996	.9984	.9954	.9891	.9777	.9590	.9308
	5	1.0000	1.0000	1.0000	.9999	.9998	.9993	.9982	.9959	.9917
	6	1.0000	1.0000	1.0000	1.0000	1.0000	1.0000	1.0000	1.0000	1.0000
7	0	.6983	.4783	.3206	.2097	.1335	.0824	.0490	.0280	.0152
	1	.9556	.8503	.7166	.5767	.4449	.3294	.2338	.1586	.1024
	2	.9962	.9743	.9262	.8520	.7564	.6471	.5323	.4199	.3164
	3	.9998	.9973	.9879	.9667	.9294	.8740	.8002	.7102	.6083
	4	1.0000	.9998	.9988	.9953	.9871	.9812	.9444	.9037	.8471
	5	1.0000	1.0000	.9999	.9996	.9987	.9962	.9910	.9812	.9643
	6	1.0000	1.0000	1.0000	1.0000	.9999	.9998	.9994	.9984	.9963
	7	1.0000	1.0000	1.0000	1.0000	1.0000	1.0000	1.0000	1.0000	1.0000

Table A-3 Binomial Distribution *(Continued)*

n	a	$\pi = .50$	.55	.60	.65	.70	.75	.80	.85	.90	.95
1	0	.5000	.4500	.4000	.3500	.3000	.2500	.2000	.1500	.1000	.0500
	1	1.0000	1.0000	1.0000	1.0000	1.0000	1.0000	1.0000	1.0000	1.0000	1.0000
2	0	.2500	.2025	.1600	.1225	.0900	.0625	.0400	.0225	.0100	.0025
	1	.7500	.6975	.6400	.5775	.5100	.4375	.3600	.2775	.1900	.0975
	2	1.0000	1.0000	1.0000	1.0000	1.0000	1.0000	1.0000	1.0000	1.0000	1.0000
3	0	.1250	.0911	.0640	.0429	.0270	.0156	.0080	.0034	.0010	.0001
	1	.5000	.4252	.3520	.2818	.2160	.1562	.1040	.0608	.0280	.0072
	2	.8750	.8336	.7840	.7254	.6570	.5781	.4880	.3859	.2710	.1426
	3	1.0000	1.0000	1.0000	1.0000	1.0000	1.0000	1.0000	1.0000	1.0000	1.0000
4	0	.0625	.0410	.0256	.0150	.0081	.0039	.0016	.0005	.0001	.0000
	1	.3125	.2415	.1792	.1265	.0837	.0508	.0272	.0120	.0037	.0005
	2	.6875	.6090	.5248	.4370	.3483	.2617	.1808	.1095	.0523	.0140
	3	.9375	.9085	.8704	.8215	.7599	.6836	.5904	.4780	.3439	.1855
	4	1.0000	1.0000	1.0000	1.0000	1.0000	1.0000	1.0000	1.0000	1.0000	1.0000
5	0	.0312	.0185	.0102	.0053	.0024	.0010	.0003	.0001	.0000	.0000
	1	.1875	.1312	.0870	.0540	.0308	.0156	.0067	.0022	.0005	.0000
	2	.5000	.4069	.3174	.2352	.1631	.1035	.0579	.0266	.0086	.0012
	3	.8125	.7438	.6630	.5716	.4718	.3672	.2627	.1648	.0815	.0226
	4	.9688	.9497	.9222	.8840	.8319	.7627	.6723	.5563	.4095	.2262
	5	1.0000	1.0000	1.0000	1.0000	1.0000	1.0000	1.0000	1.0000	1.0000	1.0000
6	0	.0156	.0083	.0041	.0018	.0007	.0002	.0001	.0000	.0000	.0000
	1	.1094	.0692	.0410	.0223	.0109	.0046	.0016	.0004	.0001	.0000
	2	.3438	.2553	.1792	.1174	.0705	.0376	.0170	.0059	.0013	.0001
	3	.6562	.5585	.4557	.3529	.2557	.1694	.0989	.0473	.0158	.0022
	4	.8906	.8364	.7667	.6809	.5798	.4661	.3446	.2235	.1143	.0328
	5	.9844	.9723	.9533	.9246	.8824	.8220	.7379	.6229	.4686	.2649
	6	1.0000	1.0000	1.0000	1.0000	1.0000	1.0000	1.0000	1.0000	1.0000	1.0000
7	0	.0078	.0037	.0016	.0006	.0002	.0001	.0000	.0000	.0000	.0000
	1	.0625	.0357	.0188	.0090	.0038	.0013	.0004	.0001	.0000	.0000
	2	.2266	.1529	.0963	.0556	.0288	.0129	.0047	.0012	.0002	.0000
	3	.5000	.3917	.2898	.1998	.1260	.0706	.0333	.0121	.0027	.0002
	4	.7734	.6836	.5801	.4677	.3529	.2436	.1480	.0738	.0257	.0038
	5	.9375	.8976	.8414	.7662	.6706	.5551	.4233	.2834	.1497	.0444
	6	.9922	.9848	.9720	.9510	.9176	.8665	.7903	.6794	.5217	.3017
	7	1.0000	1.0000	1.0000	1.0000	1.0000	1.0000	1.0000	1.0000	1.0000	1.0000

Table A-3 Binomial Distribution *(Continued)*

n	*a*	*π* = .05	.10	.15	.20	.25	.30	.35	.40	.45
8	0	.6634	.4305	.2725	.1678	.1001	.0576	.0319	.0168	.0084
	1	.9428	.8131	.6572	.5033	.3671	.2553	.1691	.1064	.0632
	2	.9942	.9619	.8948	.7969	.6785	.5518	.4278	.3154	.2201
	3	.9996	.9950	.9786	.9437	.8862	.8059	.7064	.5941	.4770
	4	1.0000	.9996	.9971	.9896	.9727	.9420	.8939	.8263	.7396
	5	1.0000	1.0000	.9998	.9988	.9958	.9887	.9747	.9502	.9115
	6	1.0000	1.0000	1.0000	.9999	.9996	.9987	.9964	.9915	.9819
	7	1.0000	1.0000	1.0000	1.0000	1.0000	.9999	.9988	.9993	.9983
	8	1.0000	1.0000	1.0000	1.0000	1.0000	1.0000	1.0000	1.0000	1.0000
9	0	.6302	.3874	.2316	.1342	.0751	.0404	.0207	.0101	.0046
	1	.9288	.7748	.5995	.4362	.3003	.1960	.1211	.0705	.0385
	2	.9916	.9470	.8591	.7382	.6007	.4628	.3373	.2318	.1495
	3	.9994	.9917	.9661	.9144	.8343	.7297	.6089	.4826	.3614
	4	1.0000	.9991	.9944	.9804	.9511	.9012	.8283	.7334	.6214
	5	1.0000	.9999	.9994	.9969	.9900	.9747	.9464	.9006	.8342
	6	1.0000	1.0000	1.0000	.9997	.9987	.9957	.9888	.9750	.9502
	7	1.0000	1.0000	1.0000	1.0000	.9999	.9996	.9986	.9962	.9909
	8	1.0000	1.0000	1.0000	1.0000	1.0000	1.0000	.9999	.9997	.9992
	9	1.0000	1.0000	1.0000	1.0000	1.0000	1.0000	1.0000	1.0000	1.0000
10	0	.5987	.3487	.1969	.1074	.0563	.0282	.0135	.0060	.0025
	1	.9139	.7361	.5443	.3758	.2440	.1493	.0860	.0464	.0233
	2	.9885	.9298	.8202	.6778	.5256	.3828	.2616	.1673	.0996
	3	.9990	.9872	.9500	.8791	.7759	.6496	.5138	.3823	.2660
	4	.9999	.9984	.9901	.9672	.9219	.8497	.7515	.6331	.5044
	5	1.0000	.9999	.9986	.9936	.9803	.9527	.9051	.8338	.7384
	6	1.0000	1.0000	.9999	.9991	.9965	.9894	.9740	.9452	.8980
	7	1.0000	1.0000	1.0000	.9999	.9996	.9984	.9952	.9877	.9726
	8	1.0000	1.0000	1.0000	1.0000	1.0000	.9999	.9995	.9983	.9955
	9	1.0000	1.0000	1.0000	1.0000	1.0000	1.0000	1.0000	.9999	.9997
	10	1.0000	1.0000	1.0000	1.0000	1.0000	1.0000	1.0000	1.0000	1.0000
11	0	.5688	.3138	.1673	.0859	.0422	.0198	.0088	.0036	.0014
	1	.8981	.6974	.4922	.3221	.1971	.1130	.0606	.0302	.0139
	2	.9848	.9104	.7788	.6174	.4552	.3127	.2001	.1189	.0652
	3	.9984	.9815	.9306	.8389	.7133	.5696	.4256	.2963	.1911
	4	.9999	.9972	.9841	.9496	.8854	.7897	.6683	.5328	.3971
	5	1.0000	.9997	.9973	.9883	.9657	.9218	.8513	.7535	.6331
	6	1.0000	1.0000	.9997	.9980	.9924	.9784	.9499	.9006	.8262
	7	1.0000	1.0000	1.0000	.9998	.9988	.9957	.9878	.9707	.9390
	8	1.0000	1.0000	1.0000	1.0000	.9999	.9994	.9980	.9941	.9852
	9	1.0000	1.0000	1.0000	1.0000	1.0000	1.0000	.9998	.9993	.9978
	10	1.0000	1 0000	1.0000	1.0000	1.0000	1.0000	1.0000	1.0000	.9998
	11	1.0000	1.0000	1.0000	1.0000	1.0000	1.0000	1.0000	1.0000	1.0000

Table A-3 Binomial Distribution *(Continued)*

n	a	$\pi = .50$	.55	.60	.65	.70	.75	.80	.85	.90	.95
8	0	.0039	.0017	.0007	.0002	.0001	.0000	.0000	.0000	.0000	.0000
	1	.0352	.0181	.0085	.0036	.0013	.0004	.0001	.0000	.0000	.0000
	2	.1445	.0885	.0498	.0253	.0113	.0042	.0012	.0002	.0000	.0000
	3	.3633	.2604	.1737	.1061	.0580	.0273	.0104	.0029	.0004	.0000
	4	.6367	.5230	.4059	.2936	.1941	.1138	.0563	.0214	.0050	.0004
	5	.8555	.7799	.6846	.5722	.4482	.3215	.2031	.1052	.0381	.0058
	6	.9648	.9368	.8936	.8309	.7447	.6329	.4967	.3428	.1869	.0572
	7	.9961	.9916	.9832	.9681	.9424	.8999	.8322	.7275	.5695	.3366
	8	1.0000	1.0000	1.0000	1.0000	1.0000	1.0000	1.0000	1.0000	1.0000	1.0000
9	0	.0020	.0008	.0003	.0001	.0000	.0000	.0000	.0000	.0000	.0000
	1	.0195	.0091	.0038	.0014	.0004	.0001	.0000	.0000	.0000	.0000
	2	.0898	.0498	.0250	.0112	.0043	.0013	.0003	.0000	.0000	.0000
	3	.2539	.1658	.0994	.0536	.0253	.0100	.0031	.0006	.0001	.0000
	4	.5000	.3786	.2666	.1717	.0988	.0489	.0196	.0056	.0009	.0000
	5	.7461	.6386	.5174	.3911	.2703	.1657	.0856	.0339	.0083	.0006
	6	.9102	.8505	.7682	.6627	.5372	.3993	.2618	.1409	.0530	.0084
	7	.9805	.9615	.9295	.8789	.8040	.6997	.5638	.4005	.2252	.0712
	8	.9980	.9954	.9899	.9793	.9596	.9249	.8658	.7684	.6126	.3698
	9	1.0000	1.0000	1.0000	1.0000	1.0000	1.0000	1.0000	1.0000	1.0000	1.0000
10	0	.0010	.0003	.0001	.0000	.0000	.0000	.0000	.0000	.0000	.0000
	1	.0107	.0045	.0017	.0005	.0001	.0000	.0000	.0000	.0000	.0000
	2	.0547	.0274	.0123	.0048	.0016	.0004	.0001	.0000	.0000	.0000
	3	.1719	.1020	.0548	.0260	.0106	.0035	.0009	.0001	.0000	.0000
	4	.3770	.2616	.1662	.0949	.0473	.0197	.0064	.0014	.0001	.0000
	5	.6230	.4956	.3669	.2485	.1503	.0781	.0328	.0099	.0016	.0001
	6	.8281	.7340	.6177	.4862	.3504	.2241	.1209	.0500	.0128	.0010
	7	.9453	.9004	.8327	.7384	.6172	.4744	.3222	.1798	.0702	.0115
	8	.9893	.9767	.9536	.9140	.8507	.7560	.6242	.4557	.2639	.0861
	9	.9990	.9975	.9940	.9865	.9718	.9437	.8926	.8031	.6513	.4013
	10	1.0000	1.0000	1.0000	1.0000	1.0000	1.0000	1.0000	1.0000	1.0000	1.0000
11	0	.0005	.0002	.0000	.0000	.0000	.0000	.0000	.0000	.0000	.0000
	1	.0059	.0022	.0007	.0002	.0000	.0000	.0000	.0000	.0000	.0000
	2	.0327	.0148	.0059	.0020	.0006	.0001	.0000	.0000	.0000	.0000
	3	.1133	.0610	.0293	.0122	.0043	.0012	.0002	.0000	.0000	.0000
	4	.2744	.1738	.0994	.0501	.0216	.0076	.0020	.0003	.0000	.0000
	5	.5000	.3669	.2465	.1487	.0782	.0343	.0117	.0027	.0003	.0000
	6	.7256	.6029	.4672	.3317	.2103	.1146	.0504	.0159	.0028	.0001
	7	.8867	.8089	.7037	.5744	.4304	.2867	.1611	.0694	.0185	.0016
	8	.9673	.9348	.8811	.7999	.6873	.5448	.3826	.2212	.0896	.0152
	9	.9941	.9861	.9698	.9394	.8870	.8029	.6779	.5078	.3026	.1019
	10	.9995	.9986	.9964	.9912	.9802	.9578	.9141	.8327	.6862	.4312
	11	1.0000	1.0000	1.0000	1.0000	1.0000	1.0000	1.0000	1.0000	1.0000	1.0000

Table A-3 Binomial Distribution *(Continued)*

n	a	π = .05	.10	.15	.20	.25	.30	.35	.40	.45
12	0	.5404	.2824	.1422	.0687	.0317	.0138	.0057	.0022	.0008
	1	.8816	.6590	.4435	.2749	.1584	.0850	.0424	.0424	.0083
	2	.9804	.8891	.7358	.5583	.3907	.2528	.1513	.0834	.0421
	3	.9978	.9744	.9078	.7946	.6488	.4925	.3467	.2253	.1345
	4	.9998	.9957	.9761	.9274	.8424	.7237	.5833	.4382	.3044
	5	1.0000	.9995	.9954	.9806	.9456	.8822	.7873	.6652	.5269
	6	1.0000	.9999	.9993	.9961	.9857	.9614	.9154	.8418	.7393
	7	1.0000	1.0000	.9999	.9994	.9972	.9905	.9745	.9427	.8883
	8	1.0000	1.0000	1.0000	.9999	.9996	.9983	.9944	.9847	.9644
	9	1.0000	1.0000	1.0000	1.0000	1.0000	.9998	.9992	.9972	.9921
	10	1.0000	1.0000	1.0000	1.0000	1.0000	1.0000	.9999	.9997	.9989
	11	1.0000	1.0000	1.0000	1.0000	1.0000	1.0000	1.0000	1.0000	.9999
	12	1.0000	1.0000	1.0000	1.0000	1.0000	1.0000	1.0000	1.0000	1.0000
13	0	.5133	.2542	.1209	.0550	.0238	.0097	.0037	.0013	.0004
	1	.8646	.6213	.3983	.2336	.1267	.0637	.0296	.0126	.0049
	2	.9755	.8661	.7296	.5017	.3326	.2025	.1132	.0579	.0269
	3	.9969	.9658	.9033	.7473	.5843	.4206	.2783	.1686	.0929
	4	.9997	.9935	.9740	.9009	.7940	.6543	.5005	.3530	.2279
	5	1.0000	.9991	.9947	.9700	.9198	.8346	.7159	.5744	.4268
	6	1.0000	.9999	.9987	.9930	.9757	.9376	.8705	.7712	.6437
	7	1.0000	1.0000	.9998	.9988	.9944	.9818	.9538	.9023	.8212
	8	1.0000	1.0000	1.0000	.9998	.9990	.9960	.9874	.9679	.9302
	9	1.0000	1.0000	1.0000	1.0000	.9999	.9993	.9975	.9922	.9797
	10	1.0000	1.0000	1.0000	1.0000	1.0000	.9999	.9997	.9987	.9959
	11	1.0000	1.0000	1.0000	1.0000	1.0000	1.0000	1.0000	.9999	.9995
	12	1.0000	1.0000	1.0000	1.0000	1.0000	1.0000	1.0000	1.0000	1.0000
	13	1.0000	1.0000	1.0000	1.0000	1.0000	1.0000	1.0000	1.0000	1.0000
14	0	.4877	.2288	.1028	.0440	.0178	.0068	.0024	.0008	.0002
	1	.8470	.5846	.3567	.1979	.1010	.0475	.0205	.0081	.0029
	2	.9699	.8416	.6479	.4481	.2811	.1608	.0839	.0398	.0170
	3	.9958	.9559	.8535	.6982	.5213	.3552	.2205	.1243	.0632
	4	.9996	.9908	.9533	.8702	.7415	.5842	.4227	.2793	.1672
	5	1.0000	.9985	.9885	.9561	.8883	.7805	.6405	.4859	.3373
	6	1.0000	.9998	.9978	.9884	.9617	.9067	.8164	.6925	.5461
	7	1.0000	1.0000	.9997	.9976	.9897	.9685	.9247	.8499	.7414
	8	1.0000	1.0000	1.0000	.9996	.9978	.9917	.9757	.9417	.8811
	9	1.0000	1.0000	1.0000	1.0000	.9997	.9983	.9940	.9825	.9574
	10	1.0000	1.0000	1.0000	1.0000	1.0000	.9998	.9989	.9961	.9886
	11	1.0000	1.0000	1.0000	1.0000	1.0000	1.0000	.9999	.9994	.9978
	12	1.0000	1.0000	1.0000	1.0000	1.0000	1.0000	1.0000	.9999	.9997
	13	1.0000	1.0000	1.0000	1.0000	1.0000	1.0000	1.0000	1.0000	1.0000
	14	1.0000	1.0000	1.0000	1.0000	1.0000	1.0000	1.0000	1.0000	1.0000

Table A-3 Binomial Distribution *(Continued)*

n	a	$\pi = .50$	.55	.60	.65	.70	.75	.80	.85	.90	.95
12	0	.0002	.0001	.0000	.0000	.0000	.0000	.0000	.0000	.0000	.0000
	1	.0032	.0011	.0003	.0001	.0000	.0000	.0000	.0000	.0000	.0000
	2	.0193	.0079	.0028	.0008	.0002	.0000	.0000	.0000	.0000	.0000
	3	.0730	.0356	.0153	.0056	.0017	.0004	.0001	.0000	.0000	.0000
	4	.1938	.1117	.0573	.0255	.0095	.0028	.0006	.0001	.0000	.0000
	5	.3872	.2607	.1582	.0846	.0386	.0143	.0039	.0007	.0001	.0000
	6	.6128	.4731	.3348	.2127	.1178	.0544	.0194	.0046	.0005	.0000
	7	.8062	.6956	.5618	.4167	.2763	.1576	.0726	.0239	.0043	.0002
	8	.9270	.8655	.7747	.6533	.5075	.3512	.2054	.0922	.0256	.0022
	9	.9807	.9579	.9166	.8487	.7472	.6093	.4417	.2642	.1109	.0196
	10	.9968	.9917	.9804	.9576	.9150	.8416	.7251	.5565	.3410	.1184
	11	.9998	.9992	.9978	.9943	.9862	.9683	.9313	.8578	.7176	.4596
	12	1.0000	1.0000	1.0000	1.0000	1.0000	1.0000	1.0000	1.0000	1.0000	1.0000
13	0	.0001	.0000	.0000	.0000	.0000	.0000	.0000	.0000	.0000	.0000
	1	.0017	.0005	.0001	.0000	.0000	.0000	.0000	.0000	.0000	.0000
	2	.0112	.0041	.0013	.0003	.0001	.0000	.0000	.0000	.0000	.0000
	3	.0461	.0203	.0078	.0025	.0007	.0001	.0000	.0000	.0000	.0000
	4	.1334	.0698	.0321	.0126	.0040	.0010	.0002	.0000	.0000	.0000
	5	.2905	.1788	.0977	.0462	.0182	.0056	.0012	.0002	.0000	.0000
	6	.5000	.3563	.2288	.1295	.0624	.0243	.0070	.0013	.0001	.0000
	7	.7095	.5732	.4256	.2841	.1654	.0802	.0300	.0053	.0009	.0000
	8	.8666	.7721	.6470	.4995	.3457	.2060	.0991	.0260	.0065	.0003
	9	.9539	.9071	.8314	.7217	.5794	.4157	.2527	.0967	.0342	.0031
	10	.9888	.9731	.9421	.8868	.7975	.6674	.4983	.2704	.1339	.0245
	11	.9983	.9951	.9874	.9704	.9363	.8733	.7664	.6017	.3787	.1354
	12	.9999	.9996	.9987	.9963	.9903	.9762	.9450	.8791	.7458	.4867
	13	1.0000	1.0000	1.0000	1.0000	1.0000	1.0000	1.0000	1.0000	1.0000	1.0000
14	0	.0000	.0000	.0000	.0000	.0000	.0000	.0000	.0000	.0000	.0000
	1	.0009	.0003	.0001	.0000	.0000	.0000	.0000	.0000	.0000	.0000
	2	.0065	.0022	.0006	.0001	.0000	.0000	.0000	.0000	.0000	.0000
	3	.0287	.0114	.0039	.0011	.0002	.0000	.0000	.0000	.0000	.0000
	4	.0898	.0462	.0175	.0060	.0017	.0003	.0000	.0000	.0000	.0000
	5	.2120	.1189	.0583	.0243	.0083	.0022	.0004	.0000	.0000	.0000
	6	.3953	.2586	.1501	.0753	.0315	.0103	.0024	.0003	.0000	.0000
	7	.6047	.4539	.3075	.1836	.0933	.0383	.0116	.0022	.0002	.0000
	8	.7880	.6627	.5141	.3595	.2195	.1117	.0439	.0115	.0015	.0000
	9	.9102	.8328	.7207	.5773	.4158	.2585	.1298	.0467	.0092	.0004
	10	.9713	.9368	.8757	.7795	.6448	.4787	.3018	.1465	.0441	.0042
	11	.9935	.9830	.9602	.9161	.8392	.7189	.5519	.3521	.1584	.0301
	12	.9991	.9971	.9919	.9795	.9525	.8990	.8021	.6433	.4154	.1530
	13	.9999	.9998	.9992	.9976	.9932	.9822	.9560	.8972	.7712	.5123
	14	1.0000	1.0000	1.0000	1.0000	1.0000	1.0000	1.0000	1.0000	1.0000	1.0000

Table A-3 Binomial Distribution *(Continued)*

n	a	π = .05	.10	.15	.20	.25	.30	.35	.40	.45
15	0	.4633	.2059	.0874	.0352	.0134	.0047	.0016	.0005	.0001
	1	.8290	.5490	.3186	.1671	.0802	.0353	.0142	.0052	.0017
	2	.9638	.8159	.6042	.3980	.2361	.1268	.0617	.0271	.0107
	3	.9945	.9444	.8227	.6482	.4613	.2969	.1727	.0905	.0424
	4	.9994	.9873	.9383	.8358	.6865	.5155	.3519	.2173	.1204
	5	.9999	.9978	.9832	.9389	.8516	.7216	.5643	.4032	.2608
	6	1.0000	.9997	.9964	.9819	.9434	.8689	.7548	.6098	.4522
	7	1.0000	1.0000	.9994	.9958	.9827	.9500	.8868	.7869	.6535
	8	1.0000	1.0000	.9999	.9992	.9958	.9848	.9578	.9050	.8182
	9	1.0000	1.0000	1.0000	.9999	.9992	.9963	.9876	.9662	.9231
	10	1.0000	1.0000	1.0000	1.0000	.9999	.9993	.9972	.9907	.9745
	11	1.0000	1.0000	1.0000	1.0000	1.0000	.9999	.9995	.9981	.9937
	12	1.0000	1.0000	1.0000	1.0000	1.0000	1.0000	.9999	.9997	.9989
	13	1.0000	1.0000	1.0000	1.0000	1.0000	1.0000	1.0000	1.0000	.9999
	14	1.0000	1.0000	1.0000	1.0000	1.0000	1.0000	1.0000	1.0000	1.0000
	15	1.0000	1.0000	1.0000	1.0000	1.0000	1.0000	1.0000	1.0000	1.0000
16	0	.4401	.1853	.0743	.0281	.0100	.0033	.0010	.0003	.0001
	1	.8108	.5147	.2839	.1407	.0635	.0261	.0098	.0033	.0010
	2	.9571	.7892	.5614	.3518	.1971	.0994	.0451	.0183	.0066
	3	.9930	.9316	.7899	.5981	.4050	.2459	.1339	.0651	.0281
	4	.9991	.9830	.9209	.7982	.6302	.4499	.2892	.1666	.0853
	5	.9999	.9967	.9765	.9183	.8103	.6598	.4900	.3288	.1976
	6	1.0000	.9995	.9944	.9733	.9204	.8247	.6881	.5272	.3660
	7	1.0000	.9999	.9989	.9930	.9729	.9256	.8406	.7161	.5629
	8	1.0000	1.0000	.9998	.9985	.9925	.9743	.9329	.8577	.7441
	9	1.0000	1.0000	1.0000	.9998	.9984	.9929	.9771	.9417	.8759
	10	1.0000	1.0000	1.0000	1.0000	.9997	.9984	.9938	.9809	.9514
	11	1.0000	1.0000	1.0000	1.0000	1.0000	.9997	.9987	.9951	.9851
	12	1.0000	1.0000	1.0000	1.0000	1.0000	1.0000	.9998	.9991	.9965
	13	1.0000	1.0000	1.0000	1.0000	1.0000	1.0000	1.0000	.9999	.9994
	14	1.0000	1.0000	1.0000	1.0000	1.0000	1.0000	1.0000	1.0000	.9999
	15	1.0000	1.0000	1.0000	1.0000	1.0000	1.0000	1.0000	1.0000	1.0000
	16	1.0000	1.0000	1.0000	1.0000	1.0000	1.0000	1.0000	1.0000	1.0000

Table A-3 Binomial Distribution *(Continued)*

n	a	$\pi = .50$	.55	.60	.65	.70	.75	.80	.85	.90	.95
15	0	.0000	.0000	.0000	.0000	.0000	.0000	.0000	.0000	.0000	.0000
	1	.0005	.0001	.0000	.0000	.0000	.0000	.0000	.0000	.0000	.0000
	2	.0037	.0011	.0003	.0001	.0000	.0000	.0000	.0000	.0000	.0000
	3	.0176	.0063	.0019	.0005	.0001	.0000	.0000	.0000	.0000	.0000
	4	.0592	.0255	.0093	.0028	.0007	.0001	.0000	.0000	.0000	.0000
	5	.1509	.0769	.0338	.0124	.0037	.0008	.0001	.0000	.0000	.0000
	6	.3036	.1818	.0950	.0422	.0152	.0042	.0008	.0001	.0000	.0000
	7	.5000	.3465	.2131	.1132	.0500	.0173	.0042	.0006	.0000	.0000
	8	.6964	.5478	.3902	.2452	.1311	.0566	.0181	.0036	.0003	.0000
	9	.8491	.7392	.5968	.4357	.2784	.1484	.0611	.0168	.0022	.0001
	10	.9408	.8796	.7827	.6481	.4845	.3135	.1642	.0617	.0127	.0006
	11	.9824	.9576	.9095	.8273	.7031	.5387	.3518	.1773	.0556	.0055
	12	.9963	.9893	.9729	.9383	.8732	.7639	.6020	.3958	.1841	.0362
	13	.9995	.9983	.9948	.9858	.9647	.9198	.8329	.6814	.4510	.1710
	14	1.0000	.9999	.9995	.9984	.9953	.9866	.9648	.9126	.7941	.5367
	15	1.0000	1.0000	1.0000	1.0000	1.0000	1.0000	1.0000	1.0000	1.0000	1.0000
16	0	.0000	.0000	.0000	.0000	.0000	.0000	.0000	.0000	.0000	.0000
	1	.0003	.0001	.0000	.0000	.0000	.0000	.0000	.0000	.0000	.0000
	2	.0021	.0006	.0001	.0000	.0000	.0000	.0000	.0000	.0000	.0000
	3	.0106	.0035	.0009	.0002	.0000	.0000	.0000	.0000	.0000	.0000
	4	.0384	.0149	.0049	.0013	.0003	.0000	.0000	.0000	.0000	.0000
	5	.1051	.0486	.0191	.0062	.0016	.0003	.0000	.0000	.0000	.0000
	6	.2272	.1241	.0583	.0229	.0071	.0016	.0002	.0000	.0000	.0000
	7	.4018	.2559	.1423	.0671	.0257	.0075	.0015	.0002	.0000	.0000
	8	.5982	.4371	.2839	.1594	.0744	.0271	.0070	.0011	.0001	.0000
	9	.7228	.6340	.4728	.3119	.1753	.0796	.0267	.0056	.0005	.0000
	10	.8949	.8024	.6712	.5100	.3402	.1897	.0817	.0235	.0033	.0001
	11	.9616	.9147	.8334	.7108	.5501	.3698	.2018	.0791	.0170	.0009
	12	.9894	.9719	.9349	.8661	.7541	.5950	.4019	.2101	.0684	.0070
	13	.9979	.9934	.9817	.9549	.9006	.8729	.6482	.4386	.2108	.0429
	14	.9997	.9990	.9967	.9902	.9739	.9365	.8593	.7161	.4853	.1892
	15	1.0000	.9999	.9997	.9990	.9967	.9900	.9719	.9257	.8147	.5599
	16	1.0000	1.0000	1.0000	1.0000	1.0000	1.0000	1.0000	1.0000	1.0000	1.0000

Table A-3 Binomial Distribution *(Continued)*

n	a	$\pi = .05$	$.10$	$.15$	$.20$	$.25$	$.30$	$.35$	$.40$	$.45$
17	0	.4181	.1668	.0631	.0225	.0075	.0023	.0007	.0002	.0000
	1	.7922	.4818	.2525	.1182	.0501	.0193	.0067	.0021	.0006
	2	.9497	.7618	.5198	.3096	.1637	.0774	.0327	.0123	.0041
	3	.9912	.9174	.7556	.5489	.3530	.2019	.1028	.0464	.0184
	4	.9988	.9779	.9013	.7582	.5739	.3887	.2348	.1260	.0596
	5	.9999	.9953	.9681	.8943	.7653	.5968	.4197	.2639	.1471
	6	1.0000	.9992	.9917	.9623	.8929	.7752	.6188	.4478	.2902
	7	1.0000	.9999	.9983	.9891	.9598	.8954	.7872	.6405	.4743
	8	1.0000	1.0000	.9997	.9974	.9876	.9597	.9006	.8011	.6626
	9	1.0000	1.0000	1.0000	.9995	.9969	.9873	.9617	.9081	.8166
	10	1.0000	1.0000	1.0000	.9999	.9994	.9968	.9880	.9652	.9174
	11	1.0000	1.0000	1.0000	1.0000	.9999	.9993	.9970	.9894	.9699
	12	1.0000	1.0000	1.0000	1.0000	1.0000	.9999	.9994	.9975	.9914
	13	1.0000	1.0000	1.0000	1.0000	1.0000	1.0000	.9999	.9995	.9981
	14	1.0000	1.0000	1.0000	1.0000	1.0000	1.0000	1.0000	.9999	.9997
	15	1.0000	1.0000	1.0000	1.0000	1.0000	1.0000	1.0000	1.0000	1.0000
	16	1.0000	1.0000	1.0000	1.0000	1.0000	1.0000	1.0000	1.0000	1.0000
	17	1.0000	1.0000	1.0000	1.0000	1.0000	1.0000	1.0000	1.0000	1.0000
18	0	.3972	.1501	.0536	.0180	.0056	.0016	.0004	.0001	.0000
	1	.7735	.4503	.2241	.0991	.0395	.0142	.0046	.0013	.0003
	2	.9419	.7338	.4797	.2713	.1353	.0600	.0236	.0082	.0025
	3	.9891	.9018	.7202	.5010	.3057	.1646	.0783	.0328	.0120
	4	.9985	.9718	.8794	.7164	.5187	.3327	.1886	.0942	.0411
	5	.9998	.9936	.9581	.8671	.7175	.5344	.3550	.2088	.1077
	6	1.0000	.9988	.9882	.9487	.8610	.7217	.5491	.3743	.2258
	7	1.0000	.9998	.9973	.9837	.9431	.8593	.7283	.5634	.3915
	8	1.0000	1.0000	.9995	.9957	.9807	.9404	.8609	.7368	.5778
	9	1.0000	1.0000	.9999	.9991	.9946	.9790	.9403	.8653	.7473
	10	1.0000	1.0000	1.0000	.9998	.9988	.9939	.9788	.9424	.8720
	11	1.0000	1.0000	1.0000	1.0000	.9998	.9986	.9938	.9797	.9463
	12	1.0000	1.0000	1.0000	1.0000	1.0000	.9997	.9986	.9942	.9817
	13	1.0000	1.0000	1.0000	1.0000	1.0000	1.0000	.9997	.9987	.9951
	14	1.0000	1.0000	1.0000	1.0000	1.0000	1.0000	1.0000	.9998	.9990
	15	1.0000	1.0000	1.0000	1.0000	1.0000	1.0000	1.0000	1.0000	.9999
	16	1.0000	1.0000	1.0000	1.0000	1.0000	1.0000	1.0000	1.0000	1.0000
	17	1.0000	1.0000	1.0000	1.0000	1.0000	1.0000	1.0000	1.0000	1.0000
	18	1.0000	1.0000	1.0000	1.0000	1.0000	1.0000	1.0000	1.0000	1.0000

Table A-3 Binomial Distribution *(Continued)*

n	a	π = .50	.55	.60	.65	.70	.75	.80	.85	.90	.95
17	0	.0000	.0000	.0000	.0000	.0000	.0000	.0000	.0000	.0000	.0000
	1	.0001	.0000	.0000	.0000	.0000	.0000	.0000	.0000	.0000	.0000
	2	.0012	.0003	.0001	.0000	.0000	.0000	.0000	.0000	.0000	.0000
	3	.0064	.0019	.0005	.0001	.0000	.0000	.0000	.0000	.0000	.0000
	4	.0245	.0086	.0025	.0006	.0001	.0000	.0000	.0000	.0000	.0000
	5	.0717	.0301	.0106	.0030	.0007	.0001	.0000	.0000	.0000	.0000
	6	.1662	.0826	.0348	.0120	.0032	.0006	.0001	.0000	.0000	.0000
	7	.3145	.1834	.0919	.0383	.0127	.0031	.0005	.0000	.0000	.0000
	8	.5000	.3374	.1989	.0994	.0403	.0124	.0026	.0003	.0000	.0000
	9	.6855	.5257	.3595	.2128	.1046	.0402	.0109	.0017	.0001	.0000
	10	.8338	.7098	.5522	.3812	.2248	.1071	.0377	.0083	.0008	.0000
	11	.9283	.8529	.7361	.5803	.4032	.2347	.1057	.0319	.0047	.0001
	12	.9755	.9404	.8740	.7652	.6113	.4261	.2418	.0987	.0221	.0012
	13	.9936	.9816	.9536	.8972	.7981	.6470	.4511	.2444	.0826	.0088
	14	.9988	.9959	.9877	.9673	.9226	.8363	.6904	.4802	.2382	.0503
	15	.9999	.9994	.9979	.9933	.9807	.9499	.8818	.7475	.5182	.2078
	16	1.0000	1.0000	.9998	.9993	.9977	.9925	.9775	.9369	.8332	.5819
	17	1.0000	1.0000	1.0000	1.0000	1.0000	1.0000	1.0000	1.0000	1.0000	1.0000
18	0	.0000	.0000	.0000	.0000	.0000	.0000	.0000	.0000	.0000	.0000
	1	.0001	.0000	.0000	.0000	.0000	.0000	.0000	.0000	.0000	.0000
	2	.0007	.0001	.0000	.0000	.0000	.0000	.0000	.0000	.0000	.0000
	3	.0038	.0010	.0002	.0000	.0000	.0000	.0000	.0000	.0000	.0000
	4	.0154	.0049	.0013	.0003	.0000	.0000	.0000	.0000	.0000	.0000
	5	.0481	.0183	.0058	.0014	.0003	.0000	.0000	.0000	.0000	.0000
	6	.1189	.0537	.0203	.0062	.0014	.0002	.0000	.0000	.0000	.0000
	7	.2403	.1280	.0576	.0212	.0061	.0012	.0002	.0000	.0000	.0000
	8	.4073	.2527	.1347	.0597	.0210	.0054	.0009	.0001	.0000	.0000
	9	.5927	.4222	.2632	.1391	.0596	.0193	.0043	.0005	.0000	.0000
	10	.7597	.6085	.4366	.2717	.1407	.0569	.0163	.0027	.0002	.0000
	11	.8811	.7742	.6257	.4509	.2783	.1390	.0513	.0118	.0012	.0000
	12	.9519	.8923	.7912	.6450	.4656	.2825	.1329	.0419	.0064	.0002
	13	.9846	.9589	.9058	.8114	.6673	.4813	.2836	.1206	.0282	.0015
	14	.9962	.9880	.9672	.9217	.8354	.6943	.4990	.2798	.0982	.0109
	15	.9993	.9975	.9918	.9764	.9400	.8647	.7287	.5203	.2662	.0581
	16	.9999	.9997	.9987	.9954	.9858	.9605	.9009	.7759	.5497	.2265
	17	1.0000	1.0000	.9999	.9996	.9984	.9944	.9820	.9464	.8499	.6028
	18	1.0000	1.0000	1.0000	1.0000	1.0000	1.0000	1.0000	1.0000	1.0000	1.0000

Table A-3 Binomial Distribution *(Continued)*

n	a	$\pi = .05$	.10	.15	.20	.25	.30	.35	.40	.45
19	0	.3774	.1351	.0456	.0144	.0042	.0011	.0003	.0001	.0000
	1	.7547	.4203	.1985	.0829	.0310	.0104	.0031	.0008	.0002
	2	.9335	.7054	.4413	.2369	.1113	.0462	.0170	.0055	.0015
	3	.9869	.8850	.6841	.4551	.2631	.1332	.0591	.0230	.0077
	4	.9980	.9648	.8556	.6733	.4654	.2822	.1500	.0696	.0280
	5	.9998	.9914	.9463	.8369	.6678	.4739	.2968	.1629	.0777
	6	1.0000	.9983	.9837	.9324	.8251	.6655	.4812	.3081	.1727
	7	1.0000	.9997	.9959	.9767	.9225	.8180	.6656	.4878	.3169
	8	1.0000	1.0000	.9992	.9933	.9713	.9161	.8145	.6675	.4940
	9	1.0000	1.0000	.9999	.9984	.9911	.9674	.9125	.8139	.6710
	10	1.0000	1.0000	1.0000	.9997	.9977	.9895	.9653	.9115	.8159
	11	1.0000	1.0000	1.0000	1.0000	.9995	.9972	.9886	.9648	.9129
	12	1.0000	1.0000	1.0000	1.0000	.9999	.9994	.9969	.9884	.9658
	13	1.0000	1.0000	1.0000	1.0000	1.0000	.9999	.9993	.9969	.9891
	14	1.0000	1.0000	1.0000	1.0000	1.0000	1.0000	.9999	.9994	.9972
	15	1.0000	1.0000	1.0000	1.0000	1.0000	1.0000	1.0000	.9999	.9995
	16	1.0000	1.0000	1.0000	1.0000	1.0000	1.0000	1.0000	1.0000	.9999
	17	1.0000	1.0000	1.0000	1.0000	1.0000	1.0000	1.0000	1.0000	1.0000
	18	1.0000	1.0000	1.0000	1.0000	1.0000	1.0000	1.0000	1.0000	1.0000
	19	1.0000	1.0000	1.0000	1.0000	1.0000	1.0000	1.0000	1.0000	1.0000
20	0	.3585	.1216	.0388	.0115	.0032	.0008	.0002	.0000	.0000
	1	.7358	.3917	.1756	.0692	.0243	.0076	.0021	.0005	.0001
	2	.9245	.6769	.4049	.2061	.0913	.0355	.0121	.0036	.0009
	3	.9841	.8670	.6477	.4114	.2252	.1071	.0444	.0160	.0049
	4	.9974	.9568	.8298	.6296	.4148	.2375	.1182	.0510	.0189
	5	.9997	.9887	.9327	.8042	.6172	.4164	.2454	.1256	.0553
	6	1.0000	.9976	.9781	.9133	.7858	.6080	.4166	.2500	.1299
	7	1.0000	.9996	.9941	.9679	.8982	.7723	.6010	.4159	.2520
	8	1.0000	.9999	.9987	.9900	.9591	.8867	.7624	.5956	.4143
	9	1.0000	1.0000	.9998	.9974	.9861	.9520	.8782	.7553	.5914
	10	1.0000	1.0000	1.0000	.9994	.9961	.9829	.9468	.8725	.7507
	11	1.0000	1.0000	1.0000	.9999	.9991	.9949	.9804	.9435	.8692
	12	1.0000	1.0000	1.0000	1.0000	.9998	.9987	.9940	.9790	.9420
	13	1.0000	1.0000	1.0000	1.0000	1.0000	.9997	.9985	.9935	.9786
	14	1.0000	1.0000	1.0000	1.0000	1.0000	1.0000	.9997	.9984	.9936
	15	1.0000	1.0000	1.0000	1.0000	1.0000	1.0000	1.0000	.9997	.9985
	16	1.0000	1.0000	1.0000	1.0000	1.0000	1.0000	1.0000	1.0000	.9997
	17	1.0000	1.0000	1.0000	1.0000	1.0000	1.0000	1.0000	1.0000	1.0000
	18	1.0000	1.0000	1.0000	1.0000	1.0000	1.0000	1.0000	1.0000	1.0000
	19	1.0000	1.0000	1.0000	1.0000	1.0000	1.0000	1.0000	1.0000	1.0000
	20	1.0000	1.0000	1.0000	1.0000	1.0000	1.0000	1.0000	1.0000	1.0000

Table A-3 Binomial Distribution *(Continued)*

n	a	$\pi = .50$	.55	.60	.65	.70	.75	.80	.85	.90	.95
19	0	.0000	.0000	.0000	.0000	.0000	.0000	.0000	.0000	.0000	.0000
	1	.0000	.0000	.0000	.0000	.0000	.0000	.0000	.0000	.0000	.0000
	2	.0004	.0001	.0000	.0000	.0000	.0000	.0000	.0000	.0000	.0000
	3	.0022	.0005	.0001	.0000	.0000	.0000	.0000	.0000	.0000	.0000
	4	.0096	.0028	.0006	.0001	.0000	.0000	.0000	.0000	.0000	.0000
	5	.0318	.0109	.0031	.0007	.0001	.0000	.0000	.0000	.0000	.0000
	6	.0835	.0342	.0116	.0031	.0006	.0001	.0000	.0000	.0000	.0000
	7	.1796	.0871	.0352	.0114	.0028	.0005	.0000	.0000	.0000	.0000
	8	.3238	.1841	.0885	.0347	.0105	.0023	.0003	.0000	.0000	.0000
	9	.5000	.3290	.1861	.0875	.0326	.0089	.0016	.0001	.0000	.0000
	10	.6762	.5060	.3325	.1855	.0839	.0287	.0067	.0008	.0000	.0000
	11	.8204	.6831	.5122	.3344	.1820	.0775	.0233	.0041	.0003	.0000
	12	.9165	.8273	.6919	.5188	.3345	.1749	.0676	.0163	.0017	.0000
	13	.9682	.9223	.8371	.7032	.5261	.3322	.1631	.0537	.0086	.0002
	14	.9904	.9720	.9304	.8500	.7178	.5346	.3267	.1444	.0352	.0020
	15	.9978	.9923	.9770	.9409	.8668	.7369	.5449	.3159	.1150	.0132
	16	.9996	.9985	.9945	.9830	.9538	.8887	.7631	.5587	.2946	.0665
	17	1.0000	.9998	.9992	.9969	.9896	.9690	.9171	.8015	.5797	.2453
	18	1.0000	1.0000	.9999	.9997	.9989	.9958	.9856	.9544	.8649	.6226
	19	1.0000	1.0000	1.0000	1.0000	1.0000	1.0000	1.0000	1.0000	1.0000	1.0000
20	0	.0000	.0000	.0000	.0000	.0000	.0000	.0000	.0000	.0000	.0000
	1	.0000	.0000	.0000	.0000	.0000	.0000	.0000	.0000	.0000	.0000
	2	.0002	.0000	.0000	.0000	.0000	.0000	.0000	.0000	.0000	.0000
	3	.0013	.0003	.0000	.0000	.0000	.0000	.0000	.0000	.0000	.0000
	4	.0059	.0015	.0003	.0000	.0000	.0000	.0000	.0000	.0000	.0000
	5	.0207	.0064	.0016	.0003	.0000	.0000	.0000	.0000	.0000	.0000
	6	.0577	.0214	.0065	.0015	.0003	.0000	.0000	.0000	.0000	.0000
	7	.1316	.0580	.0210	.0060	.0013	.0002	.0000	.0000	.0000	.0000
	8	.2517	.1308	.0565	.0196	.0051	.0009	.0001	.0000	.0000	.0000
	9	.4119	.2493	.1275	.0532	.0171	.0039	.0006	.0000	.0000	.0000
	10	.5881	.4086	.2447	.1218	.0480	.0139	.0026	.0002	.0000	.0000
	11	.7483	.5857	.4044	.2376	.1133	.0409	.0100	.0013	.0001	.0000
	12	.8684	.7480	.5841	.3990	.2277	.1018	.0321	.0059	.0004	.0000
	13	.9423	.8701	.7500	.5834	.3920	.2142	.0867	.0219	.0024	.0000
	14	.9793	.9447	.8744	.7546	.5836	.3828	.1958	.0673	.0113	.0003
	15	.9941	.9811	.9490	.8818	.7625	.5852	.3704	.1702	.0432	.0026
	16	.9987	.9951	.9840	.9556	.8929	.7748	.5886	.3523	.1330	.0159
	17	.9998	.9991	.9964	.9879	.9645	.9087	.7939	.5951	.3231	.0755
	18	1.0000	.9999	.9995	.9979	.9924	.9757	.9308	.8244	.6083	.2642
	19	1.0000	1.0000	1.0000	.9998	.9992	.9968	.9885	.9612	.8784	.6415
	20	1.0000	1.0000	1.0000	1.0000	1.0000	1.0000	1.0000	1.0000	1.0000	1.0000

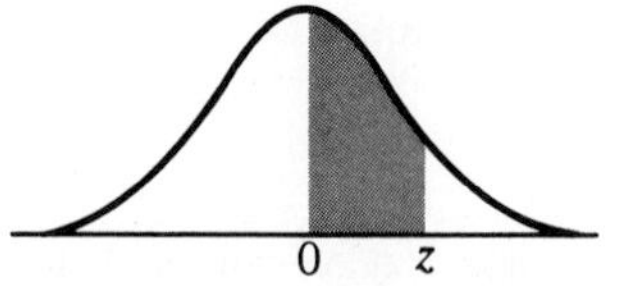

Table A-4 Normal Curve Areas

z	.00	.01	.02	.03	.04	.05	.06	.07	.08	.09
0.0	.0000	.0040	.0080	.0120	.0160	.0199	.0239	.0279	.0319	.0359
0.1	.0398	.0438	.0478	.0517	.0557	.0596	.0636	.0675	.0714	.0753
0.2	.0793	.0832	.0871	.0910	.0948	.0987	.1026	.1064	.1103	.1141
0.3	.1179	.1217	.1255	.1293	.1331	.1368	.1406	.1443	.1480	.1517
0.4	.1554	.1591	.1628	.1664	.1700	.1736	.1772	.1808	.1844	.1879
0.5	.1915	.1950	.1985	.2019	.2054	.2088	.2123	.2157	.2190	.2224
0.6	.2257	.2291	.2324	.2357	.2389	.2422	.2454	.2486	.2517	.2549
0.7	.2580	.2611	.2642	.2673	.2704	.2734	.2764	.2794	.2823	.2852
0.8	.2881	.2910	.2939	.2967	.2995	.3023	.3051	.3078	.3106	.3133
0.9	.3159	.3186	.3212	.3238	.3264	.3289	.3315	.3340	.3365	.3389
1.0	.3413	.3438	.3461	.3485	.3508	.3531	.3554	.3577	.3599	.3621
1.1	.3643	.3665	.3686	.3708	.3729	.3749	.3770	.3790	.3810	.3830
1.2	.3849	.3869	.3888	.3907	.3925	.3944	.3962	.3980	.3997	.4015
1.3	.4032	.4049	.4066	.4082	.4099	.4115	.4131	.4147	.4162	.4177
1.4	.4192	.4207	.4222	.4236	.4251	.4265	.4279	.4292	.4306	.4319
1.5	.4332	.4345	.4357	.4370	.4382	.4394	.4406	.4418	.4429	.4441
1.6	.4452	.4463	.4474	.4484	.4495	.4505	.4515	.4525	.4535	.4545
1.7	.4554	.4564	.4573	.4582	.4591	.4599	.4608	.4616	.4625	.4633
1.8	.4641	.4649	.4656	.4664	.4671	.4678	.4686	.4693	.4699	.4706
1.9	.4713	.4719	.4726	.4732	.4738	.4744	.4750	.4756	.4761	.4767
2.0	.4772	.4778	.4783	.4788	.4793	.4798	.4803	.4808	.4812	.4817
2.1	.4821	.4826	.4830	.4834	.4838	.4842	.4846	.4850	.4854	.4857
2.2	.4861	.4864	.4868	.4871	.4875	.4878	.4881	.4884	.4887	.4890
2.3	.4893	.4896	.4898	.4901	.4904	.4906	.4909	.4911	.4913	.4916
2.4	.4918	.4920	.4922	.4925	.4927	.4929	.4931	.4932	.4934	.4936
2.5	.4938	.4940	.4941	.4943	.4945	.4946	.4948	.4949	.4951	.4952
2.6	.4953	.4955	.4956	.4957	.4959	.4960	.4961	.4962	.4963	.4964
2.7	.4965	.4966	.4967	.4968	.4969	.4970	.4971	.4972	.4973	.4974
2.8	.4974	.4975	.4976	.4977	.4977	.4978	.4979	.4979	.4980	.4981
2.9	.4981	.4982	.4982	.4983	.4984	.4984	.4985	.4985	.4986	.4986
3.0	.4987	.4987	.4987	.4988	.4988	.4989	.4989	.4989	.4990	.4990

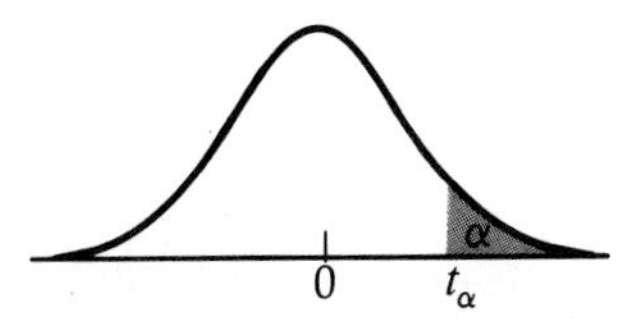

Table A-5 Critical Values of t

$d.f.$	$t_{.100}$	$t_{.050}$	$t_{.025}$	$t_{.010}$	$t_{.005}$	$d.f.$
1	3.078	6.314	12.706	31.821	63.657	1
2	1.886	2.920	4.303	6.965	9.925	2
3	1.638	2.353	3.182	4.541	5.841	3
4	1.533	2.132	2.776	3.747	4.604	4
5	1.476	2.015	2.571	3.365	4.032	5
6	1.440	1.943	2.447	3.143	3.707	6
7	1.415	1.895	2.365	2.998	3.499	7
8	1.397	1.860	2.306	2.896	3.355	8
9	1.383	1.833	2.262	2.821	3.250	9
10	1.372	1.812	2.228	2.764	3.169	10
11	1.363	1.796	2.201	2.718	3.106	11
12	1.356	1.782	2.179	2.681	3.055	12
13	1.350	1.771	2.160	2.650	3.012	13
14	1.345	1.761	2.145	2.624	2.977	14
15	1.341	1.753	2.131	2.602	2.947	15
16	1.337	1.746	2.120	2.583	2.921	16
17	1.333	1.740	2.110	2.567	2.898	17
18	1.330	1.734	2.101	2.552	2.878	18
19	1.328	1.729	2.093	2.539	2.861	19
20	1.325	1.725	2.086	2.528	2.845	20
21	1.323	1.721	2.080	2.518	2.831	21
22	1.321	1.717	2.074	2.508	2.819	22
23	1.319	1.714	2.069	2.500	2.807	23
24	1.318	1.711	2.064	2.492	2.797	24
25	1.316	1.708	2.060	2.485	2.787	25
26	1.315	1.706	2.056	2.479	2.779	26
27	1.314	1.703	2.052	2.473	2.771	27
28	1.313	1.701	2.048	2.467	2.763	28
29	1.311	1.699	2.045	2.462	2.756	29
inf.	1.282	1.645	1.960	2.326	2.576	inf.

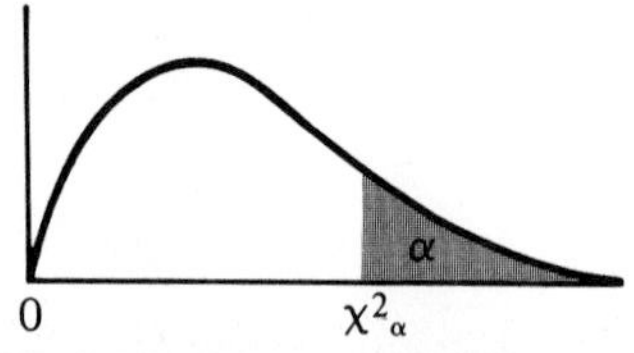

Table A-6 Critical Values of χ^2

d.f.	$\chi^2 0.995$	$\chi^2 0.990$	$\chi^2 0.975$	$\chi^2 0.950$	$\chi^2 0.900$
1	0.0000393	0.0001571	0.0009821	0.0039321	0.0157908
2	0.0100251	0.0201007	0.0506356	0.102587	0.210720
3	0.0717212	0.114832	0.215795	0.351846	0.584375
4	0.206990	0.297110	0.484419	0.710721	1.063623
5	0.411740	0.554300	0.831211	1.145476	1.61031
6	0.675727	0.872085	1.237347	1.63539	2.20413
7	0.989265	1.239043	1.68987	2.16735	2.83311
8	1.344419	1.646482	2.17973	2.73264	3.48954
9	1.734926	2.087912	2.70039	3.32511	4.16816
10	2.15585	2.55821	3.24697	3.94030	4.86518
11	2.60321	3.05347	3.81575	4.57481	5.57779
12	3.07382	3.57056	4.40379	5.22603	6.30380
13	3.56503	4.10691	5.00874	5.89186	7.04150
14	4.07468	4.66043	5.62872	6.57063	7.78953
15	4.60094	5.22935	6.26214	7.26094	8.54675
16	5.14224	5.81221	6.90766	7.96164	9.31223
17	5.69724	6.40776	7.56418	8.67176	10.0852
18	6.26481	7.01491	8.23075	9.39046	10.8649
19	6.84398	7.63273	8.90655	10.1170	11.6509
20	7.43386	8.26040	9.59083	10.8508	12.4426
21	8.03366	8.89720	10.28293	11.5913	13.2396
22	8.64272	9.54249	10.9823	12.3380	14.0415
23	9.26042	10.19567	11.6885	13.0905	14.8479
24	9.88623	10.8564	12.4011	13.8484	15.6587
25	10.5197	11.5240	13.1197	14.6114	16.4734
26	11.1603	12.1981	13.8439	15.3791	17.2919
27	11.8076	12.8786	14.5733	16.1513	18.1138
28	12.4613	13.5648	15.3079	16.9279	18.9392
29	13.1211	14.2565	16.0471	17.7083	19.7677
30	13.7867	14.9535	16.7908	18.4926	20.5992
40	20.7065	22.1643	24.4331	26.5093	29.0505
50	27.9907	29.7067	32.3574	34.7642	37.6886
60	35.5346	37.4848	40.4817	43.1879	46.4589
70	43.2752	45.4418	48.7576	51.7393	55.3290
80	51.1720	53.5400	57.1532	60.3915	64.2778
90	59.1963	61.7541	65.6466	69.1260	73.2912
100	67.3276	70.0648	74.2219	77.9295	82.3581

Table A-6 Critical Values of χ^2 (Continued)

$\chi^2_{0.100}$	$\chi^2_{0.050}$	$\chi^2_{0.025}$	$\chi^2_{0.010}$	$\chi^2_{0.005}$	*d.f.*
2.70554	3.84146	5.02389	6.63490	7.87944	1
4.60517	5.99147	7.37776	9.21034	10.5966	2
6.25139	7.81473	9.34840	11.3449	12.8381	3
7.77944	9.48773	11.1433	13.2767	14.8602	4
9.23635	11.0705	12.8325	15.0863	16.7496	5
10.6446	12.5916	14.4494	16.8119	18.5476	6
12.0170	14.0671	16.0128	18.4753	20.2777	7
13.3616	15.5073	17.5346	20.0902	21.9550	8
14.6837	16.9190	19.0228	21.6660	23.5893	9
15.9871	18.3070	20.4831	23.2093	25.1882	10
17.2750	19.6751	21.9200	24.7250	26.7569	11
18.5494	21.0261	23.3367	26.2170	28.2995	12
19.8119	22.3621	24.7356	27.6883	29.8194	13
21.0642	23.6848	26.1190	29.1413	31.3193	14
22.3072	24.9958	27.4884	30.5779	32.8013	15
23.5418	26.2962	28.8454	31.9999	34.2672	16
24.7690	27.5871	30.1910	33.4087	35.7185	17
25.9894	28.8693	31.5264	34.8053	37.1564	18
27.2036	30.1435	32.8523	36.1908	38.5822	19
28.4120	31.4104	34.1696	37.5662	39.9968	20
29.6151	32.6705	35.4789	38.9321	41.4010	21
30.8133	33.9244	36.7807	40.2894	42.7956	22
32.0069	35.1725	38.0757	41.6384	44.1813	23
33.1963	36.4151	39.3641	42.9798	45.5585	24
34.3816	37.6525	40.6465	44.3141	46.9278	25
35.5631	38.8852	41.9232	45.6417	48.2899	26
36.7412	40.1133	43.1944	46.9630	49.6449	27
37.9159	41.3372	44.4607	48.2782	50.9933	28
39.0875	42.5569	45.7222	49.5879	52.3356	29
40.2560	43.7729	46.9792	50.8922	53.6720	30
51.8050	55.7585	59.3417	63.6907	66.7659	40
63.1671	67.5048	71.4202	76.1539	79.4900	50
74.3970	79.0819	83.2976	88.3794	91.9517	60
85.5271	90.5312	95.0231	100.425	104.215	70
96.5782	101.879	106.629	112.329	116.321	80
107.565	113.145	118.136	124.116	128.299	90
118.498	124.342	129.561	135.807	140.169	100

Table A-7 Critical Values of F_{ν_1, ν_2} for $\alpha = .05$

υ_1 = Degrees of freedom for numerator; υ_2 = Degrees of freedom for denominator

υ_2	1	2	3	4	5	6	7	8	9	10	12	15	20	24	30	40	60	120	∞
1	161	200	216	225	230	234	237	239	241	242	244	246	248	249	250	251	252	253	254
2	18.5	19.0	19.2	19.2	19.3	19.3	19.4	19.4	19.4	19.4	19.4	19.4	19.4	19.5	19.5	19.5	19.5	19.5	19.5
3	10.1	9.55	9.28	9.12	9.01	8.94	8.89	8.85	8.81	8.79	8.74	8.70	8.66	8.64	8.62	8.59	8.57	8.55	8.53
4	7.71	6.94	6.59	6.39	6.26	6.16	6.09	6.04	6.00	5.96	5.91	5.86	5.80	5.77	5.75	5.72	5.69	5.66	5.63
5	6.61	5.79	5.41	5.19	5.05	4.95	4.88	4.82	4.77	4.74	4.68	4.62	4.56	4.53	4.50	4.46	4.43	4.40	4.37
6	5.99	5.14	4.76	4.53	4.39	4.28	4.21	4.15	4.10	4.06	4.00	3.94	3.87	3.84	3.81	3.77	3.74	3.70	3.67
7	5.59	4.74	4.35	4.12	3.97	3.87	3.79	3.73	3.68	3.64	3.57	3.51	3.44	3.41	3.38	3.34	3.30	3.27	3.23
8	5.32	4.46	4.07	3.84	3.69	3.58	3.50	3.44	3.39	3.35	3.28	3.22	3.15	3.12	3.08	3.04	3.01	2.97	2.93
9	5.12	4.26	3.86	3.63	3.48	3.37	3.29	3.23	3.18	3.14	3.07	3.01	2.94	2.90	2.86	2.83	2.79	2.75	2.71
10	4.96	4.10	3.71	3.48	3.33	3.22	3.14	3.07	3.02	2.98	2.91	2.85	2.77	2.74	2.70	2.66	2.62	2.58	2.54
11	4.84	3.98	3.59	3.36	3.20	3.09	3.01	2.95	2.90	2.85	2.79	2.72	2.65	2.61	2.57	2.53	2.49	2.45	2.40
12	4.75	3.89	3.49	3.26	3.11	3.00	2.91	2.85	2.80	2.75	2.69	2.62	2.54	2.51	2.47	2.43	2.38	2.34	2.30
13	4.67	3.81	3.41	3.18	3.03	2.92	2.83	2.77	2.71	2.67	2.60	2.53	2.46	2.42	2.38	2.34	2.30	2.25	2.21
14	4.60	3.74	3.34	3.11	2.96	2.85	2.76	2.70	2.65	2.60	2.53	2.46	2.39	2.35	2.31	2.27	2.22	2.18	2.13
15	4.54	3.68	3.29	3.06	2.90	2.79	2.71	2.64	2.59	2.54	2.48	2.40	2.33	2.29	2.25	2.20	2.16	2.11	2.07
16	4.49	3.63	3.24	3.01	2.85	2.74	2.66	2.59	2.54	2.49	2.42	2.35	2.28	2.24	2.19	2.15	2.11	2.06	2.01
17	4.45	3.59	3.20	2.96	2.81	2.70	2.61	2.55	2.49	2.45	2.38	2.31	2.23	2.19	2.15	2.10	2.06	2.01	1.96
18	4.41	3.55	3.16	2.93	2.77	2.66	2.58	2.51	2.46	2.41	2.34	2.27	2.19	2.15	2.11	2.06	2.02	1.97	1.92
19	4.38	3.52	3.13	2.90	2.74	2.63	2.54	2.48	2.42	2.38	2.31	2.23	2.16	2.11	2.07	2.03	1.98	1.93	1.88
20	4.35	3.49	3.10	2.87	2.71	2.60	2.51	2.45	2.39	2.35	2.28	2.20	2.12	2.08	2.04	1.99	1.95	1.90	1.84
21	4.32	3.47	3.07	2.84	2.68	2.57	2.49	2.42	2.37	2.32	2.25	2.18	2.10	2.05	2.01	1.96	1.92	1.87	1.81
22	4.30	3.44	3.05	2.82	2.66	2.55	2.46	2.40	2.34	2.30	2.23	2.15	2.07	2.03	1.98	1.94	1.89	1.84	1.78
23	4.28	3.42	3.03	2.80	2.64	2.53	2.44	2.37	2.32	2.27	2.20	2.13	2.05	2.01	1.96	1.91	1.86	1.81	1.76
24	4.26	3.40	3.01	2.78	2.62	2.51	2.42	2.36	2.30	2.25	2.18	2.11	2.03	1.98	1.94	1.89	1.84	1.79	1.73
25	4.24	3.39	2.99	2.76	2.60	2.49	2.40	2.34	2.28	2.24	2.16	2.09	2.01	1.96	1.92	1.87	1.82	1.77	1.71
30	4.17	3.32	2.92	2.69	2.53	2.42	2.33	2.27	2.21	2.16	2.09	2.01	1.93	1.89	1.84	1.79	1.74	1.68	1.62
40	4.08	3.23	2.84	2.61	2.45	2.34	2.25	2.18	2.12	2.08	2.00	1.92	1.84	1.79	1.74	1.69	1.64	1.58	1.51
60	4.00	3.15	2.76	2.53	2.37	2.25	2.17	2.10	2.04	1.99	1.92	1.84	1.75	1.70	1.65	1.59	1.53	1.47	1.39
120	3.92	3.07	2.68	2.45	2.29	2.18	2.09	2.02	1.96	1.91	1.83	1.75	1.66	1.61	1.55	1.50	1.43	1.35	1.25
∞	3.84	3.00	2.60	2.37	2.21	2.10	2.01	1.94	1.88	1.83	1.75	1.67	1.57	1.52	1.46	1.39	1.32	1.22	1.00

Table A-8 Critical Values of F_{ν_1, ν_2} for $\alpha = .01$

0 F .01

υ_1 = Degrees of freedom for numerator

υ_2 = Degrees of freedom for denominator

υ_2	1	2	3	4	5	6	7	8	9	10	12	15	20	24	30	40	60	120	∞
1	4,052	5,000	5,403	5,625	5,764	5,859	5,928	5,982	6,023	6,056	6,106	6,157	6,209	6,235	6,261	6,287	6,313	6,339	6,366
2	98.5	99.0	99.2	99.2	99.3	99.3	99.4	99.4	99.4	99.4	99.4	99.4	99.4	99.5	99.5	99.5	99.5	99.5	99.5
3	34.1	30.8	29.5	28.7	28.2	27.9	27.7	27.5	27.3	27.2	27.1	26.9	26.7	26.6	26.5	26.4	26.3	26.2	26.1
4	21.2	18.0	16.7	16.0	15.5	15.2	15.0	14.8	14.7	14.5	14.4	14.2	14.0	13.9	13.8	13.7	13.7	13.6	13.5
5	16.3	13.3	12.1	11.4	11.0	10.7	10.5	10.3	10.2	10.1	9.89	9.72	9.55	9.47	9.38	9.29	9.20	9.11	9.02
6	13.7	10.9	9.78	9.15	8.75	8.47	8.26	8.10	7.98	7.87	7.72	7.56	7.40	7.31	7.23	7.14	7.06	6.97	6.88
7	12.2	9.55	8.45	7.85	7.46	7.19	6.99	6.84	6.72	6.62	6.47	6.31	6.16	6.07	5.99	5.91	5.82	5.74	5.65
8	11.3	8.65	7.59	7.01	6.63	6.37	6.18	6.03	5.91	5.81	5.67	5.52	5.36	5.28	5.20	5.12	5.03	4.95	4.86
9	10.6	8.02	6.99	6.42	6.06	5.80	5.61	5.47	5.35	5.26	5.11	4.96	4.81	4.73	4.65	4.57	4.48	4.40	4.31
10	10.0	7.56	6.55	5.99	5.64	5.39	5.20	5.06	4.94	4.85	4.71	4.56	4.41	4.33	4.25	4.17	4.08	4.00	3.91
11	9.65	7.21	6.22	5.67	5.32	5.07	4.89	4.74	4.63	4.54	4.40	4.25	4.10	4.02	3.94	3.86	3.78	3.69	3.60
12	9.33	6.93	5.95	5.41	5.06	4.82	4.64	4.50	4.39	4.30	4.16	4.01	3.86	3.78	3.70	3.62	3.54	3.45	3.36
13	9.07	6.70	5.74	5.21	4.86	4.62	4.44	4.30	4.19	4.10	3.96	3.82	3.66	3.59	3.51	3.43	3.34	3.25	3.17
14	8.86	6.51	5.56	5.04	4.70	4.46	4.28	4.14	4.03	3.94	3.80	3.66	3.51	3.43	3.35	3.27	3.18	3.09	3.00
15	8.68	6.36	5.42	4.89	4.56	4.32	4.14	4.00	3.89	3.80	3.67	3.52	3.37	3.29	3.21	3.13	3.05	2.96	2.87
16	8.53	6.23	5.29	4.77	4.44	4.20	4.03	3.89	3.78	3.69	3.55	3.41	3.26	3.18	3.10	3.02	2.93	2.84	2.75
17	8.40	6.11	5.19	4.67	4.34	4.10	3.93	3.79	3.68	3.59	3.46	3.31	3.16	3.08	3.00	2.92	2.83	2.75	2.65
18	8.29	6.01	5.09	4.58	4.25	4.01	3.84	3.71	3.60	3.51	3.37	3.23	3.08	3.00	2.92	2.84	2.75	2.66	2.57
19	8.19	5.93	5.01	4.50	4.17	3.94	3.77	3.63	3.52	3.43	3.30	3.15	3.00	2.92	2.84	2.76	2.67	2.58	2.49
20	8.10	5.85	4.94	4.43	4.10	3.87	3.70	3.56	3.46	3.37	3.23	3.09	2.94	2.86	2.78	2.69	2.61	2.52	2.42
21	8.02	5.78	4.87	4.37	4.04	3.81	3.64	3.51	3.40	3.31	3.17	3.03	2.88	2.80	2.72	2.64	2.55	2.46	2.36
22	7.95	5.72	4.82	4.31	3.99	3.76	3.59	3.45	3.35	3.26	3.12	2.98	2.83	2.75	2.67	2.58	2.50	2.40	2.31
23	7.88	5.66	4.76	4.26	3.94	3.71	3.54	3.41	3.30	3.21	3.07	2.93	2.78	2.70	2.62	2.54	2.45	2.35	2.26
24	7.82	5.61	4.72	4.22	3.90	3.67	3.50	3.36	3.26	3.17	3.03	2.89	2.74	2.66	2.58	2.49	2.40	2.31	2.21
25	7.77	5.57	4.68	4.18	3.86	3.63	3.46	3.32	3.22	3.13	2.99	2.85	2.70	2.62	2.53	2.45	2.36	2.27	2.17
30	7.56	5.39	4.51	4.02	3.70	3.47	3.30	3.17	3.07	2.98	2.84	2.70	2.55	2.47	2.39	2.30	2.21	2.11	2.01
40	7.31	5.18	4.31	3.83	3.51	3.29	3.12	2.99	2.89	2.80	2.66	2.52	2.37	2.29	2.20	2.11	2.02	1.92	1.80
60	7.08	4.98	4.13	3.65	3.34	3.12	2.95	2.82	2.72	2.63	2.50	2.35	2.20	2.12	2.03	1.94	1.84	1.73	1.60
120	6.85	4.79	3.95	3.48	3.17	2.96	2.79	2.66	2.56	2.47	2.34	2.19	2.03	1.95	1.86	1.76	1.66	1.53	1.38
∞	6.63	4.61	3.78	3.32	3.02	2.80	2.64	2.51	2.41	2.32	2.18	2.04	1.88	1.79	1.70	1.59	1.47	1.32	1.00

Table A-9 Values for Studentized Range q_{ν_1, ν_2} for $\alpha = .05$

ν_2 \ ν_1	2	3	4	5	6	7	8	9	10	11	12	13	14	15	16	17	18	19	20
1	18.0	27.0	32.8	37.1	40.4	43.1	45.4	47.4	49.1	50.6	52.0	53.2	54.3	55.4	56.3	57.2	58.0	58.8	59.6
2	6.09	8.3	9.8	10.9	11.7	12.4	13.0	13.5	14.0	14.4	14.7	15.1	15.4	15.7	15.9	16.1	16.4	16.6	16.8
3	4.50	5.91	6.82	7.50	8.04	8.48	8.85	9.18	9.46	9.72	9.95	10.15	10.35	10.52	10.69	10.84	10.98	11.11	11.24
4	3.93	5.04	5.76	6.29	6.71	7.05	7.35	7.60	7.83	8.03	8.21	8.37	8.52	8.66	8.79	8.91	9.03	9.13	9.23
5	3.64	4.60	5.22	5.67	6.03	6.33	6.58	6.80	6.99	7.17	7.32	7.47	7.60	7.72	7.83	7.93	8.03	8.12	8.21
6	3.46	4.34	4.90	5.31	5.63	5.89	6.12	6.32	6.49	6.65	6.79	6.92	7.03	7.14	7.24	7.34	7.43	7.51	7.59
7	3.34	4.16	4.68	5.06	5.36	5.61	5.82	6.00	6.16	6.30	6.43	6.55	6.66	6.76	6.85	6.94	7.02	7.09	7.17
8	3.26	4.04	4.53	4.89	5.17	5.40	5.60	5.77	5.92	6.05	6.18	6.29	6.39	6.48	6.57	6.65	6.73	6.80	6.87
9	3.20	3.95	4.42	4.76	5.02	5.24	5.43	5.60	5.74	5.87	5.98	6.09	6.19	6.28	6.36	6.44	6.51	6.58	6.64
10	3.15	3.88	4.33	4.65	4.91	5.12	5.30	5.46	5.60	5.72	5.83	5.93	6.03	6.11	6.20	6.27	6.34	6.40	6.47
11	3.11	3.82	4.26	4.57	4.82	5.03	5.20	5.35	5.49	5.61	5.71	5.81	5.90	5.99	6.06	6.14	6.20	6.26	6.33
12	3.08	3.77	4.20	4.51	4.75	4.95	5.12	5.27	5.40	5.51	5.62	5.71	5.80	5.88	5.95	6.03	6.09	6.15	6.21
13	3.06	3.73	4.15	4.45	4.69	4.88	5.05	5.19	5.32	5.43	5.53	5.63	5.71	5.79	5.86	5.93	6.00	6.05	6.11
14	3.03	3.70	4.11	4.41	4.64	4.83	4.99	5.13	5.25	5.36	5.46	5.55	5.64	5.72	5.79	5.85	5.92	5.97	6.03
15	3.01	3.67	4.08	4.37	4.60	4.78	4.94	5.08	5.20	5.31	5.40	5.49	5.58	5.65	5.72	5.79	5.85	5.90	5.96
16	3.00	3.65	4.05	4.33	4.56	4.74	4.90	5.03	5.15	5.26	5.35	5.44	5.52	5.59	5.66	5.72	5.79	5.84	5.90
17	2.98	3.63	4.02	4.30	4.52	4.71	4.86	4.99	5.11	5.21	5.31	5.39	5.47	5.55	5.61	5.68	5.74	5.79	5.84
18	2.97	3.61	4.00	4.28	4.49	4.67	4.82	4.96	5.07	5.17	5.27	5.35	5.43	5.50	5.57	5.63	5.69	5.74	5.79
19	2.96	3.59	3.98	4.25	4.47	4.65	4.79	4.92	5.04	5.14	5.23	5.32	5.39	5.46	5.53	5.59	5.65	5.70	5.75
20	2.95	3.58	3.96	4.23	4.45	4.62	4.77	4.90	5.01	5.11	5.20	5.28	5.36	5.43	5.49	5.55	5.61	5.66	5.71
24	2.92	3.53	3.90	4.17	4.37	4.54	4.68	4.81	4.92	5.01	5.10	5.18	5.25	5.32	5.38	5.44	5.50	5.54	5.59
30	2.89	3.49	3.84	4.10	4.30	4.46	4.60	4.72	4.83	4.92	5.00	5.08	5.15	5.21	5.27	5.33	5.38	5.43	5.48
40	2.86	3.44	3.79	4.04	4.23	4.39	4.52	4.63	4.74	4.82	4.91	4.98	5.05	5.11	5.16	5.22	5.27	5.31	5.36
60	2.83	3.40	3.74	3.98	4.16	4.31	4.44	4.55	4.65	4.73	4.81	4.88	4.94	5.00	5.06	5.11	5.16	5.20	5.24
120	2.80	3.36	3.69	3.92	4.10	4.24	4.36	4.48	4.56	5.64	4.72	4.78	4.84	4.90	4.95	5.00	5.05	5.09	5.13
∞	2.77	3.31	3.63	3.86	4.03	4.17	4.29	4.39	4.47	4.55	4.62	4.68	4.74	4.80	4.85	4.89	4.93	4.97	5.01

Table A-10 Critical Values for the Mann-Whitney U

n	p	m = 2	3	4	5	6	7	8	9	10	11	12	13	14	15	16	17	18	19	20
	.001	0	0	0	0	0	0	0	0	0	0	0	0	0	0	0	0	0	0	0
	.005	0	0	0	0	0	0	0	0	0	0	0	0	0	0	0	0	0	1	1
2	.01	0	0	0	0	0	0	0	0	0	0	0	1	1	1	1	1	1	2	2
	.025	0	0	0	0	0	0	1	1	1	1	2	2	2	2	2	3	3	3	3
	.05	0	0	0	1	1	1	2	2	2	2	3	3	4	4	4	4	5	5	5
	.10	0	1	1	2	2	2	3	3	4	4	5	5	5	6	6	7	7	8	8
	.001	0	0	0	0	0	0	0	0	0	0	0	0	0	0	0	1	1	1	1
	.005	0	0	0	0	0	0	0	1	1	1	2	2	2	3	3	3	3	4	4
3	.01	0	0	0	0	0	1	1	2	2	2	3	3	3	4	4	5	5	5	6
	.025	0	0	0	1	2	2	3	3	4	4	5	5	6	6	7	7	8	8	9
	.05	0	1	1	2	3	3	4	5	5	6	6	7	8	8	9	10	10	11	12
	.10	1	2	2	3	4	5	6	6	7	8	9	10	11	11	12	13	14	15	16
	.001	0	0	0	0	0	0	0	0	1	1	1	2	2	2	3	3	4	4	4
	.005	0	0	0	0	1	1	2	2	3	3	4	4	5	6	6	7	7	8	9
4	.01	0	0	0	1	2	2	3	4	4	5	6	6	7	9	8	9	10	10	11
	.025	0	0	1	2	3	4	5	5	6	7	8	9	10	11	12	12	13	14	15
	.05	0	1	2	3	4	5	6	7	8	9	10	11	12	13	15	16	17	18	19
	.10	1	2	4	5	6	7	8	10	11	12	13	14	16	17	18	19	21	22	23

Table A-10 Critical Values for the Mann-Whitney *U* (Continued)

n	*p*	*m* = 2	3	4	5	6	7	8	9	10	11	12	13	14	15	16	17	18	19	20
	.001	0	0	0	0	0	0	1	2	2	3	3	4	4	5	6	6	7	8	8
	.005	0	0	0	1	2	2	3	4	5	6	7	8	8	9	10	11	12	13	14
5	.01	0	0	1	2	3	4	5	6	7	8	9	10	11	12	13	14	15	16	17
	.025	0	1	2	3	4	6	7	8	9	10	12	13	14	15	16	18	19	20	21
	.05	1	2	3	5	6	7	9	10	12	13	14	16	17	19	20	21	23	24	26
	.10	2	3	5	6	8	9	11	13	14	16	18	19	21	23	24	26	28	29	31
	.001	0	0	0	0	0	0	2	3	4	5	5	6	7	8	9	10	11	12	13
	.005	0	0	1	2	3	4	5	6	7	8	10	11	12	13	14	16	17	18	19
6	.01	0	0	2	3	4	5	7	8	9	10	12	13	14	16	17	19	20	21	23
	.025	0	2	3	4	6	7	9	11	12	14	15	17	18	20	22	23	25	26	28
	.05	1	3	4	6	8	9	11	13	15	17	18	20	22	24	26	27	29	31	33
	.10	2	4	6	8	10	12	14	16	18	20	22	24	26	28	30	32	35	37	39
	.001	0	0	0	0	1	2	3	4	6	7	8	9	10	11	12	14	15	16	17
	.005	0	0	1	2	4	5	7	8	10	11	13	14	16	17	19	20	22	23	25
7	.01	0	1	2	4	5	7	8	10	12	13	15	17	18	20	22	24	25	27	29
	.025	0	2	4	6	7	9	11	13	15	17	19	21	23	25	27	29	31	33	35
	.05	1	3	5	7	9	12	14	16	18	20	22	25	27	29	31	34	36	38	40
	.10	2	5	7	9	12	14	17	19	22	24	27	29	32	34	37	39	42	44	47
	.001	0	0	0	1	2	3	5	6	7	9	10	12	13	15	16	18	19	21	22
	.005	0	0	2	3	5	7	8	10	12	14	16	18	19	21	23	25	27	29	31
8	.01	0	1	3	5	7	8	10	12	14	16	18	21	23	25	27	29	31	33	35
	.025	1	3	5	7	9	11	14	16	18	20	23	25	27	30	32	35	37	39	42
	.05	2	4	6	9	11	14	16	19	21	24	27	29	32	34	37	40	42	45	48
	.10	3	6	8	11	14	17	20	23	25	28	31	34	37	40	43	46	49	52	55

Table A-10 Critical Values for the Mann-Whitney *U* (*Continued*)

n	*p*	*m* = 2	3	4	5	6	7	8	9	10	11	12	13	14	15	16	17	18	19	20
	.001	0	0	0	2	3	4	6	8	9	11	13	15	16	18	20	22	24	26	27
	.005	0	1	2	4	6	8	10	12	14	17	19	21	23	25	28	30	32	34	37
9	.01	0	2	4	6	8	10	12	15	17	19	22	24	27	29	32	34	37	39	41
	.025	1	3	5	8	11	13	16	18	21	24	27	29	32	35	38	40	43	46	49
	.05	2	5	7	10	13	16	19	22	25	28	31	34	37	40	43	46	49	52	55
	.10	3	6	10	13	16	19	23	26	29	32	36	39	42	46	49	53	56	59	63
	.001	0	0	1	2	4	6	7	9	11	13	15	18	20	22	24	26	28	30	33
	.005	0	1	3	5	7	10	12	14	17	19	22	25	27	30	32	35	38	40	43
10	.01	0	2	4	7	9	12	14	17	20	23	25	28	31	34	37	39	42	45	48
	.025	1	4	6	9	12	15	18	21	24	27	30	34	37	40	43	46	49	53	56
	.05	2	5	8	12	15	18	21	25	28	32	35	38	42	45	49	52	56	59	63
	.10	4	7	11	14	18	22	25	29	33	37	40	44	48	52	55	59	63	67	71
	.001	0	0	1	3	5	7	9	11	13	16	18	21	23	25	28	30	33	35	38
	.005	0	1	3	6	8	11	14	17	19	22	25	28	31	34	37	40	43	46	49
11	.01	0	2	5	8	10	13	16	19	23	26	29	32	35	38	42	45	48	51	54
	.025	1	4	7	10	14	17	20	24	27	31	34	38	41	45	48	52	56	59	63
	.05	2	6	9	13	17	20	24	28	32	35	39	43	47	51	55	58	62	66	70
	.10	4	8	12	16	20	24	28	32	37	41	45	49	53	58	62	66	70	74	79
	.001	0	0	1	3	5	8	10	13	15	18	21	24	26	29	32	35	38	41	43
	.005	0	2	4	7	10	13	16	19	22	25	28	32	35	38	42	45	48	52	55
12	.01	0	3	6	9	12	15	18	22	25	29	32	36	39	43	47	50	54	57	61
	.025	2	5	8	12	15	19	23	27	30	34	38	42	46	50	54	58	62	66	70
	.05	3	6	10	14	18	22	27	31	35	39	43	48	52	56	61	65	69	73	78
	.10	5	9	13	18	22	27	31	36	40	45	50	54	59	64	68	73	78	82	87

Table A-10 Critical Values for the Mann-Whitney *U* (Continued)

n	p	$m = 2$	3	4	5	6	7	8	9	10	11	12	13	14	15	16	17	18	19	20
	.001	0	0	2	4	6	9	12	15	18	21	24	27	30	33	36	39	43	46	49
	.005	0	2	4	8	11	14	18	21	25	28	32	35	39	43	46	50	54	58	61
13	.01	1	3	6	10	13	17	21	24	28	32	36	40	44	48	52	56	60	64	68
	.025	2	5	9	13	17	21	25	29	34	38	42	46	51	55	60	64	68	73	77
	.05	3	7	11	16	20	25	29	34	38	43	48	52	57	62	66	71	76	81	85
	.10	5	10	14	19	24	29	34	39	44	49	54	59	64	69	75	80	85	90	95
	.001	0	0	2	4	7	10	13	16	20	23	26	30	33	37	40	44	47	51	55
	.005	0	2	5	8	12	16	19	23	27	31	35	39	43	47	51	55	59	64	68
14	.01	1	3	7	11	14	18	23	27	31	35	39	44	48	52	57	61	66	70	74
	.025	2	6	10	14	18	23	27	32	37	41	46	51	56	60	65	70	75	79	84
	.05	4	8	12	17	22	27	32	37	42	47	52	57	62	67	72	78	83	88	93
	.10	5	11	16	21	26	32	37	42	48	53	59	64	70	75	81	86	92	98	103
	.001	0	0	2	5	8	11	15	18	22	25	29	33	37	41	44	48	52	56	60
	.005	0	3	6	9	13	17	21	25	30	34	38	43	47	52	56	61	65	70	74
15	.01	1	4	8	12	16	20	25	29	34	38	43	48	52	57	62	67	71	76	81
	.025	2	6	11	15	20	25	30	35	40	45	50	55	60	65	71	76	81	86	91
	.05	4	8	13	19	24	29	34	40	45	51	56	62	67	73	78	84	89	95	101
	.10	6	11	17	23	28	34	40	46	52	58	64	69	75	81	87	93	99	105	111
	.001	0	0	3	6	9	12	16	20	24	28	32	36	40	44	49	53	57	61	66
	.005	0	3	6	10	14	19	23	28	32	37	42	46	51	56	61	66	71	75	80
16	.01	1	4	8	13	17	22	27	32	37	42	47	52	57	62	67	72	77	83	88
	.025	2	7	12	16	22	27	32	38	43	48	54	60	65	71	76	82	87	93	99
	.05	4	9	15	20	26	31	37	43	49	55	61	66	72	78	84	90	96	102	108
	.10	6	12	18	24	30	37	43	49	55	62	68	75	81	87	94	100	107	113	120

Table A-10 Critical Values for the Mann-Whitney *U* (*Continued*)

n	p	$m = 2$	3	4	5	6	7	8	9	10	11	12	13	14	15	16	17	18	19	20
	.001	0	1	3	6	10	14	18	22	26	30	35	39	44	48	53	58	62	67	71
	.005	0	3	7	11	16	20	25	30	35	40	45	50	55	61	66	71	76	82	87
17	.01	1	5	9	14	19	24	29	34	39	45	50	56	61	67	72	78	83	89	94
	.025	3	7	12	18	23	29	35	40	46	52	58	64	70	76	82	88	94	100	106
	.05	4	10	16	21	27	34	40	46	52	58	65	71	78	84	90	97	103	110	116
	.10	7	13	19	26	32	39	46	53	59	66	73	80	86	93	100	107	114	121	128
	.001	0	1	4	7	11	15	19	24	28	33	38	43	47	52	57	62	67	72	77
	.005	0	3	7	12	17	22	27	32	38	43	48	54	59	65	71	76	82	88	93
18	.01	1	5	10	15	20	25	31	37	42	48	54	60	66	71	77	83	89	95	101
	.025	3	8	13	19	25	31	37	43	49	56	62	68	75	81	87	94	100	107	113
	.05	5	10	17	23	29	36	42	49	56	62	69	76	83	89	96	103	110	117	124
	.10	7	14	21	28	35	42	49	56	63	70	78	85	92	99	107	114	121	129	136
	.001	0	1	4	8	12	16	21	26	30	35	41	46	51	56	61	67	72	78	83
	.005	1	4	8	13	18	23	29	34	40	46	52	58	64	70	75	82	88	94	100
19	.01	2	5	10	16	21	27	33	39	45	51	57	64	70	76	83	89	95	102	108
	.025	3	8	14	20	26	33	39	46	53	59	66	73	79	86	93	100	107	114	120
	.05	5	11	18	24	31	38	45	52	59	66	73	81	88	95	102	110	117	124	131
	.10	8	15	22	29	37	44	52	59	67	74	82	90	98	105	113	121	129	136	144
	.001	0	1	4	8	13	17	22	27	33	38	43	49	55	60	66	71	77	83	89
	.005	1	4	9	14	19	25	31	37	43	49	55	61	68	74	80	87	93	100	106
20	.01	2	6	11	17	23	29	35	41	48	54	61	68	74	81	88	94	101	108	115
	.025	3	9	15	21	28	35	42	49	56	63	70	77	84	91	99	106	113	120	128
	.05	5	12	19	26	33	40	48	55	63	70	78	85	93	101	108	116	124	131	139
	.10	8	16	23	31	39	47	55	63	71	79	87	95	103	111	120	128	136	144	152

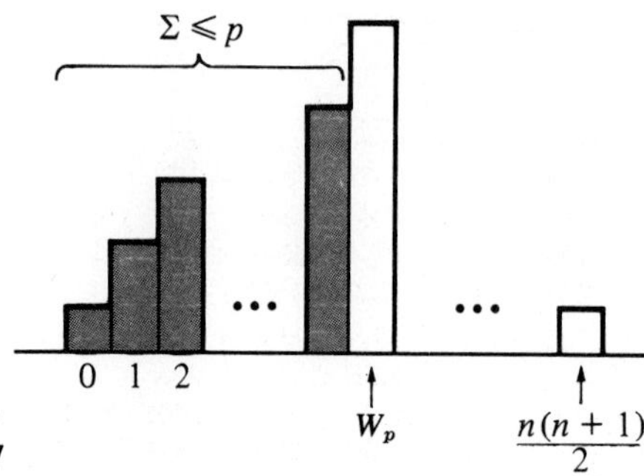

Table A-11 Critical Values for the Wilcoxon Signed Ranks Statistic W

	$w_{.005}$	$w_{.01}$	$w_{.025}$	$w_{.05}$	$w_{.10}$	$w_{.20}$	$w_{.30}$	$w_{.40}$	$w_{.50}$	$\frac{n(n+1)}{2}$
$n = 4$	0	0	0	0	1	3	3	4	5	10
5	0	0	0	1	3	4	5	6	7.5	15
6	0	0	1	3	4	6	8	9	10.5	21
7	0	1	3	4	6	9	11	12	14	28
8	1	2	4	6	9	12	14	16	18	36
9	2	4	6	9	11	15	18	20	22.5	45
10	4	6	9	11	15	19	22	25	27.5	55
11	6	8	11	14	18	23	27	30	33	66
12	8	10	14	18	22	28	32	36	39	78
13	10	13	18	22	27	33	38	42	45.5	91
14	13	16	22	26	32	39	44	48	52.5	105
15	16	20	26	31	37	45	51	55	60	120
16	20	24	30	36	43	51	58	63	68	136
17	24	28	35	42	49	58	65	71	76.5	153
18	28	33	41	48	56	66	73	80	85.5	171
19	33	38	47	54	63	74	82	89	95	190
20	38	44	53	61	70	82	91	98	105	210

Table A-12 Logarithms

N	0	1	2	3	4	5	6	7	8	9
10	0000	0043	0086	0128	0170	0212	0253	0294	0334	0374
11	0414	0453	0492	0531	0569	0607	0645	0682	0719	0755
12	0792	0828	0864	0899	0934	0969	1004	1038	1072	1106
13	1139	1173	1206	1239	1271	1303	1335	1367	1399	1430
14	1461	1492	1523	1553	1584	1614	1644	1673	1703	1732
15	1761	1790	1818	1847	1875	1903	1931	1959	1987	2014
16	2041	2068	2095	2122	2148	2175	2201	2227	2253	2279
17	2304	2330	2355	2380	2405	2430	2455	2480	2504	2529
18	2553	2577	2601	2625	2648	2672	2695	2718	2742	2765
19	2788	2810	2833	2856	2878	2900	2923	2945	2967	2989
20	3010	3032	3054	3075	3096	3118	3139	3160	3181	3201
21	3222	3243	3263	3284	3304	3324	3345	3365	3385	3404
22	3424	3444	3464	3483	3502	3522	3541	3560	3579	3598
23	3617	3636	3655	3674	3692	3711	3729	3747	3766	3784
24	3802	3820	3838	3856	3874	3892	3909	3927	3945	3962
25	3979	3997	4014	4031	4048	4065	4082	4099	4116	4133
26	4150	4166	4183	4200	4216	4232	4249	4265	4281	4298
27	4314	4330	4346	4362	4378	4393	4409	4425	4440	4456
28	4472	4487	4502	4518	4533	4548	4564	4579	4594	4609
29	4624	4639	4654	4669	4683	4698	4713	4728	4742	4757
30	4771	4786	4800	4814	4829	4843	4857	4871	4886	4900
31	4914	4928	4942	4955	4969	4983	4997	5011	5024	5038
32	5051	5065	5079	5092	5105	5119	5132	5145	5159	5172
33	5185	5198	5211	5224	5237	5250	5263	5276	5289	5302
34	5315	5328	5340	5353	5366	5378	5391	5403	5416	5428
35	5441	5453	5465	5478	5490	5502	5514	5527	5539	5551
36	5563	5575	5587	5599	5611	5623	5635	5647	5658	5670
37	5682	5694	5705	5717	5729	5740	5752	5763	5775	5786
38	5798	5809	5821	5832	5843	5855	5866	5877	5888	5899
39	5911	5922	5933	5944	5955	5966	5977	5988	5999	6010
40	6021	6031	6042	6053	6064	6075	6085	6096	6107	6117
41	6128	6138	6149	6160	6170	6180	6191	6201	6212	6222
42	6232	6243	6253	6263	6274	6284	6294	6304	6314	6325
43	6335	6345	6355	6365	6375	6385	6395	6405	6415	6425
44	6435	6444	6454	6464	6474	6484	6493	6503	6513	6522
45	6532	6542	6551	6561	6571	6580	6590	6599	6609	6618
46	6628	6637	6646	6656	6665	6675	6684	6693	6702	6712
47	6721	6730	6739	6749	6758	6767	6776	6785	6794	6803
48	6812	6821	6830	6839	6848	6857	6866	6875	6884	6893
49	6902	6911	6920	6928	6937	6946	6955	6964	6972	6981
50	6990	6998	7007	7016	7024	7033	7042	7050	7059	7067
51	7076	7084	7093	7101	7110	7118	7126	7135	7143	7152
52	7160	7168	7177	7185	7193	7202	7210	7218	7226	7235
53	7243	7251	7259	7267	7275	7284	7292	7300	7308	7316
54	7324	7332	7340	7348	7356	7364	7372	7380	7388	7396

Table A-12 Logarithms *(Continued)*

N	0	1	2	3	4	5	6	7	8	9
55	7404	7412	7419	7427	7435	7443	7451	7459	7466	7474
56	7482	7490	7497	7505	7513	7520	7528	7536	7543	7551
57	7559	7566	7574	7582	7589	7597	7604	7612	7619	7627
58	7634	7642	7649	7657	7664	7672	7679	7686	7694	7701
59	7709	7716	7723	7731	7738	7745	7752	7760	7767	7774
60	7782	7789	7796	7803	7810	7818	7825	7832	7839	7846
61	7853	7860	7868	7875	7882	7889	7896	7903	7910	7917
62	7924	7931	7938	7945	7952	7959	7966	7973	7980	7987
63	7993	8000	8007	8014	8021	8028	8035	8041	8048	8055
64	8062	8069	8075	8082	8089	8096	8102	8109	8116	8122
65	8129	8136	8142	8149	8156	8162	8169	8176	8182	8189
66	8195	8202	8209	8215	8222	8228	8235	8241	8248	8254
67	8261	8267	8274	8280	8287	8293	8299	8306	8312	8319
68	8325	8331	8338	8344	8351	8357	8363	8370	8376	8382
69	8388	8395	8401	8407	8414	8420	8426	8432	8439	8445
70	8451	8457	8463	8470	8476	8482	8488	8494	8500	8506
71	8513	8519	8525	8531	8537	8543	8549	8555	8561	8567
72	8573	8579	8585	8591	8597	8603	8609	8615	8621	8627
73	8633	8639	8645	8651	8657	8663	8669	8675	8681	8686
74	8692	8698	8704	8710	8716	8722	8727	8733	8739	8745
75	8751	8756	8762	8768	8774	8779	8785	8791	8797	8802
76	8808	8814	8820	8825	8831	8837	8842	8848	8854	8859
77	8865	8871	8876	8882	8887	8893	8899	8904	8910	8915
78	8921	8927	8932	8938	8943	8949	8954	8960	8965	8971
79	8976	8982	8987	8993	8998	9004	9009	9015	9020	9025
80	9031	9036	9042	9047	9053	9058	9063	9069	9074	9079
81	9085	9090	9096	9101	9106	9112	9117	9122	9128	9133
82	9138	9143	9149	9154	9159	9165	9170	9175	9180	9186
83	9191	9196	9201	9206	9212	9217	9222	9227	9232	9238
84	9243	9248	9253	9258	9263	9269	9274	9279	9284	9289
85	9294	9299	9304	9309	9315	9320	9325	9330	9335	9340
86	9345	9350	9355	9360	9365	9370	9375	9380	9385	9390
87	9395	9400	9405	9410	9415	9420	9425	9430	9435	9440
88	9445	9450	9455	9460	9465	9469	9474	9479	9484	9489
89	9494	9499	9504	9509	9513	9518	9523	9528	9533	9538
90	9542	9547	9552	9557	9562	9566	9571	9576	9581	9586
91	9590	9595	9600	9605	9609	9614	9619	9624	9628	9633
92	9638	9643	9647	9652	9657	9661	9666	9671	9675	9680
93	9685	9689	9694	9699	9703	9708	9713	9717	9722	9727
94	9731	9736	9741	9745	9750	9754	9759	9763	9768	9773
95	9777	9782	9786	9791	9795	9800	9805	9809	9814	9818
96	9823	9827	9832	9836	9841	9845	9850	9854	9859	9863
97	9868	9872	9877	9881	9886	9890	9894	9899	9903	9908
98	9912	9917	9921	9926	9930	9934	9939	9943	9948	9952
99	9956	9961	9965	9969	9974	9978	9983	9987	9991	9996

Selected Answers

CHAPTER 1

1. *a.* $+7$; -3; $+21$; $+10/1 = 10$; $-27/3 = -9$
   b. 12; $-12/1 = -12$; 139,056; $-8{,}742{,}278$; $121/11 = 11$

2. *a.* 1/2 or 2/4 or 3/6 or 4/8 or 5/10, etc.
   b. 1/3 or 2/6 or 3/9 or 4/12, etc.
   c. 4/1 or 8/2 or 12/3, etc.
   d. 6/20 or 9/30, etc.
   e. 8/10 or 16/20 or 24/30, etc.
   f. 247/100

3. *a.* 1/2
   b. $-5/3$
   c. 5/6
   d. 4/3
   e. 24/11
   f. $4/1 = 4$
   g. 1/6
   h. $13a/19b$
   i. b/a
   j. 2/3
   k. $2/1 = 2$
   l. 29/4

4. *a.* 7/5
   b. $9/3 = 3$
   c. $6/9 = 2/3$
   d. $-5/7$
   e. 25/6
   f. $-7/12$
   g. $(x - \mu)/\sigma$
   h. $(x - \bar{x})/\sigma$
   i. 8/27
   j. 2/9
   k. 64/63
   l. 10

5. *a.* 6
   b. 24
   c. 5
   d. $3x + 3y + 3$

6. *a.* 3
   b. 4.5
   c. 13/3
   d. −2
   e. −4
   f. −4
   g. 5
   h. −100/9
   i. 11.645
   j. 10.98

7. *a.* >
   b. >
   c. >
   d. <

8. *a.* 1.96
   b. $(x - \mu)/\sigma$
   c. $(x - \mu)/\sigma$
   d. 1.96

9. *a.* $(x - \mu)/\sigma$
   b. $(x - \mu)/\sigma$
   c. 1.645
   d. $(x - \mu)/\sigma$

10. *a.* >
    b. >
    c. >
    d. <
    e. >
    f. <
    g. >
    h. <
    i. <
    j. <
    k. >
    l. <
    m. <
    n. >

11. *a.* 46
    b. −37
    c. 99
    d. −96
    e. 25
    f. 20/10 = 2
    g. 0.8
    h. 2318
    i. 400

12. *a.* 24
    b. 35
    c. 5
    d. 1
    e. 1980
    f. 37

13. *a.*

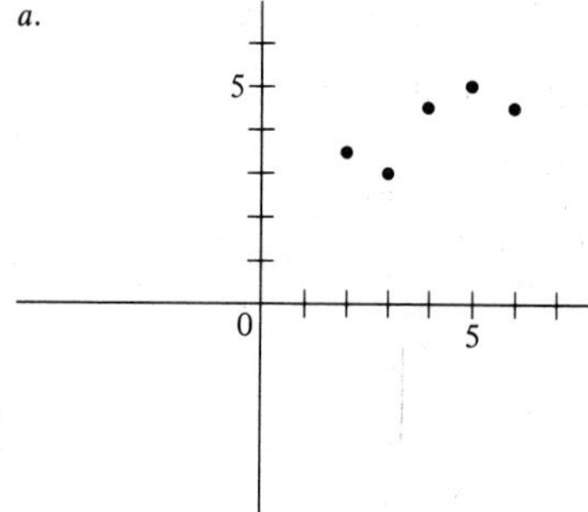

b.

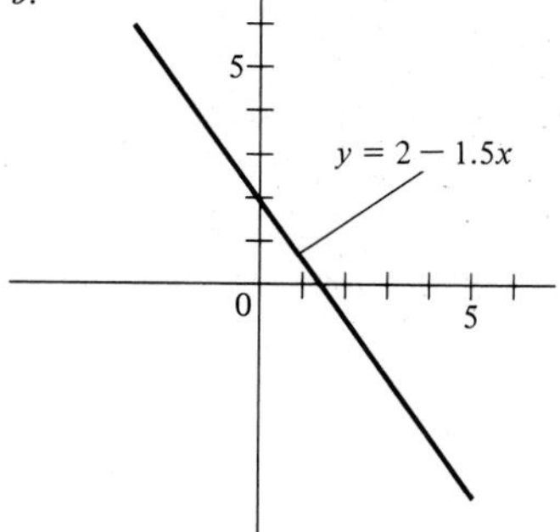

14.

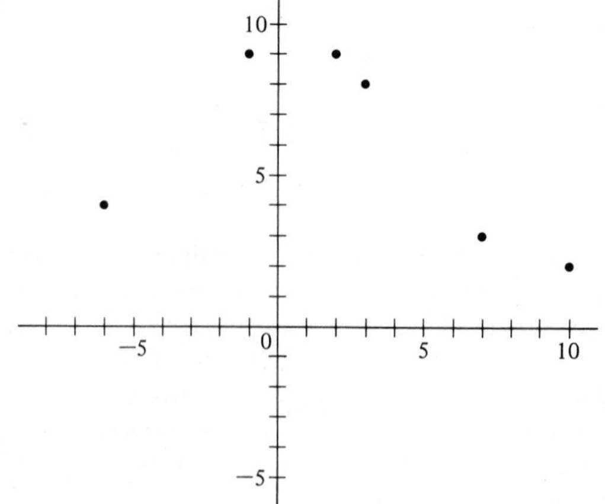

15. *a.* .25 *d.* .0025
b. 4.41 *e.* 15,876
c. .1225 *f.* .054756

16. *a.* 80
b. 166

17. *a.* 20
b. 20
c. 77

18. *a.* 4.6583 *d.* 699.29
b. 0.23664 *e.* 0.018166
c. 22.113 *f.* 0.078740

CHAPTER 2

3.

	Stock A	Stock B	Stock C
a.	$\mu = 100$	$\mu = 100$	$\mu = 100$
b.	$\sigma^2 = 21.625$	$\sigma^2 = 45.125$	$\sigma^2 = 47.0$
c.	$\sigma = 4.650$	$\sigma = 6.718$	$\sigma = 6.856$

d. All three stocks have the same mean, but stock A has the smallest variance of the three. Recommend stock A.

4. *a.* 22, 27, 32, 37, 42, 47, 52, 57, 62

b. $\mu \simeq 238{,}471/5698$
$\simeq 41.851702$

d. If you redefine the classes to correspond to the three groups, you find fewer workers in the (35–49) group than either of the other two.

6.

	Machine A	Machine B
b.	$\mu = 100$	$\mu = 100$
c.	$\sigma^2 = 1.12$	$\sigma^2 = 0.83$

10. Both countries A and B are possibilities since their populations are relatively large. However, the large difference between their means and medians would suggest that both countries contain some households with relatively large annual incomes. Their large variances tend to confirm this conclusion. Country C is a definite yes, as at least 1,350,000 households have an annual income of at least $402. Country D is a definite no since no more than 1 million households have annual income in excess of $390. Country E is certainly possible as at least 950,000 households have annual income at or above $410.

CHAPTER 3

2. Two possible indices:

   a. $I = \frac{P_{t,i}}{P_{o,i}} \times 100$, where 1968 is the base year

   b. $I = \frac{P_{t,i}}{P_{t-1,i}} \times 100$.

5. Using a quantity-weighted price index, the adjusted sales figure for 1977 is $1,012,727.20.

6. Mr. O'Dell's sales have grown in each year except 1972 and 1977.

CHAPTER 4

2. *Pr* ("overfill" ∪ "underfill") = .016

3. *Pr* (oil|positive seismographic survey) = .778

4. *Pr* (interviewee joins the company) = .08; company should interview 5625 applicants

5. *Pr* (at least one) = .952
   Pr (any two) = .464
   Pr (all three) = .192

8. Assuming independence, *Pr* (defective) = .0038965

10. *Pr* (at least 15,000) = 1.00
    Pr (at least 25,000) = .80
    Pr (at least 35,000) = .60
    Pr (at least 45,000) = .40
    Pr (at least 55,000) = .20

CHAPTER 5

2. Yes; 210

3. $340,000

4. Northwood Center as its expected payoff is $300,000 greater than that for Western Hills.

6. $4

CHAPTER 6

3. $Pr\ (x \geq 2) = .8208$; so $n(.8208) = 5{,}663{,}520$

5. $Pr(x \leq 4) = .1662$

6. $Pr\ (x \geq 111.23) = .10$, so need seating for 112

7. $Pr\ (x \geq 108.42) = .20$

8. The relevant probabilities are: Marshall, .1702; Ornales, .8244; Greer, .5951; Washington, .9612; Jones, .3523

11. $Pr\ (x \geq \$50.00) = .0509$

13. $Pr\ (x < 10\%) = .3696$

14. $Pr\ (x < 18 \text{ months}) - Pr\ (x < 12 \text{ months}) = .0685$

CHAPTER 7

2. *a.* 55% $.55 \pm 0.0563$
   b. 67.5% $.675 \pm 0.0545$
   c. 60% $.60 \pm 0.0564$

3. $16{,}217 \pm 1{,}159$
   No, as 20,000 is not in this confidence interval.

5. $\hat{\tau} = 2{,}368{,}000$ oz
   $2{,}368{,}000 \pm 175{,}137.27$

7.

"Virtually positive"	n
$Pr = .99$	21,722
$Pr = .95$	12,585
$Pr = .90$	8,865

10. $n = 254$

CHAPTER 8

2. H_0: $\pi \leq .28$
   Reject H_0 for $z \geq +2.33$.
   As the computed z value is 1.27, which is less than the critical z value of $+2.33$, accept H_0 and conclude that the rejection rate for Travis is not greater than the national average.

3. H_0: $\pi_T - \pi_N \geq .03$
   Reject H_0 for $z \leq -1.645$
   As the computed z value is $+2.17$, which is greater than the critical z value of -1.645, accept H_0 and conclude that local advertising cost is justified.

4. H_0: $\mu_s - \mu_u \leq 0$
   Reject H_0 for $z \geq +2.33$.

   As the computed z value is +3.16, which is greater than the critical z value of +2.33, reject H_0 and conclude that the mean respondent evaluation of the nutrition for the successful brand is larger than that for the unsuccessful brand.

5. H_0: $\mu \leq 15.7$
   Reject H_0 for $z \geq +1.645$.

   As the computed z value is +1.16, which is less than the critical z value of +1.645, accept H_0 and conclude that Ms. Warner's company is performing at the national average.

8. H_0: $\sigma^2 \geq 0.015$
   Reject H_0 for $\chi^2 \leq 77.9295$

   As the computed χ^2 value is 133, accept H_0 and conclude that the machine would not be acceptable.

14. H_0: $\sigma^2_{\text{with}} \geq \sigma^2_{\text{without}}$
    Reject H_0 for F $\geq$ 5.91 when
    $F = s^2_{\text{without}}/s^2_{\text{with}}$

    As the computed F value is 8.07, reject H_0 and conclude that the adapter does decrease the variance of dispensed fluid ounces.

CHAPTER 9

2. The expected payoff for bidding is $275.

3. The expected costs are:
   With campaign, $6,625,000
   Without campaign, $6,250,000

6. For Watson, −$2,950,000
   For Ellsworth, $1,550,000

8. *a.* The expected payoff for adding the flight is $980,000.
   b. The value of the information is $56,000.

9. The expected payoff from signing is $7,780,000.

10. The expected profits are: Add third shift, −$160,000; do not add, $390,000.

CHAPTER 10

2. *b.* $\hat{y} = -6.719 + 5.112x$
   c. H_0: $\beta_1 = 0$
   H_1: $\beta_1 > 0$
   t computed = 48.69

   The decision rule is to reject H_0 for $t \geq 1.812$. As $48.69 > 1.812$, reject H_0 and conclude that x and y are linearly related.

d. $r = 0.99788$
e. 197.761 ± 31.697

4. *a*. $\hat{y} = -3012.6 + 0.9734x$
H_0: $\beta_1 = 0$
H_1: $\beta_1 > 0$
t computed $= 23.06$
The decision rule is to reject H_0 for $t \geq 1.943$. As $23.06 > 1.943$, reject H_0 and conclude that x and y are linearly related.
d. $r = 0.9965$
e. $23{,}804.6 \pm 1{,}145.0$

6. *a*. 72,400
b. 16,004,567
c. 73,945.6
d. 1,614,704.9
e. Use the regression estimator as it has a much smaller variance.

8. Yes, it looks as though x and y are linearly related. The plot of the observations appear linear and the regression coefficient $b_1 = 0.09572$ is significant at any alpha level given our tables.

CHAPTER 11

3. *b*. $y = \beta_0 + \beta_1 x + \beta_2 x^2 + \epsilon$
c. $\hat{y} = 45.49979 + 13.63340x - 1.24243x^2$
d. H_0: $\beta_1 = \beta_2 = 0$
MSR/MSE $= 36.058$

Decision rule; reject H_0 for $F_{2,17} \geq 3.59$; thus reject H_0 and conclude that threat score and attitude-change score are not independent.

4. *a*. Response variable is the performance score. GPA, percent of expenses earned, and number of social organizations are potential explanatory variables.
c. $y = \beta_0 + \beta_1 x_1 + \beta_2 x_2 + \beta_3 x_3 + \epsilon$

where x_1 = GPA
x_2 = % expenses earned
x_3 = number of social organizations

d. $\hat{y} = -27.28114 + 28.75150x_1 + 0.33507x_2 + 3.21120x_3$
e.

Variable	Computed t
x_1	4.162174
x_2	2.611743
x_3	2.015688

Critical t value is 2.080. Thus, it appears as though x_3 should be deleted from the model.

9. Retail sales is the only one of the three potential independent variables that seems to explain total industry sales. The computed t values for housing starts, government expenditures, and retail sales are respectively, 1.3575, 0.6949, and 19.9887. The critical t value for $\alpha = .05$ is 2.306. Since total retail sales explains 98.43 percent of the variation in total industry sales, we might recommend that Hardin, Inc. use the model
$\hat{y} = 442.5313 + 17.9690x$

where $\hat{y}$ = estimated total industry sales for a territory
x = total retail sales for the corresponding territory

CHAPTER 12

2. *a.* **Anova**

Source	df	SS	MS
Among	3	78.3948	26.1316
Within	16	31.1221	1.9451
Total	19	109.5169	

$F = 26.1316/1.9451$
$= 13.43$

Critical value for $F_{3,16} = 3.24$. As $13.43 > 3.24$ we reject H_0 and conclude that not all store layouts produce equal mean purchase per customer.

b. The four populations are normally distributed with equal variances.

3. *a.* **Anova**

Source	df	SS	MS
Among	3	531692	177230.66
Within	12	42422	3535.17
Total	15	574114	

$F = 177230.66/3535.17$
$= 50.13$

Critical value for $F_{3,12} = 3.49$ as $50.13 > 3.49$ we reject H_0 and conclude that the four machines do not all produce the same mean output per hour.

b. $T\sqrt{\text{MSW}} = 2.1\sqrt{3535.17}$
$= 124.86$

All $|\bar{x}_j - \bar{x}_j'|$ exceed 124.86 except for $|\bar{x}_B - \bar{x}_D|$, which is 113.0.
Therefore, we conclude that

$\mu_A \neq \mu_B$
$\mu_A \neq \mu_C$
$\mu_A \neq \mu_D$
$\mu_B \neq \mu_C$
$\mu_B = \mu_D$
$\mu_C \neq \mu_D$

c. On the basis of *b* we conclude that all machines (B, C, and D) are better than machine A. Although the difference between $\bar{x}_B$ and $\bar{x}_D$ was not large enough to reject the hypothesis that $\mu_B = \mu_D$, if we had to choose a brand to buy, it would be brand D.

7. *a*. Two-way ANOVA without replications. One factor is oil type and the other factor is vehicle type.

   b. **Anova**

Source	df	SS	MS
Oils	2	2297.556	1148.778
Vehicles	2	3072.222	1536.111
Error	4	173.778	43.445
Total	8	5543.556	

   c. $F = \text{MSA/MSE}$
   $= 1148.778/43.445$
   $= 26.445$

   As the critical value of $F = 6.94$, reject H_0 and conclude that there is a difference in the mean maintenance cost for the three oils.

   d. Applying the Tukey method we can conclude that both premium oil and super premium oil are "better" (produce lower mean maintenance cost) than standard oil. However, the difference between the sample means for premium and super premium oils is not "significant."

11. **Anova**

Source	df	SS	MS
Among	4	728.12	182.03
Within	120	107489.60	895.747
Total	124	108217.72	

   $F = 182.03/895.747$
   $= 0.2032$

   At $\alpha = 0.05$, $F_{4,120} = 2.45$. As $0.2032 < 2.45$ we would accept H_0 and conclude no difference in the five means. Recommendation would be to market the version that is cheapest to produce.

CHAPTER 13

2. *a*. Value of χ^2 with one df = .826; "accept"
   b. $z = .9088$; "accept"

4. Testing whether Fort Worth's proportion of sales remains the same with the campaign; value of χ^2 with one df = 3.905. Reject H_0 for $\alpha = .05$.

5. Value of χ^2 with four df's = 9.946. Reject for $\alpha = .10$.

8. Value of χ^2 with three df's = 1.807. "Accept" for $\alpha = .10$.

10. Value of χ^2 with two df's = 1.39; "accept"

CHAPTER 14

2. $Pr(x \geq 15) = .0207 < \alpha = .10$; reject

3. $Pr(x \leq 2) = .0547 > \alpha = .01$; "accept"

5. $U = 228.5$; for $n = 20$, $m = 15$, reject for any $\alpha \geq .005$.

8. $U = 294$; using normal approximation, "accept" for any $\alpha \leq .10$.

9. $W = 26.5$; for $n = 7$, reject for $\alpha \geq .025$.

11. $W = 4$; for $n = 6$, "accept" for $\alpha \leq .10$.

CHAPTER 15

4. *a.* $T = 2608.4 + 59.4t$, where $t = 7$ represents 19X7.

   b.

Jan.	.82	May	1.06	Sept.	1.09
Feb.	.82	June	1.11	Oct.	1.09
Mar.	.84	July	1.04	Nov.	1.06
Apr.	.94	Aug.	1.05	Dec.	1.08

   d.

19X8				19X9			
Jan.	210.7	July	267.3	Jan.	214.8	July	272.4
Feb.	210.7	Aug.	269.8	Feb.	214.8	Aug.	275.0
Mar.	215.9	Sept.	280.1	Mar.	220.0	Sept.	285.5
Apr.	241.6	Oct.	280.1	Apr.	246.2	Oct.	285.5
May	272.4	Nov.	272.4	May	277.6	Nov.	277.6
June	285.3	Dec.	277.6	June	290.7	Dec.	282.9

5.

19X7		19X8	
Apr.	478	Jan.	414
May	459	Feb.	399
June	407	Mar.	483
July	393		
Aug.	460		
Sept.	470		
Oct.	494		
Nov.	525		
Dec.	599		

8. (Sales) = 2.8669 + 1.1105 (housing starts)
   $F_{1,34} = 155.948$.

Index